The Joy of Science

The Joy of Science

An Examination of How Scientists Ask and Answer Questions Using the Story of Evolution as a Paradigm

Edited by

Richard A. Lockshin, Ph.D.

Department of Biological Sciences, St. John's University

 Springer

A C.I.P. Catalogue record for this book is available from the Library of Congress.

ISBN 978-1-4020-6098-4 (HB)
ISBN 978-1-4020-6099-1 (e-book)

Published by Springer,
P.O. Box 17, 3300 AA Dordrecht, The Netherlands.

www.springer.com

Printed on acid-free paper

CONTENTS

PREFACE

Scientists have great passion. What could be more exhilarating than to go to work every day feeling as if you were once again a nine-year-old called up to he stage to help the magician with his trick? To be a researcher is to always be in the position of having the chance to see how the trick works. No wonder that many researchers feel that each new day is the most exciting day to be a scientist.

It therefore is not surprising that scientists have such trouble communicating with non-scientists. It is difficult for the scientist to understand a life not focused on the desire to understand. But the differences are not that. Everyone wants to understand; that is one of the factors that make us human. The difference is more that scientists limit their definition of comprehension to specific rules of logic and evidence. These rules apply and are used in everyday life, but often with less rigor or restrictions on evidence.

The structure of this book is therefore tripartite. On the first level, we wish to demonstrate that, far from being arcane or inaccessible, the scientific approach is simply a variant of normal, common experience and judgment, easily accessible to any educated person. The second goal is to explain the structure of scientific thinking, which we will describe as the requirement for evidence, logic, and falsification (experimental testing). The third goal is to illustrate the scientific method by looking at the story of the development of the idea of evolution.

Evolution is a branch of scientific inquiry that is distinguished by its minimal level of laboratory experimentation, as least in its early period. Nevertheless, the story of evolution seems for several reasons to be an excellent choice to examine the nature of scientific inquiry. First, it is, almost without doubt, the most important idea of the 19th and 20th centuries. Second, it is often misunderstood. Third, understanding the story does not require an extensive technical background. Finally, it is very multidisciplinary.

This latter point may be confusing to some – what do Einstein's Theory of relativity, X-rays of molecules, or the physics of flight have to do with evolution? But all knowledge is interconnected, and the best science (and the best ideas generally) come when thoughts range across disciplines. If you are unfamiliar with, or uncomfortable with, this approach, try it! It is much easier than you think, and making the connection between history and biology, or between any two disciplines, makes our understanding of both much richer and deeper. Furthermore, the facts will make more sense and be easier to remember. If you understand, you don't have to memorize, because the facts will be obvious. This is why the questions at the ends of the chapters are essay style. Isolated facts are the basis for a trivia contest, while connected facts are the gateways to understanding.

Finally, for those concerned about using this book for teaching or learning within the confines of a course: all knowledge is connected, and it would be possible in taking a topic as global as evolution to expand into every realm of science and theology. I have found it useful in my teaching to allow the curiosity of students to redefine the directions I take, and the book reflects some of these directions. It is not necessary to address evolution through an excursion into molecular biology, but molecular biology is relevant, interesting, and currently in the headlines. I therefore have included excursions such as these into the text, but I highly encourage teachers and others planning a course to omit these excursions, as they see fit, or to use them as supplementary materials. I have also included several comments on the relationship of history and culture to the development of science. Since the book is written for those who do not intend to major in sciences, these comments should help these students to connect the various trains of developing thought and culture to the growing science as well as providing launchpads for teachers more comfortable with these subjects.

It is possible to use this book for a one-semester or two-semester course. Each of the chapters may be treated briefly or in more detail—for instance, in developing the story of quantitation and statistics in Chapter 32 or following in greater or lesser detail the excursion into molecular biology in Chapters 14–16. It will also be possible to spend more time on such issues as the distinction among the various historical eras, the modern classification of animals and plants, or the relationship between ecology and evolution. If possible, it would be best to use this book in the setting of small classes in which discussion is encouraged.

For further resources, more technical sources and interesting web pages are listed at the end of most chapters. Of course, nothing beats reading Darwin's original books, *The Origin of Species, The Descent of Man, and Voyage of the Beagle*, or any of several books and essays by Stephen Jay Gould, Ernst Mayr, or other more recent giants of the field. A more popular summary, written by a science reporter, is Carl Zimmer's *Evolution: The Triumph of an Idea*, Harper Collins, 2001. It was written in conjunction with a PBS series on Evolution, which is likewise available from the Public Broadcasting System (http://www.pbs.org). Some of the references that you will find in this book are to Wikipedia (http://www.wikipedia.org). They are used because they are readily accessible–the function of Wikipedia. However, readers should appreciate that most articles are written by graduate students, who may have good understanding but rarely a historical perspective, and the articles are usually not written by established authorities. Most of the articles, however, contain appended references that are generally reliable.

Finally, there are of course many people to whom I am indebted for assistance in the preparation of this book. Many readers will recognize my indebtedness to many excellent writers in this field such as Steven Jay Gould (several writings, but especially *The Mismeasurement of Man*) and Jared Diamond (*Guns, Germs, and Steel* and *Collapse*). I attempt to summarize some of their arguments. Hopefully, readers will be encouraged to read the more voluminous but exciting and challenging full works. In addition to the many teachers and lecturers from whom I have profited at all stages of my career and the administrators at St. John's University who encouraged and supported the development of the course from which this book is derived. Among the friends who have read and commented—with excellent suggestions—on various sections and drafts, and offered many worthwhile books and readings, I count (in alphabetical order) Mitchell Baker, Dan Brovey, Andrew Greller, and Michael Lockshin. My colleague, friend, and wife, Zahra Zakeri, has offered many cogent criticisms and, of course, has been most helpful and tolerant of my endless searches, writings, and musings. I dedicate this book to her None of these individuals has any responsibility for any weaknesses, errors, or other problems.

PART 1

HOW SCIENCE WORKS

CHAPTER 1

SCIENCE IS AN ELF

Evidence, Logic, and Falsification as the criterion for scientific decision-making. A question beginning with the interrogative "Why" is not a good scientific question. The art of structuring a question so that it can be tested. The controlled experiment

WHY BOTHER WITH SCIENCE?

This book has several goals. In the first instance it is about how scientists evaluate information and draw conclusions. Understanding this process is a requirement for modern life and it is an important aspect of every part of our lives. Thomas Jefferson is reputed to have said, "An informed citizenry is the bulwark of a democracy..." Today, to be a participant in the community of "informed citizenry," one must be able to interpret scientific information. It is difficult if not impossible to function effectively in society without some knowledge of the scientific process.

Every day the newspaper or television brings forth a large issue of some concern to each of us, but how prepared are you, really, to evaluate the arguments that global warming is real, will affect your way of life, will threaten coastlines, is responsible for severe hurricanes? Can you truly compare moral vs scientific arguments concerning stem cells, correction of genetic defects, medical manipulation of fertility (to achieve conception or prevent it), or maintenance of life by use of machines? Should you vote to protect wetlands, to prevent future floods, to maintain a fishing industry, or to allow resting places for migratory birds? Or are wetlands simply breeders for mosquitoes and places that could be profitably developed for housing or commercial purposes? Can you participate in a meaningful discussion of the dangers of nuclear reactors, or the merits or disadvantages of genetically engineered foods? On a more personal level, can you evaluate different potential diets, or interpret an advertisement for a medication? Can you read and understand the information inserts in medicine?

Ultimately, each of these discussions, and many more, depend on highly technical details that are not readily presented to the non-scientist. On the other hand, all scientists are expected to present their data in a manner that a layman can understand. Much scientific research is supported by your tax dollars through government-sponsored research programs. Each proposal for research is presented to a scientific board for evaluation, but the proposal typically also contains a summary that is expected to be meaningful to a congressman or congresswoman who will vote on the subsidy for the overall program, and meaningful to interested citizens who would like to know how their money is spent. That means you.

The goal of the scientist in this abstract is not to teach a lay audience the highly technical details of a complex proposal but to make the goals, limitations, and potential of the proposed research clear enough that you will understand the purpose and agree that it is a good idea and has the potential of producing knowledge of interest and value to you. Thus the first goal of this book and this course is to prepare you for this role as a citizen. What we hope to achieve is to give you a sense of how scientific data are collected and evaluated, so that you will be able to interpret the information inundating you. Thus throughout this book we will be emphasizing the scientific method.

EVOLUTION

We have chosen the approach of illustrating the scientific method through the study of evolution. We have chosen evolution for several reasons. First and foremost, evolution is the most important idea of the 19th Century and the most influential of the 20th Century. (Scientists almost never speak in absolutes, and almost inevitably qualify or restrict any statement that they make. I was therefore tempted to state, "evolution is arguably the most important idea..." but in this case there seems to be little reason to deny these claims.) Second, unlike, for instance, astrophysics or molecular biology, one needs relatively little technical background or familiarity with very abstruse and abstract topics to understand what is going on. For these reasons the topic seemed a logical choice.

SCIENCE IS AN ELF

Evolution, like astrophysics, lacks one essential of laboratory science, the ability to readily design and carry out experiments. It is possible to make predictions, which are in a sense thought experiments, and in some instances it is possible to design and conduct experiments, and we will address these issues as best we can. In all other senses, evolution is in every way a full science and illustrates the logic and construction of scientific thinking. That is, it depends fully on three elements that I define as an "ELF" principle: Evidence, Logic, and Falsification. A scientific idea must be based on **evidence,** whether obtained by observation or experiment. The evidence suggests a link between two phenomena. A scientist will attempt to understand the link by establishing that one phenomenon causes another, or in other words he or she will form a hypothesis of cause and result. For instance, every year as spring approaches the sun gets higher in the sky and the days get longer. This is the **evidence**—both the length of the day and the mean temperature—that we can observe and measure. A reasonable hypothesis would be that the increased sunlight warmed the earth, rather than that the warming of the earth caused the days to get longer. This is the **logic** of the hypothesis, associating the heat that one feels in sunlight with the larger issue of gradually-increased warmth. Finally, the scientist will wish to test the hypothesis. The way that a hypothesis is tested is to try to disprove it: Can I create or envisage a situation in which the days will get longer but the earth will NOT get warmer? If so, does this disprove my hypothesis, or can I explain the seeming contradiction in a manner that still preserves the hypothesis? This is the **falsification** step (See Table 1.1). We will discuss these steps in considerable detail in the next chapter, and then use the principles throughout the book.

This means of analyzing information is not only not very difficult, it is something that humans do every day of their lives. Hunting-stage humans must have done it by observing, "if animal tracks from here go toward the setting sun (west), but when I am two days walk toward the setting sun, the animal tracks go toward the rising sun (east) then the animals must be heading towards a water hole between here and two days' walk west of here," (Fig. 1.1) or, "if that fat plant (cactus or succulent) contained water to drink, perhaps this fat plant also contains water" (Fig. 1.2). These are basically examples of classical syllogisms:

"If all antelope go to water in the evening
And if all antelopes here go west in the evening
Then there is water to the west."

Table 1.1. Evidence, Logic, Falsification

Evidence	Logic	Falsification
Weather gets warmer as days get longer	Sunlight warms the earth	Prevent all sunlight and warmed air from reaching an object
The lamp does not light when switched on	Perhaps it is unplugged	Verify that it is plugged in; plug it in. If it is plugged in, or plugging it in does not work, the hypothesis is falsified and we have to go to another hypothesis (bulb is burned out?)
Animals go west at twilight	Animals go to water	Follow animals, or determine when they return that they have drunk water
Cactus type A contains water; cacti type B and C have similar fat appearance	Fat plants contain water	Open cactus type B and C to see if they contain water
See bus leave stop; buses run every half hour	I walk 3 miles/hour and want to go 1 mile; walking is faster than waiting for next bus	Walk the distance; time yourself; observe if another bus passes

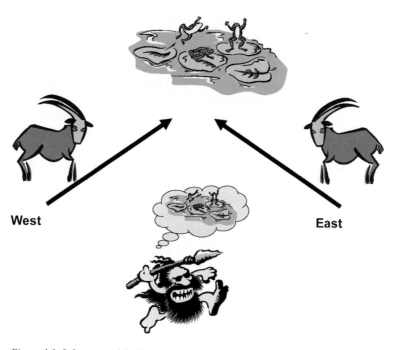

Figure 1.1. Inference and logic in a simple decision. The hunter-gatherer knows that antelopes seek water in the evening. When the antelope comes from the west, it heads toward the northeast. When antelopes come from a position several kilometers to the east, they head toward the northwest. Our hunter infers that water can be found somewhere at the intersection of these two tracks, or toward the north

When you buy a pen, and you say to yourself, "I really like that pen, but it costs five times more than this pen, and I usually lose pens in three days, so I had better buy the cheap one," you are using scientific logic, prediction, and evaluation; if you choose the more expensive pen, in spite of the evidence, you are conducting the experiment, "If my motivation—budgetary or desire—is strong enough, I will remember where I put the pen and gain the pleasure of owning it." Or again, suppose a candidate for mayor announces a platform of being "against crime in the streets". You are likely to say, "That's nice, what are you going to do?" If the candidate says, "I'll put all the criminals in jail," you are likely to say, "How are you going to do that?" If the candidate continues, "I'll arrest them all," you are likely very soon to wonder, "Is what the candidate suggests practical? Is he or she going to be threatening or harassing specific groups of innocent citizens? Can we afford the plan, whether it is better lighting, more police, more judges, more jails? Will the plan demand too much information about my life? If it includes restrictions on access to guns, knives, spray paint cans, box cutters, is this a good idea? How much will it restrict my life?" In other words, the candidate has hypothesized that a specific number of habitual criminals are the primary cause of crime (as opposed,

Figure 1.2. Generalization. Our hunter-gatherer is aware that, in dry lands, some plants with thick stems and leaves (which may be cacti, shown here, or succulents such as jade plants) store water in the stems and leaves. He is thirsty when he encounters a new type of plant, which has some resemblance to the cacti that he knows. He generalizes the first information to deduce that the new plant also stores water in its leaves, and thus finds something to drink

for instance, to poverty, lack of employment, insufficient care and protection of objects, lack of activities for teenagers and young adults, or other causes) and has proposed the experiment that isolating these individuals will eliminate the problem. You are asking for evidence that you will test against your own logic. You may well apply a form of falsification to the candidate's hypothesis: "Arrest rates differ from city to city and state to state. Do states with higher arrest rates, or more aggressive prosecution of criminals, have lower rates of crime? Do other factors, such as numbers of young men, play a role? How about the availability of employment, or of youth centers?" Collecting any or all of this information would in essence be an experiment in the same sense that a laboratory scientist designs an experiment. In other words,

IT AIN'T ROCKET SCIENCE (OR, IT IS UNDERSTANDABLE)

You apply the logic of science (hopefully) every day of your life. A local fast-food chain offers a huge ice-cream sundae that contains "only five calories"; you wonder if that's true (what the evidence is; how logically it can be sweet without sugar).

One mild day in winter, a friend remarks that the mildness is due to global warming; it crosses your mind that last week was a record low temperature. On television, an ad touts a "miracle brush" that can remove spilled dry paint with a single swipe; you are very skeptical and look very closely at the ad to judge if what is being shown actually happened. You notice, on another ad, for a weight-loss regime, that the actors in the "after" pictures are smiling, flexing their muscles, holding in their stomachs, and are turned so that their least flattering parts are hard to see, whereas in the "before pictures" they are not smiling and are making no efforts to hide their flab. Even a decision whether or not to walk to the next bus stop rather than wait, or to take a taxi rather than wait for the bus, is based on a hypothesis about the time on the route and your fatigue or energy.

This point cannot be made too strongly: The logic of science, and the structure of science, is simply human logic. It requires the same skills that we use on a daily basis, and is no more complex than that. There are only three things that seem to be difficult about science: its use of mathematics, its large and complex vocabulary, and the abstractness of many of its concepts. None of these presents an insuperable barrier to the student who wants to understand how science works.

MATHEMATICS AND TECHNICAL TERMS

Working scientists need to understand mathematics because quantification is a very important aspect of what we do. For obvious reasons, we need to know more than the fact that a volcano is a volcano. We need to know if it will erupt, which is a calculation based on the location of its magma (molten rock, lava), its past history and the history of similar volcanoes, what the earth is doing under the volcano, etc. If the volcano is not completely dead, we need to know when it is likely to erupt, and how severe the eruption is likely to be. All of these require extensive calculations, but even a non-mathematical person is likely to understand a scientist who says, "The molten rock moved this week from a half mile beneath the cone to within 600 feet of the cone, and the surface temperature at the cone rose $50°$ F. We consider the volcano dangerous."

Likewise, statistics is a large part of medical and sociological research. New medical treatments, and the licensing or banning of drugs, are based on comparisons of groups done by elaborate mathematical procedures. These procedures are based on analyses designed to eliminate inadvertent bias (smokers might also be heavy drinkers; a group of aspirin users might on average be considerably fatter than the non-aspirin users to whom they are compared; vegetarians may differ in lifestyle from non-vegetarians in more ways than in diet). The non-statistician needs to know how reliable others judge these statistics to be, and what the implications are, not the mathematics of how it is done.

Scientists use technical terms such as magma because they need other scientists to understand exactly what they mean. This is an important distinction from casual speech (though not from careful writing in any discipline). Listen to how many times "y'know" as in "It's like y'know, cool, man" translates to, "I haven't explained this coherently. I hope that you can fill in the gaps." This is not to say that common language is wrong or is not appropriate; it just does not have a place in scientific communication. One summer I worked in a factory, and a fellow worker liked to engage me in conversation. Unfortunately most of his conversation consisted of one obscenity, used as a noun, verb, and adjective: "That bleepin' son of a bleep of a bleepin' son of a bleep bleeped me!" My participation in the first part of most conversations usually consisted of non-committal responses as I trolled (in frustration) for the meaning of what he said. Was he talking about our boss? Politicians? His friends around home? His wife? Had someone insulted him? Short-changed him?

The vocabulary does not need to be daunting. Scientists use complex vocabulary partly because sometimes what the words describe have no counterparts in common language—no Biblical or other early writer truly imagined a molecule structured like DNA—but mostly because of a need for precision. Scientific language strives for a precision that assures that any worker throughout the world, on seeing a specific word, will have the same mental image. This is very different, and sometimes much drier, than common or poetic language. A poet may describe a lovely woman as having diaphanous skin and hair like gossamer, but the beauty of the poetry is that these phrases conjure an image rather than paint a picture; the language evokes an image unique to each reader, based on that reader's experiences and desires. Each reader will imagine a different woman and different circumstances, collecting impressions from his or her experience, and hopefully each reader will generate a different very personal but equally compelling and pleasurable image. Poetry frequently loses its value as it becomes more specific, as a film based on a very romantic novel may prove disappointing if the hero or heroine in the film is very different from the person one imagined. This is nothing like a police report, giving height, weight, hair shape, length, and color, age, skin color, shape of eyes, nose, lips, etc. …not very exciting, but everyone will have same image. Again: Which of the following passages better *evokes* autumn? Alternatively, if you had never heard of the word "autumn" (for instance, if you spoke Tibetan and were learning English) which would give you a better and more precise idea of the term?

"SEASON of mists and mellow fruitfulness!
Close bosom-friend of the maturing sun;
Conspiring with him how to load and bless
With fruit the vines that round the thatch-eaves run;
To bend with apples the moss'd cottage-trees,
And fill all fruit with ripeness to the core;
To swell the gourd, and plump the hazel shells
With a sweet kernel; to set budding more,
And still more, later flowers for the bees,
Until they think warm days will never cease,
For Summer has o'er-brimm'd their clammy cells.

Keats, Ode to Autumn
Compare Keats' poem to this description of autumn:

"The season starting at the fall equinox (normally September 21 in the Northern Hemisphere, March 21 in the Southern Hemisphere) and ending with the winter solstice (December 21 and June 21, respectively). In popular use, the dates are often constrained by holidays, as in the U.S., between Labor Day and either Halloween or Thanksgiving; or are defined by climate, as in northern North America many people consider that autumn ends with the first killing frost or the first snowfall."

A scientific report is far more similar to a police report than to poetry—the goal is that everyone have as close to the same image as possible.

Common spoken English does not have this requirement. When an Englishman refers to a robin or to robin redbreast, he is describing a very different bird from the thrush that Americans call a robin (because the first English in the new world thought that the bird was the same). To prevent confusion, scientists would use a 300 year old tradition, from a time in which all educated persons spoke Latin, and would refer to the European bird by the Latin name of *Erithacus rubecula* and the American bird as *Turdus migratorius*. (The two-name system functions like the first or given name and last or family name system by which people in western societies are known. In the case of Latin names for animals, the capitalized first word is the equivalent of the family name. For the American bird, the name simply translates to "the thrush that migrates". We will discuss the definition of a species and the terminology in Chapter 11, begining on page 157.) Likewise we know a turkey by the name of a country because of confusion with a large bird from that country. If one asks for "regular" coffee, in some parts of the United States one will get black coffee and in others coffee with milk. We also use several words to describe the same thing: a long sandwich with several types of meat, cheese, and lettuce may be called a submarine sandwich, a hero (sometimes even jiro), or a hoagie, depending on the region of the country.

As a more specific illustration of the point, let's look at the word "significant," which has several meanings. One, its original meaning, was "giving a sign," as in "To the Greeks, it was significant that the general saw a meteorite the night before the big battle". Another common meaning is "important," "large," or "considerable," as "the loss was not significant", and there are several variants of these, as in "significant other," referring to a person with whom one is romantically involved. In biomedical sciences, the word has *only* a statistical sense: A difference between two groups that would occur so rarely by chance alone that the difference most likely supports the hypothesis of a relationship. For instance, if one measured lung cancers among 100 gum chewers and 100 non-chewers, and found 2 cancers in the first group and 3 in the second, the chances are that a repeat of the same assessment would the next time find 3 cancers in the first group and 2 in the second. There was no real difference, only a minor one dependent on chance. On the other hand, if one compared lung cancers among smokers, and found 10 cancers among the smokers and 1 among the non-smokers, the chances are that a repeat of the assessment would find a similar difference the next time, supporting the hypothesis that smoking can cause lung cancer. Statisticians can mathematically determine the probability that the results would be repeated, and biomedical scientists would call the difference between smokers and non-smokers significant. This is the only sense in which the word would be used by a scientist. In a scientific paper, "significant" NEVER means "important" or "meaningful". We will explore the precise meaning of the word "significant" in page 126.

SCIENTIFIC THEORIES AND HYPOTHESES

In common conversation, a theory is a guess as to how something works: "My theory is that the thermostat turns on the pump that circulates the water." To a scientist, a theory is not a guess but a hypothesis—that is, a logical inference as to how something works, or about the relationship of two phenomena—that has been tested many times, and each time supported by the test. When scientists review applications by other scientists for the support of their research, they often ask, "Is the work hypothesis-driven?" meaning, "Has this scientist created a model, based on preliminary evidence, as to how this works?" When many scientists have done this, and attempted many times to disprove their argument (falsification), and all the scientists come to the same conclusion, the hypothesis earns the title of theory. In general, calling something a theory means that it is logical, (logic); in many situations it is a plausible explanation of the relationship of two phenomena (evidence); and that many attempts to disprove it have failed (falsification). Thus other scientists can with some confidence consider the hypothesis sufficiently valid to base further, extrapolated, work on the assumption that the hypothesis is true. This is as close as we get to a higher level of certainty, a law. For a law, for instance, the law of gravity, we are sufficiently confident that all bodies produce and respond to gravity that we can base everything from planning the orbits of space ships to calculating tides to building very exotic medical and analytical machinery to aspects of atomic physics on the assumption that the "law" of gravity will apply, and we would be genuinely astonished if it did not. Although the terminology is a bit fuzzy at the borders, we do not have quite this level of confidence in a theory. We are only quite certain that a theory is true. A theory, and even a law, can potentially always be disproved, if an experiment or an observation can contradict it and no reasonable explanation can place the result into a category of interesting but comprehensible exceptions. The essence of science is testability, and thus everything is tentative pending the next experiment. It is quite humbling, and it is a source of considerable friction between scientists and public understanding. To a scientist, "the theory of evolution" means that the idea is well thought out, based on lots of evidence, and not disproved by any of myriad experiments—but there is always the outside chance that something that we have not imagined may someday disprove it, in the sense that we cannot predict that a lake will not suddenly appear in the middle of Arizona. To a scientist, the "theory of evolution" does NOT mean "a rather casual guess by a bunch of people who have not thought of other possibilities". And in any case, science addresses only the mechanics of how things work (which therefore can be tested) and never addresses the untestable.

ABSTRACT CONCEPTS

Finally, science demands abstractions, because in most instances the subject of the science is something that is not part of common experience. For instance, we cannot see a molecule. We can create an image of it, using specialized technology such as electron microscopy or atomic force microscopy, and we can view the image, or we can use various complex machines to detect the presence of molecules and determine their properties. What scientists do is to use their training about what these machines do so that they can construct mental images of the molecules as if they were 1,000,000 times bigger. In brief, the ability to think abstractly is the ability to make the abstract concrete. Throughout this book, we will attempt to help you, the reader, imagine some of these abstract and seemingly difficult concepts.

SCIENTIFIC PRESENTATION: FIGURES, GRAPHS, AND TABLES

There you have it! Science, its logic, and its findings can be understood by all students. One task that you should undertake, however, is something that is often neglected but is important, and which will considerably simplify your effort: look at the tables and figures that appear in the subsequent chapters. To working scientists, figures are not sidebars or attempts to render the text more fun. Well designed figures summarize important points, indicate relationships, and suggest further expansion of an idea. A figure such as that in Fig 1.3 can contain the ideas and relationships that would take pages to explain, and if one can grasp how it does so, one has saved oneself all of this memorization. Note how long it takes to explain in words what is shown in the graph, and how much clearer the graph is than the verbal explanation.

This figure illustrates the cost of printing magazines. One can read it as follows: Before a single magazine is printed, there is a cost of approximately $20,000 (point A). (This cost presumably includes the cost of the conception and design of the magazine, the collection of articles and pictures, the machines, and the building in which the press is housed, as well as salaries and incidentals. For someone trying to handle the budget, it would be important to know this initial cost.) For the first 20,000 magazines published, the real cost per magazine is approximately 2 dollars per magazine (calculated from the slope of the line between points A and B; the initial cost at 0 magazines is $20,000, and the cost at 20,000 magazines is $60,000. $60,000-$20,000 = $40,000, which is divided by the 20,000 magazines produced. The effective cost for the first 20,000 magazines, including the initial cost, is $60,000/20,000 or $3/magazine. After the first 20,000 magazines have been printed, presumably some basic costs have been met, and the cost per magazine falls to $5 per magazine. (Between 20,000 magazines and 40,000 magazines (point C), the cost has risen from $60,000 to $75,000, or $15,000 for 20,000 magazines.) The effective cost per magazine is $75,000/40,000 or $1.875/magazine. Beyond 40,000 magazines, presumably all background costs have been satisfied, and the

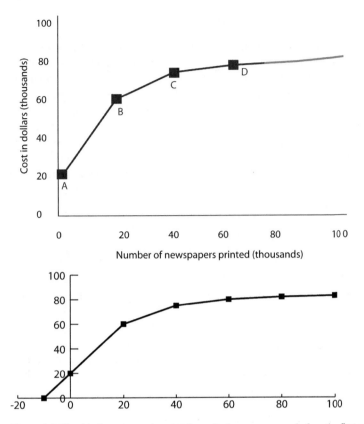

Figure 1.3. Graphical representation. As shown in the upper curve, before the first magazine is printed, $20,000 must be spent (point A: $20,000 for 0 magazines). This money represents paper stock, ink, staples, and costs for the building and personnel. If 20,000 copies are printed, the cost rises to $60,000 (point B), representing the baseline cost plus an incremental cost of $2.00/magazine ($60,000–$20,000 = $40,000/20,000 magazines). These costs presumably include the electricity, transport and delivery, and extra hours of labor. Between 20,000 and 40,000 copies (B to C), the incremental cost drops to $0.75/magazine ($15,000/20,000 copies), presumably because delivery costs do not increase much once the first shipment has been made. Between 40,000 and 60,000 magazines (C to D), the cost rises only $5,000, or $0.25 per magazine, for a total cost per magazine of $1.33/magazine ($80,000/60,000 magazines). Obviously there is higher profit in a larger production run if all the magazines can be sold. It is possible to extrapolate beyond point D (grey line) to see what higher numbers would per print run would cost, but extrapolating backwards to see what number of magazines would cost nothing, as shown in the lower curve (–10,000 copies), would be meaningless. Note how much simpler it is to read this information from the graph than to have to listen to an explanation

only remaining costs are supplies and salaries for the extra time, as the cost per magazine has fallen to approximately $0.25 per magazine between points C and D, for an effective cost of $80.000/60,000 or $1.33 per magazine at 60,000 magazines. One could even use the graph to predict the cost should the press decide to publish, for instance, 100,000 magazines.

CURIOSITY AND VALID SCIENTIFIC QUESTIONS

The last point to understand about how science works can be summarized in a single word: curiosity. All children are curious (just listen to conversations between 2- to 6-year old children and their parents) and some retain this curiosity throughout life, so that everything evokes a question: "How did this mountain get here? Why are male birds more brightly colored than female birds? How do insects survive freezing in the winter? Why do leaves turn color?" This curiosity can be summed up in an aphorism that is worth keeping in mind throughout this course: *"Phenomena are questions"*. In other words, there is a mechanism to explain how cells migrate to their proper places in an embryo, how the body fights an infection, how trees move water as high as 300 feet, how a bird or a whale finds its way half way around the world, or how the world is constituted such that flightless, ostrich-like birds are found in Australia, New Zealand, South America, and Africa, but not in Europe, Asia, or North America.

There are two related modifications to this last statement. The first is that science is about the mechanics of how things work. For this reason a question beginning "Why" is almost never a legitimate scientific question. Science is about the how, not the why, and a good question suggests a means of testing the how. It is rarely possible to test a "why". This is also why the scientific method presents far less confrontation with religion than many assume. A question beginning "Why", when it is not meaningless, is a religious question rather than a scientific one. For instance, "Why are rabbits brown?" may have a religious answer ("Because God made them brown so that they could hide"), but the scientific question could be any of several: "What is the selective advantage of brown color?" "What is the mechanism of inheritance of brown as opposed to other colors?" "What developmental mechanism arranges for pigment to appear on the back, but not the belly, of the rabbit?" "In what cells is the brown pigment?" (The cells are called melanocytes, or "black cells".) "How do the melanocytes carrying the brown pigment get into the skin?" "What is the biochemical pathway by which the pigment is synthesized?" "What is the biochemical structure of the pigment?" "How does a pigment molecule absorb light?" These questions can be carried deep into sub-atomic physics, and all are legitimate scientific questions, because, at least indirectly, they suggest possible mechanisms that can be tested. For this reason they differ from the non-scientific question "Why are rabbits brown?"

In summary, science is not an incomprehensible subject. The scientific method is an approach to understanding that is identical to the approach we use to understand any aspect of our lives. It differs only in that it has a series of specific codes and disciplines that allow a very structured means of asking questions, together with specific rules concerning what constitutes a meaningful answer. Understanding how these rules operate can demystify the world of science.

This book is primarily about the rules of science–how science works–as illustrated through examples of experiments, thought processes, and incidents in the lives of scientists, taking as a primary intellectual issue the development of a major theory. The subject of our inquiry and analysis will be evolution or, more properly, natural selection. The story of evolution encompasses three major steps that were accomplished in the mid 19th Century. The first was that thinkers had to conclude that the world was much older than the biblical approximation of 6000 years. Second, they had to accept the idea that species of plants and animals could evolve, or change with time (descent with modification). (This step also required a firm sense of what was meant by the term 'species', which itself depended considerably on new and confusing findings as Europeans explored the New World.) Finally, they then had to accept Darwin's contention that this descent with modification was directed by the non-random survival of certain favored individuals in an intense competition for food, protection, nesting resources, and mates (natural selection). As we shall see, none of these ideas was particularly new or original in the mid 19th Century, but it was the connecting of all of these ideas that revolutionized the world. It was known, for instance, that farmers could improve crops by using only the seeds of plants displaying the desired characteristics. Breeders of dogs were aware that numerous variations of dogs were produced by selective mating. Thus species could vary considerably. Within a few hundred years, one could breed dogs to produce dachshunds, great danes, and bulldogs. What was not obvious, however, was how it would be possible to generate all the varieties of plants and animals in the world in 6000 years. Furthermore, the exploration of the new world had produced new conundrums or puzzles, based on the realization that different continents contained different animals and plants—a finding not readily obvious from the story of Noah's Ark. We shall address each of these issues in turn. We will, however, branch into other, related subjects where appropriate. After all, all subjects are related in some sense: the exploration that led Europeans to reach the Americas would not have been possible without advances in astronomy and physics, and the history of 16th C Europe would have been very different without the struggles to acquire the riches of the Americas. Donne[1] would never have marveled at a woman, "Oh my America, my new found land!" It is sometimes very confusing, but ultimately exhilarating, to see these connections. Look for them. There are many rewards. First, you will be thinking like a scientist. Facts will become richer and more meaningful. Most of all, you will see that you will reduce the amount of tedious rote memorization you have to do because, once you see the connection, one fact necessarily leads to the next. For instance, Spain did not become a major power in Europe until it could draw on the resources of the New World. Once you know the date 1492, you can approximate the dates of the next events in Europe. No more getting the dates wrong by 200 years.

One last note: a well-developed, mature college-level vocabulary helps greatly to clarify issues. Therefore we will not attempt to "talk down" to you, the student, by using a less mature, less specific vocabulary. We will, however, in introducing a less common word attempt to explain it in passing. In doing so, we will use the trick of the King James Version of the Bible. The authors of that translation

---

[1] John Donne, To His Mistress Going to Bed

were faced with the problem that some of the English, mostly peasants, spoke Saxon, derived from the Germanic languages, while the upper class spoke French. They therefore repeated many terms, giving a Saxon and a French version of the same statement:

Gen.4
[1] And Adam knew **Eve** his wife; and she conceived (FRENCH), and bare (GERMANIC) Cain, and said, I have gotten a man from the LORD.
[14] Behold, thou hast driven me out this day from the face of the earth; and from thy face shall I be hid; and I shall be a fugitive (FRENCH) and a vagabond (GERMANIC, FROM LATIN) in the earth; and it shall come to pass, that

REFERENCES

http://www.journaloftheoretics.com/Editorials/Vol-1/el-3.htm

Brody, Baruch A., and Grandy, Richard E., *Readings in the Philosophy of Science,* 2nd edition, Prentice Hall, Englewood Cliffs, NJ, 1989

Latour, Bruno, *Science in Action, How to Follow Scientists and Engineers through Society*, Harvard University Press, Cambridge, MA, 1987.

Bacon, Francis *Novum Organum (The New Organon),* 1620. Bacon's work described many of the accepted principles, underscoring the importance of theory, empirical results, data gathering, experiment, and independent corroboration.

Holton, Gerald, *Thematic Origins of Scientific Thought, Kepler to Einstein*, 1st edition 1973, revised edition, Harvard University Press, Cambridge, MA, 1988.

Ziman, John (2000). *Real Science: what it is, and what it means.* Cambridge, UK: Cambridge University Press.

Popper, Karl R., *The Logic of Scientific Discovery*, 1934, 1959.

Kuhn, Thomas S., *The Structure of Scientific Revolutions*, University of Chicago Press, Chicago, IL, 1962. 2nd edition 1970. 3rd edition 1996.

STUDY QUESTIONS

1. Give an example from daily life and from a scientific/technical situation in which you can identify evidence, logic, and falsification. Clearly indicate which aspects are evidence, which logic, and which falsification.
2. From television or daily news, choose an example of a claim being made for a political or scientific issue and dissect the claim into evidence, logic, and falsification.
3. A medical report notes that there was a significant difference in survival between patients who walked at least one mile per day and patients who did not walk much. Explain what this statement means.
4. A weight-loss treatment is advertised using testimonials from satisfied customers. How would you evaluate the advertisement?
5. Why do we give Latin names to animals and plants?
6. Give examples of theories, laws, and hypotheses.
7. Choose a graph from a newspaper or news magazine and write a verbal description of the information found in the graph. Pay special attention to the

logical relationship between the data on the abscissa (horizontal or X axis) and the data on the ordinate (vertical or Y axis).

8. Describe three questions that you have asked at some point concerning how something works. What evidence would you need to answer your question?

9. What words did you not know when you read this chapter? What is the meaning of these words?

PART 2

**ORIGIN OF THE THEORY OF EVOLUTION:
TIME AND CHANGE**

CHAPTER 2

THE ORIGIN OF THE EARTH AND OF SPECIES OF ANIMALS AND PLANTS AS SEEN BEFORE THE ENLIGHTENMENT

WHERE DID I COME FROM? THE EARLIEST INTERPRETATIONS

All societies have faced the issue of "Where did I come from?" and have usually assigned a divine cause for creation. Few have pondered the issue more deeply, owing to two factors: First, for all societies, the world is tolerably constant. Second, in western society, the influence of Aristotle, Plato, and the Old Testament, which heavily relied on the assumption of constancy, mitigated against further exploration and analysis, even when logical contradictions were acknowledged.

As is discussed in Chapter 4, page 45, at a given time and place the world appears to be constant. One summer may be warmer or a winter colder than another, and there may be other modest changes in climate or the precise bed of a river, but by and large the old-timers can remember hotter summers or heavier snowfalls. The biological world also appears to be constant and discrete. To take an oversimplified but illustrative example, any reasonably observant person realizes that there are different kinds of birds in his or her neighborhood. In a northeastern urban or suburban neighborhood, for instance, there are pigeons, robins, cardinals, gull, sparrows, mockingbirds, crows, and Canada geese. There are also several others, such as owls and hawks, but they might not be commonly noticed by the casual observer. The point is that one does not mistake one species for another. Pigeons might have many colors, but they are certainly not robins. A female cardinal might be greenish-brown, rather than red, but her body shape, her crown, her beak shape, and her markings make her distinguishable from any other bird in the vicinity. We do not find birds that are half-way between a pigeon and a robin, or birds that we could not with little effort classify and identify. Even if a species becomes extinct, it is known in its last stages as a rare species and, unless one is specifically attempting to document its existence, its disappearance is simply perceived as a lack of a recent sighting until, in a few years, it is forgotten. With these observations, there is little reason for assuming that the world is not as it has always been, other than by the divine placement of humans into the scene. The concept of change does not become obvious until one has a long historical (written) record of the world. Furthermore this record must sufficiently preserve earlier writings and later generations must be able to read them, so that the differences between then and now become apparent.

Ancient Greece constructed a world image that did not depend on divine creation. Thus it was that earlier Greeks, Anaximander, Anaximenes, Empedocles, and Democritus, argued that humans arose from the earth or a primordial moist element, being engendered by the sun's warmth and spontaneously arising as maggots appeared to do in rotting flesh. In general, perhaps by noting the obvious biology and by understanding a hierarchical world in which animation (life) was superior to inanimation (rocks), movement (animals) superior to immobility (plants) and thinking (humans) superior to reactive behavior (animals) they perceived a creation in which plants preceded animals and animals preceded humans. They recognized but did not however address the logical problem of the state of the first human. If the first human appears as a baby, it must be cared for, but then who (or what) cares for it? With divine creation, it is possible to accept the idea of the first humans appearing on earth as adults, but for the Greeks this was a conundrum. As far as we know, they did not pursue this problem with great enthusiasm or to any depth, partly because of the rising influence of Plato and his student Aristotle. Plato felt that each object in the universe was an imperfect representation of an ideal type or archetype, and that the universe consisted of more- or less-successful approaches to that archetype. Note, however, the inference: if there are archetypes, then by definition the archetypes do not change. Therefore, in the biological world, species do not change. A robin is a better or worse approximation of the ideal of "robin-ness" but that ideal, or archetypical, robin persists and will remain in all generations as the goal of robins. Aristotle carried this idea further in attempting to systematize or classify all forms of animate and inanimate nature, in his Scale of Life (page 55). As with Plato, each species was an attempt to replicate an absolute ideal, but beyond that, the archetype of each species occupied a particular rank in nature. Thus animals were above plants, vertebrates were above invertebrates, birds and reptiles (which had perfect or shelled eggs) above fish (which had imperfect or soft eggs), mammals above reptiles and birds, and humans above mammals. He counted over 500 links in the chain, or species. His classifications improved on the earlier versions such as with or without feet or wings. All this is well and good, but it ultimately gets complicated, as the Aristotelian scale allowed no ambiguity or ties in rank. Thus, for instance, a peach tree had to be above or below a cherry tree, a trout had to be above or below a bass, a cat had to be above or below a dog, a sheep above or below a goat. Life could be created, but new organisms would join their appropriate rank. So, by this argument, not only was there no possibility of change of a single species, there was no possibility of movement among species. It was not possible for a goat to pass a sheep, or vice versa. In this world view, evolution is an absurdity. Each species is fixed in its type, and fixed in relationship to every other species. Together with the Judeo-Christian view of Creation, as expressed in Genesis, this view dominated western culture for two thousand years.

In Genesis, the world was created at one time. Thus all species were formed at that period, and by this argument again, there was no logical means by which species could change or evolve from one type into another. Fish, frogs, reptiles, birds, and mammals first appeared during creation and have been present on earth since then.

Thus, between the teachings of Genesis (written approximately 450 B.C. recounting tales of 1000 years earlier) and the teachings of Plato (427–c347 B.C.) and Aristotle (384–322 B.C.) from a logical standpoint as well as from the evidence at hand, species were fixed and there was little reason to worry about, or even concern oneself with, the relationship of one organism to another. The similarity of monkeys to humans had been noticed, as had the similarity of organs and bones among different vertebrates, but these resemblances were considered to be examples of God's choice or God's wisdom, rather than peculiarities of the world that deserved attention and analysis.

WHERE DID I COME FROM? INTIMATIONS THAT NOT ALL WAS STABLE

By the 17th C, however, the European world had changed. The world now had a long tradition of literacy, coupled with printing presses that made knowledge accessible to a much larger population, and explorers were describing the strangeness of the new continents that they were exploring. Philosophers, who at that time were not distinguished from scientists, were pondering the meaning of all of the new knowledge, and the evidence that life in 17th C Europe was very different from what had been described in the Bible and in Aristotle. They were susceptible to the concept of change. In terms of social structure, economic structure, political order, and even values and mores, the world today (17th C) is different from what it once was. There have been periods of wealth and poverty, pestilence and health, democracy and tyranny; and what had been a rural society (Germany, England, Ireland) became a society with great cities. Islam had appeared in the 8th C and grew strong enough to compete with Christianity, and the religions of the Orient and of the New World were very different.

Thus the world could be restructured, perhaps not in front of one's eyes, but over time. Where did it come from? What caused the restructuring? Was it possible that the natural world could change as well? Perhaps the similarity of the bones of a dog to those of a human told us more than we had suspected.

It may strike many as surprising, but many of the main elements of the story of evolution were well known long before 1859 and were the subject of popular discussion among intelligent and educated, but not professional, members of upper-class society. The Enlightenment had not truly invented but had brought to the forefront of intellectual life several attitudes that continue to pervade our society: an emphasis on material evidence and human logic, as opposed to mysticism or unquestioning faith, as the basis of rationality (Galileo & Copernicus); a powerful sense of the mechanical or physical construction of the universe (Galileo, Newton, Pascal); and a widespread but quintessentially British assumption of continuous progress in the history of the earth, leading of course to the summum bonum (maximum good) exemplified by contemporary British society. Each of the episodes that we now identify as landmarks in the history of science originated in the attempt to address a specific practical problem, and each had generated spectacular and immediate success. These successes validated the assumption that much could be learned from the physical world, and led to further inquiries about the anomalies of the earth, ranging from curiosity about the origin of mountains to efforts to understand fossils in the context of or opposed to the biblical description of the history of the earth. We will discuss these below, but to give a sense of the pragmatism that allowed natural philosophers to gain ascendancy over theologians and philosophers, we can cite a few examples: One was that motion was associated with life, and thus its laws were worthy of exploration. Furthermore, issues such as the trajectory of cannonballs and, for the purposes of armies and explorers, measuring movement around the earth provided plenty of work for those who would ultimately become physicists. Galileo was well known for his studies of trajectories and compasses. Others needs included the measurement of longitude and interpretation of disease, as is described below.

THE USES OF SCIENCE AND THE DISCOVERY OF THE MECHANICS OF THE EARTH

Both philosophical inclination and practical considerations drove a 17th C interest in movement. From the philosophical viewpoint, movement, or at least directed movement, was one of the few features that separated the living from the inanimate. Thus the difference between a dog, horse, or human one minute after death and one minute before death was manifest primarily in movement, of the chest, heart, limbs, or eyes. Thus, as the value of mechanics impressed itself more on European society (see below) attention turned to an understanding of motion as part of the deeply philosophical and even holy quest to answer the age-old question, "What is life?" Rather than address this question from purely theoretical or philosophical terms, thinkers turned to mechanics, or experimentalists, to help them understand. There was plenty of reason to view this approach with optimism.

Greek scholarship had returned to Europe via Spain, since the Islamic unlike the Christian world had never lost it and the Moslems, though eager to keep their distance from "heathens," nevertheless would communicate through intermediaries, frequently Jews. It was no accident that Maimonides, the greatest of the Jewish philosophers, and certainly a great physician and philosopher by any criteria, had a strongly Aristotelian attitude, including the argument that God's miracles worked through, and did not violate, physical laws. Likewise, Nostradamus' writings arose from an effort to reconcile a profound logic with apparent contradictions in holy writings—the effort that gave rise to the Kabala (Page 403). Thus the role of (perhaps) Archimedes (287–212 B.C.) in devising catapults and other instruments of war based on the theoretical understanding of the physics of levers, and the practical benefit of his correlating density of matter with displacement of water, so that he could tell whether gold had been removed and substituted in the king's crown, were familiar to scholars. The question became whether such approaches could contribute to various practical problems, ranging from the construction of machines to accomplish the heavy labor of building large buildings or destroying fortress walls in battle to the prediction of seasons and correct assessment of Church holidays. There was even, for the heads of state, a very urgent and large issue. By the 17th C it was very apparent that there was much wealth and resources to be gained (or plundered) from the New World and the now-accessible Asia, and that the power of a country would depend on its ability to assert pre-eminence in that exchange. After all, small and relatively weak countries such as Spain, Portugal, and Holland, were achieving considerable influence at the expense of previously much more powerful England, France, and Italy. However, trans-oceanic voyages were still hazardous and unreliable. A prince might well, at great expense, outfit a fleet to barter, plunder, or otherwise collect the wealth of another land, but if the fleet went down in a storm, was lost in a raid, or otherwise foundered, the entire investment would be lost. Such a catastrophe was far more likely if the fleet wandered off course. Wandering off course was highly likely since ship captains knew how to calculate latitude (distance north or south of the equator) by the height of the sun at noon—again a practical result of the mechanistic approach to the philosophy of what the world was—but they could not calculate longitude, the distance east or west of their home base. This lack of information would have been an inconvenience if the captains had been able simply to chart their own course, for instance in the return trip simply sailing to the appropriate latitude and then sailing due east, but they were sailing ships, and they therefore followed the prevailing winds, which flowed basically westward near the equator and eastward far from the equator, with an area of relatively little movement (the doldrums) in between. Thus they had to be able to assess their positions accurately, lest, in the worst type of scenario, a gold-laden Spanish ship returning from Mexico would find itself, while still at the latitude of England, approaching European shores. The hostile English might well capture the ship. There was obviously a premium on the ability of the fleet to home right into its base port. There was so much of a premium, in fact, that the king of Spain offered serious prize money to the person or persons who could devise an accurate means of calculating longitude. This competition motivated some of the best scholars of the time, including Galileo who, using the newly-invented dual or compound arrangement of lenses to devise a telescope, searched the skies for markers that could be used to assess longitude. He was so assiduous in his search that he even devised a means of determining longitude by noting the positions of the moons of Jupiter! This exploration, of course, led him also to realize that the moon was not a perfect sphere or component of one of the "cool, crystálline spheres" praised by John Donne, and to realize that he could even calculate the heights of the mountains on the moon. This interest in motion and in celestial mechanics even allowed him to address the issue of the motion of tides. Even though his interpretation is now considered to be incorrect, his rules for calculating or predicting the tides were of obvious value in an era in which merchant and military ships were becoming larger and bulkier, riding much lower in the water when fully loaded, but deep water harbors were not yet being dredged.

Science as a Means of Solving Problems

Another hugely important source of wealth was minerals. Coal was known and sought, though not to the extent that it would be later, but a range of other minerals, ranging from marble for construction to iron for tools and weapons to gold, silver, and gemstones for the holding and display of wealth, were high demand. Recall the myths of the desire of Midas, prior to the Christian era, for gold and the opulent garments, sewn with gold thread and encrusted with precious stones, of royalty. The function of these garments, heavy nearly to the point of immobility, was to impress upon others, even flaunt, the wealth and by inference the power of the wearer. The Spanish Conquistadors sought gold with febrile intensity, even melting down wonderful artifacts to ingots and so overloading their ships that they sank in the Caribbean. Needless to say, in this type of atmosphere, the discovery of gemstones and precious metals was too important to be left to chance. Thus there was substantial interest in understanding the characteristics of the earth so that the locations could be predicted and mines dug if necessary. Analytical observers such as Nicholaus Steno, described immediately below (page 27) and pages 40–42 in terms of his contribution to understanding the age of the earth, were sought by the courts of Europe. (It is of note that even stone-age humans had been known to dig ten feet into the ground to find flint, indicating that they understood the structure of the land, and that the Chinese by the 8th C and Avicenna and al-Biruni by the 10th C clearly described sedimentation and the meaning of fossils; and Native American legends likewise gave some suggestion of the massive sedimentation fields of central North America. All of this understanding was unknown to, lost to, or suppressed by, European scholars and theologians.)

Likewise, in medicine, new concerns demanded greater attention to the details and practical aspects of life. Plague had entered Europe in the 13th C and was still a feared disease, clearly related to urban life but of unknown origin (Chapter 27, page 359); kings and queens, entrusted with (and depending on) the welfare of their subjects, needed to understand and control it. Malaria, attributed to the fumes around Venice (literally, "malaria" means "bad air") could incapacitate even a rich, powerful, and elegant city; an expensive, vast, and well-trained army could be defeated as easily by disease as by its enemy; and, of course, in addition to individual self-interest, there was considerable motivation in terms of inheritance and control of the government in protecting (or discretely terminating) the life of the regent or his or her potential successors. Thus courts had their royal physicians, often as distinguished as Maimonides in the 13th C or William Harvey in the 17th C. Ambrose Paré in the 16th C had improved the handling of war wounds by tying off wounds rather than cauterizing them, and using what was later learned to be moderately antiseptic solutions to wash the wounds. Paracelsus, the great (and arrogant—he gave himself the name 'Paracelsus,' meaning "beside Celsus" (a famous Roman physician)) physician, recognized that diseases were carried by and caused by outside agents (confirmed by Pasteur's germ theory in 1862), and advocated the observational and experimental approach as opposed to following ancient texts. He literally threw the works of the revered Avicenna into bonfires.

Likewise Andreas Vesalius scorned a slavish following of ancient texts, leading the world to a new understanding of anatomy and the function of the body (page 406), in the same year (1543) that Copernicus rejected elaborate mathematical models of the universe in favor of simple calculations based on the idea that the sun, not the earth, was the center of rotation. By the 17th C the experimentalists, like their counterparts in physics and geology, were in charge: Francisco Redi had established that maggots on exposed meat had come from the eggs of flies (page 141), leading Harvey to the conclusion "Ex ova omnia" ("All [life] from eggs"). Harvey also showed in 1628 that the heart circulated blood in the body, leading to better insight into the importance of dehydration and bleeding. There were many other scientific activities at the time, including of course the work of Sir Isaac Newton on optics, gravity, and the laws of motion.

Europe meets the Americas

This increased respect for, and interest in, the mechanics and the tinkerers, led to the subjects of the nascent scientific research becoming a matter of interest for all educated citizenry. In fact, since all exploration costs money; money was available only in the noble and mercantile classes; and merchants were, by and large, too busy trying to earn the money to be very philosophical, scientific exploration was to a large extent a hobby or amateur (literally, lover) occupation of the more relaxed (idle?) nobility. As such, these activities were widely discussed in the upper classes. Curious findings (in the broader, original sense, meaning unusual enough to provoke wonder about their meaning or origin) were considered, marveled upon, and discussed. In the age of exploration, there were many curious findings. New animals and plants, and reports of wonders, were being brought from abroad, and explorations of the geology of Europe were forcing people to ask questions about their meaning. The level of excitement over new wonders can be appreciated in a few anecdotes: chocolate, brought from Mexico, was presumed to be a powerful aphrodisiac, and therefore sequestered to nobility; tulips, brought from Turkey, were considered so precious that there was a tulip frenzy, with rare bulbs being sold, in a stock market-like structure, at today's equivalent of hundreds of dollars per bulb; and newly-met indigenous peoples were routinely interpreted as being descendents of one of the lost tribes of Israel. As is described in Chapter 7, page 81, the realization that the rest of the world contained novel species initiated the query of how all of this fit in with the story of Genesis, but at home the new-found interest in the structure of the land meant that, instead of simply accepting phenomena, the mechanism-based scientists began to ask how the phenomena came to be. In common terms, the transition was from "Yes, those hills have funny [or pretty] stripes" to "What made those stripes in those hills, and why do they look like the stripes on the hills on the other side of the valley?" This was the basis of Steno's identification of the principles of geology, but in terms of evolution the argument is much more cogent: "I know that I can get limestone for making my mortar from the white areas of the earth, but those white areas are white because they are filled with old shells. The shells look a bit like the ones on the beach, but they are not the same and, besides, they are on the top of a mountain. What is going on?"

The Discovery of Anatomy Raises New Questions

Georges Cuvier, the director of the Musée d'Histoire Naturelle, was a master anatomist. As is described in Chapter 3, page 35, there are specific correlations among organs and structures such that it is possible to assess the lifestyle of an animal from its general appearance. Any person, and indeed any animal, can distinguish between a dangerous carnivore such as a shark or a lion and a peaceful herbivore like a zebra or a goose. Our films and our creative fiction exploit this ability, showing dangerous fictitious predators such as werewolves, zombies, and aliens from outer space with the appropriate paraphernalia of a predator: large, sharp, tearing teeth like canines, strong arms with claws or other lethal cutting weapons, and forward-facing, distance-judging eyes. A hypothetical science fiction movie showing people terrified by an invasion of cows or guinea pigs would be laughed out of a theater. Cuvier was one of the men who verbalized these intuitive judgments, but he went much further. To Cuvier, each part of the anatomy necessarily related to every other part, in the sense that, if one takes a femur (upper leg bone) from an unknown animal, the shape of the joints indicate how it attached to the pelvis and the tibia (lower leg bone) and from this one can determine if the animal was truly quadruped (four-footed) or walked upright. In fact, it was said of Cuvier, and he did not deny it, that he could reconstruct an entire animal from a single bone. For this talent, he was justly famous and, in the structure of society at the time, he and his colleague and to some extent mentor Geoffroy Saint-Hilaire associated with and were admired by such prominent literary figures as Wolfgang Goethe, Etienne Balzac, and Georges Sand. (Goethe was also an outstanding botanist, having recognized that flowers and other appendages of plants were modified leaves.) In the midst of the French revolution of 1830, Goethe was far more excited by the prospect of a debate between Cuvier and Geoffroy than by news of the war. Balzac was sufficiently interested in the debate to describe it in his introduction to *The Divine Comedy*.

The social structure however is another story, and told at greater length and in more detail elsewhere (see bibliography for books by S. J. Gould). Of interest here is what Cuvier learned from his skills and knowledge. First, he realized that the fossils in his museum and being collected at an increasing pace represented real animals, and ones that he could classify and for which he could describe lifestyles. Second, as Geoffroy would summarize in an aphorism ("There is only one animal."), all the tetrapod (four-legged) vertebrates had essentially the same bones in their limbs, whether the limbs served for swimming (whales), flying (birds or bats), walking (dogs), digging (moles) or carrying (humans). Third, many of the animals represented by the fossils were unlike anything seen on earth. Fourth, understanding Steno's principles of stratigraphy pages 40–42, the ones most like today's creatures were closest to recent times, and they never appeared in the earlier layers. Fifth, many species had finally disappeared. Geoffroy had studied the anatomy of different organisms to the extent of trying to identify in fish the homologs (parts related by ancestry) of the bones of the inner ear of mammals. To the logical and analytical Cuvier, the data had only one interpretation: the stratification of fossils told the history of animal and plant life. The creatures found on this earth had changed over time, with some types of animals completely disappearing from the record. The similarity of bones betokened a common ancestry. He stated this argument clearly in his first major book on the subject, in 1812, *Research on the fossil bones of quadrupeds, from which one reestablishes the characteristics of several species of animals that the upheavals of the earth*[2] *appear to have destroyed.* Could one state more clearly the concept of extinction and possibly evolution? Why then do we mark 1859, the year of the publication of "*Origin of the Species*," as a turning point, rather than 1812?

Cuvier saw what had happened, but he lacked two crucial points: First, he understood sequence, but he had no conception of the time that it took. In other words, if you live in a big or industrial city, you are familiar with the fact that every day a little bit of soot accumulates. You can imagine that, over the space of 1000 years, on an undisturbed space a few inches will accumulate—let's say, five inches. If you now find a soot layer four feet deep, you might reasonably conclude that the soot had been accumulating for approximately 10,000 years. However, suppose that there is a volcano not too far away. A single eruption of a volcano might produce a foot of ashfall, or two feet of ash, or four feet of ash. You surely can establish the sequence of the accumulation, but without sophisticated modern technology, can you unequivocally argue that it represents 10,000 years of accumulation, as opposed to a single day of volcanic eruption, or anything in between?

The second problem that he had was the inability, because of lack of this sense of time as well as the social context that led to his asking the specific questions, to conceptualize a new, grand theory of mechanism. In the world of Cuvier and Saint-Hilaire, the issue was much more how the fact that vertebrate bones were homologous would demonstrate the wisdom and beneficence of God. What was the genius of using the same basic plan for all vertebrates? There was surely method, but what advantage did it bring? The great debate of 1830, fervently followed by the intellectual community and continued with follow-up books and pamphlets, was not over the issue of evolution, but whether God's plan ordained specific types of creatures, each containing a modest variation on a theoretical ideal type (Geoffroy) or whether God's wisdom was displayed in the excellent fit that He had constructed from a basic sketch to serve each animal's unique needs (to swim, run, fly, walk, or dig—Cuvier). What Darwin brought to the picture was the certainty that the fossil record was a true representation of a sequence of historical events; that the species had changed rather than been replaced; that the earth was old enough to account for

---

[2] The French title that I have translated as "upheavals of the earth" is "les révolutions du globe," literally "revolutions of the globe" but the term "révolution" is more similar to the meaning "revolt" or "American Revolution" than to the concept of turning in a circle.

these changes (this information was inaccessible to Cuvier but was widely believed forty-some years later); and, above all, a MECHANISM by which it could have occurred. The mechanism, the Logic of the ELF triumvirate, was obligatory for a theory of evolution. The function of the preceding discussion is therefore to argue that the evidence of the fossil record had been available, and that its implication—the true existence of antecedent animals, and their successive replacements over time—was well accepted. Furthermore, there was extensive knowledge of the anatomy of common and exotic animals, and their relationships were puzzled over, from the obvious homologies of the bones even to bewilderment over the existence of vestigial and completely useless pelvic bones in walruses and some whales. They worried about such issues as, if the failure of the skull bones to fuse before birth in mammals is Divine provision to allow the head to be smaller and to mold during birth, thus demanding less distention of the birth canal, why were the skull bones of birds not fused before hatching? All the birds had to do was to break the shell, not push through the narrow pelvis of the mother. These issues were being hotly debated in England as well, most notably by Richard Owen, "the British Cuvier," who likewise was deeply concerned by the similarity of bones in limbs of such different functions. As he wrote in 1848, "The recognition of an ideal Exemplar for the vertebrate animals proves that the knowledge of such a being as man must have existed before man appeared. For the Divine mind which planned the Archetype also foreknew all its modifications."

Embryology was also appearing on the scene, as microscopes and techniques improved to allow the first embryologists to preserve, dissect, and observe the typically tiny, watery, and mushy early embryos of animals. What Ernst von Baer observed and correctly interpreted by 1828 was quite startling: embryonic humans had tails like other mammals, and all vertebrate embryos had gills. Human tails disappeared by failing to grow at the same rate as the rest of the embryo, ultimately being seen as the internal curved end of the spine, the coccyx. In land animals, the gills ultimately ended up as (morphed into) structures of the throat. If he had not traced their development, he would have never recognized the relationship in the adult. In any event, to von Baer it was clear that the embryo of a human contained also the embryonic stages of aquatic and tailed creatures. He considered that they were there by inheritance, but did not extend the argument. Once the story of evolution had broken, Ernst Haeckel made the connection with his famous aphorism, "Ontogeny recapitulates phylogeny," meaning that the developmental stages indicate the evolutionary line of descent.

What do the Relationships Mean?

Jean-Baptiste Lamarck is today somewhat unfairly ridiculed for one of his extrapolations of his findings, but at the beginning of the 19th C his careful observations and interpretations contributed another step on the ladder to the story of natural selection. What Lamarck saw was the marvelous fit of form to function, such that wings of birds allowed them to fly while the limbs and overall shape of porpoises were well adapted for swimming. Giraffes had long necks to feed on tall acacia trees, and ducks had webbed feet to allow them to swim. The perfection of these matches, according to Lamarck, could only be explained by (God's generosity in arranging) the adaptation of animals to their needs. Taking his cue from the obvious adaptation of individuals to changing circumstances—muscles grow in individuals who do hard physical labor, and atrophy in immobilized limbs, and plants send leaves toward the light and roots to the soil—he proposed that the adaptations of animals to their surroundings was a direct growth or other response to their situation. Furthermore, he studied fossil mollusks, which are shells often with a long and continuous history. He saw, in the series that he studied, substantial evidence for a gradual change in form and size from the archaic to the modern forms. From what he knew and saw, he proposed that animals adapted to their environments and that the adaptations would be inherited. In this latter point he was wrong, as he had no idea that the cells of inheritance, the germ cells, which produce the gametes (eggs and sperm) are independent of the body cells (somatic cells) and cannot pick up what we today call acquired characteristics. This distinction was discovered only in 1888 by August Weissmann (page 178), in direct test of Lamarck's theory, and even Darwin assumed that the body's characteristics drained into the gametes. However, the fundamental observation that species changed over time was provocative. It challenged Linnaeus' assumption that the species were fixed, instigating a controversy and opening the speculation as to exactly what would have been taken onto Noah's Ark. What was important to this story is that he put onto the table for all, including Darwin, to see the evidence that species were not fixed. He did not believe in extinction, which substantially undercut his argument. Although many argued vehemently with his theory, emphasizing such evidence of imperfect adaptation as vestigial organs and the massive teeth of sabertooth tigers, the evidence of the gradual change of at least the molluscan species was not denied. Cuvier later demonstrated that many fossils represented creatures no longer found on earth. As Pietro Corsi notes, Lamarck's was "the first major evolutionist synthesis in modern biology" (quoted in Browne).

THE SEARCH FOR MEANING AND THE DISCOVERY OF TIME

The other major limitation to a theory of evolution was time. To anatomists and interpreters of fossils such as Cuvier, the biblical accounting, as interpreted by Ussher and others, was dubious, but they had no measuring rod against which to judge the scale of events. This measuring rod, if not precisely constructed, was at least given a meaningful existence by Charles Lyell. Lyell, who combined a scientist's precision and attentiveness to detail with a persuasiveness derived from his career as a lawyer, had set out to deny the theory of catastrophism, the theory that all events and changes on earth had resulted from (bible-described) catastrophes and cataclysmic events. He argued that the great changes now recognized on earth could result from gradual changes over great periods of time. For instance, one might encounter massively folded sedimentary rocks (see Chapter 2, page 27) overlying or underlying horizontal layers. (Fig 2.1). According to Steno's rules (page 41),

Figure 2.1. Upper. Strong uplift of originally sedimentary rock. On these mountains in Alaska, the originally horizontal surfaces, distinguished as individual jagged edges, have been lifted to nearly vertical. Nevertheless, the original plane of the land can be distinguished as described in the text. In the middle of the photograph is the origin of a glacier, which gives evidence of flow. See Chapter 6. Lower. Strongly folded sedimentary rock. Folds like this indicate considerable activity and plasticity of the earth. Credits: Photograph:–Phil Stoffer, U.S. Geological Survey http://3dparks.wr.usgs.gov/goldengate2/large/ribbonchert.html

it was no longer seriously argued that the layering was not due to sedimentation, but the catastrophists argued that the sharp discontinuities indicated catastrophic events. Lyell argued, quite reasonably, that sedimentation occurred only in a time of heavy water flow. If a stream meandered quietly to the sea, it would not have much sediment to deposit in the sea. Similarly, if the water level changed, then sediment would accumulate only where the water met the sea. If the sea level dropped, the area of sediment accumulation would move farther out. If the beach front eroded, the area of sediment accumulation would move farther inland. Therefore, a line of sharp discontinuity could reflect a period during which sediment was not accumulating, suggesting a very slow process rather than a sudden one. Although Cuvier had attributed the changes in the fossil records to catastrophes, in reality all the major changes in the land could be produced by gradual processes such as those noted in current times, but these would require vast amounts of time, surely orders of magnitude greater than the biblical record.

Of course there were many other intellectual currents. Lord Kelvin's measurements of temperature (Chapter 8), Linnaeus' efforts to classify all organisms (Chapter 5) and the observations and theories of social scientists such as Malthus and Adam Smith (Chapters 7 and 10) all were part of the intellectual ferment of the 19th C and will be discussed in relation to the topics that they influenced.

REFERENCES

Browne, J, 2002, Charles Darwin, The Power of Place, Alfred A. Knopf, New York, NY.
Gould, Stephen Jay, 2002, The structure of evolutionary theory. Harvard University Press, Cambridge MA.
Sobel, Dava, 1999, Galileo's daughter, Penguin Books, New York.
http://www.victorianweb.org/science/cuvier.html (Georges Cuvier (1769–1832).

STUDY QUESTIONS

1. Making your best judgments as to how life is organized, build your own "Scale of Life". Explain the criteria by which you make the judgment, and compare your scale to those of Aristotle and Linnaeus.
2. What are the criteria by which modern "Scales of Life" are built?
3. Is it fair to call today's groupings of animals and plants a "Scale of Life"? Why or why not?
4. Look around your environment and note any evidence that the physical world is stable, has changed, or is changing. If you feel that it is changing, estimate how rapidly it is changing. Be prepared to defend your arguments in class.
5. Assume that you are talking with someone who has never left the region and has little knowledge of the geography, biology, or history of the rest of the world. How would you convince him or her that species can vary?
6. What hypotheses can you generate to explain the differences of animals and plants among the five continents?

7. In single sentences, describe the major contributions of at least five of the historical figures mentioned in this chapter.

8. To what extent was the concept of evolution prior to Darwin hindered by the failure of ELF logic?

9. Would it have been possible to develop the theory of evolution without exploring the world? Why or why not?

10. Argue for or against the proposition that the person who contributed most importantly to the development of the theory of evolution was Thomas Malthus. (The subject is discussed in Chapter 10. Considering the question at this point will help you to understand the issues.)

CHAPTER 3

THE SEASHELLS ON THE MOUNTAINTOP

INTERPRETATIONS OF MARINE AND OTHER FOSSILS. HOW DO YOU TELL WHAT A FOSSIL IS OR DID?

Most of us, meeting an unknown animal for the first time, would be able to make some assessment of how threatening it might be. Whether we realized it or not, we would note its teeth, its claws, and the strength of its legs to judge if it was a carnivore and how fast it might move. From experience, if not from direct knowledge, we would identify carnivores by their forward-facing eyes and binocular vision, allowing them to make good judgments of distance, and herbivores by their side-facing eyes and good peripheral vision, allowing them to observe predators approaching from above or behind (Fig. 3.1).

As we will discuss further in Chapter 4, there is much more that a trained biologist can read from the appearance of animals or plants. For instance, a flying animal must be light-weight, meaning that its bones must be small or hollow; its weight must be balanced for flight; and it must have strong attachments for its flight muscles (the keel or sternum of a flying bird—Fig. 3.2) Flying animals need to take advantage of the lift provided by thermal currents, so that soaring creatures, be they reptiles or birds, have similar configuration in flight (Fig. 3.3).

There are physical reasons why gills work in water and lungs work for air breathers, and so we do not see gills on land- or air-living creatures, and we can interpret the lungs of whales and porpoises as evidence that their ancestors lived on land. Hard shells and spines suggest fighting (usually among males) or protection from predators. Animals that swim tend to be shaped like fish, because the physical constraints of moving in water impose certain limits on their shape (Fig. 3.4).

Warm-blooded animals need some form of insulation, such as hair or feathers, and also they need a higher rate of blood flow. Plants with very large leaves dry out quickly and therefore need lots of water, while plants that live in very dry regions often have tiny or absent leaves and thick, water-retaining, stems. A tree with very shallow diffuse roots can grow in rocky land with thin topsoil, while trees with deep tap roots, though more stable in high wind, need much deeper soil. Thus it is possible to make many judgments about the lives of plants or animals just by looking at them.

The same is true of fossils. A fish is readily identifiable, and would not be mistaken for any other type of animal (Fig. 3.5a). Likewise, though we have never

35

Figure 3.1. Reading the function of animals from their forms and, ultimately, their skeletal systems. Herbivores (left), whose noses are often in grass, have a high need to see what is coming from behind and would ideally be suited with 360° vision. They therefore have eyes on the sides of their heads. Carnivores (right) need good depth perception, achieved when both eyes register the same image, in order to capture prey, and so they have forward-facing eyes. Carnivores also have tearing claws, teeth, or beaks. Closely-related pairs are shown, the herbivores on the left, and the carnivores on the right. From top to bottom on left: grasshopper or locust; bullfrog tadpole; mourning dove; antelope. From top to bottom on right: praying mantis; bullfrog; bald eagle; lioness. Credits: Praying mantis - © Photographer: Rogelio Hernandez | Agency: Dreamstime.com, Bullfrog - © Photographer: Loricarol Lori Froeb, Yorktown Heights NY | Agency: Dreamstime.com

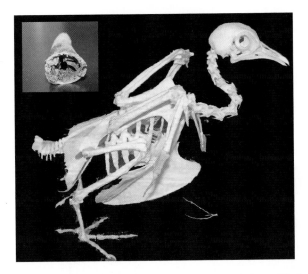

Figure 3.2. Reading the function of animals from their forms. The large keel or breastbone of a pigeon gives a location for attachment of the flight muscles, indicating that this creature is a good flyer. Other indicators of course are the wings and the fact that the weight is well balanced for flight in that there is a beak, but no heavy teeth, in the mouth. Inset: even the bones are very lightweight and even hollow. (Birds use spaces in their skeletons both to lighten the density of the bone and to increase the efficiency of air circulation so that their lungs can work more efficiently, both when they are working hard and when they are at high altitude.)

seen a living trilobite (Fig. 3.5b) we can recognize a certain similarity to that of living horseshoe crabs (Fig. 3.5b) and we can certainly conclude, based on its general body plan, legs, mouthparts, and gill structures, that it was a creature that crawled along ocean bottoms and lived a life fairly similar to that of the horseshoe crab.

This brings us to the main argument, which was an issue that was known to Aristotle and which became more important from the 17th to the mid-19th centuries. There are fossils, unequivocally of marine animals, near the tops of mountains (Fig. 3.6).

Why are they there? Over the centuries, several explanations were posited: The sea was once that high; the fossils are evidence that life can be generated out of rock; the animals were deposited there during Noah's flood; God put them there; the Devil put them there to confuse and challenge humans. Some thinkers were more analytical. Ovid wrote in *Metamorphoses* (Book XV), "Nothing lasts long under the same form. I have seen what once was solid earth changed into sea, and lands created out of what once was ocean. Seashells lie far away from ocean's waves, and ancient anchors have been found on mountain tops." In one of the most remarkable and illustrative stories of the history of science, a group unfettered by commitment to a specific theology or philosophy had clearly worked out an understanding of rock strata 500 years before the first Europeans dared suggest

PELICAN

ALBATROSS

MAGNIFICENT
FRIGATE BIRD

PTERODACTYL

CONDOR

Figure 3.3. Reading the function of animals from their forms. The physics of flight dictates that only certain types of shapes can efficiently collect the updrafts of thermal currents to soar well. Illustrated are the silhouettes of four birds known as excellent soaring birds: the pelican, magnificent frigate bird, and albatross, all of which soar over oceans; and the condor, which habitually soars away from its mountain nest at elevations of 10,000–16,000 feet. Darwin watched condors soar for over an hour without once flapping their wings. Also illustrated is a reconstruction of a fossil pterodactyl, a flying reptile. From its shape it appears to have been similar to a soaring seabird. The figures are not drawn to scale

it. Aristotle had thought that the world was eternal. In Syria, a mystical sect of Shiite Muslims, the "Brothers of Purity," had a motto, "Shun no science, scorn no book, nor cling fanatically to a single creed." In an encyclopedia they wrote, they clearly described the erosion of mountains and hills by rivers, the carrying of the pebbles and rocks to the sea, the conversion of the larger particles to sand by

Figure 3.4. Reading the function of animals from their forms. Fast-moving swimming creatures are constrained by the physics of movement through water to a limited range of forms. A. A fish (goldfish); B: Extinct reptile (ichthyosaur); C: bird (penguin); D: mammal (sea lion); E: mammal (orca or killer whale); F: artifact (submarine). Credits: Ichthyosaur - Traced from an 1863 image on Wikipedia, Penguin - Penguin swimming 1441337: © Photographer: Steinar Figved | Agency: Dreamstime.com, Submarine - Image provided by Dreamstime.com

wave action, the depositing of sediments of sand and clay into sedimentary layers, and the eventual uplifting of these layers into new hills and mountains. The great Uzbeki/Persian physician and scholar Avicenna (980–1037) was also familiar with the ideas, as were the Jewish scholars of medieval Spain, but Christian Europe was either indifferent or frankly hostile to the idea that the world was of great age.

WHY ARE SEASHELLS ON THE TOPS OF MOUNTAINS?

In 1569 the traveler Jan Van Gorp found shells in the Tridentine Alps, though he refused to believe that they could be real; and in Bolivia Albaro Alonzo Barba reported his astonishment at finding seashells ("Cockles [the type of clamlike shell commonly found on seashores] . . . with the smallest Lineaments of those shells

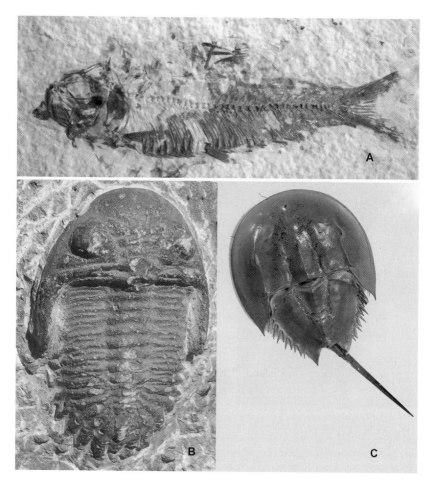

Figure 3.5. Interpreting fossils. A. There is no difficulty in identifying this fossil as a fish similar in most respects to modern fish. The physical structure of this trilobite (B) is very similar to that of a horseshoe crab (C), thus suggesting that, like horseshoe crabs, this animal scuttled along ocean bottoms. See also discussion on page 162

drawn in great Perfection") at 13,000 feet above sea level. Although these reports were known, there was no known means of getting the fossils there, and they were usually interpreted to be peculiar crystallizations of stones or evidence that life could be generated in stones. The question of whether or not life could be spontaneously generated persisted until the mid 19th C (see descriptions of Redi's experiments on page 141 and Pasteur's contest with Pouchet on page 143).

In the mid 17th C, it was sharks' teeth that interested Nicolai Steno, a Florentine monk of Danish origin. In several locations, but especially in the island of Malta, there were peculiar formations known as glossopetrae, literally tongue stones. Some were quite large, as big as a human hand, and there were various theories as to what

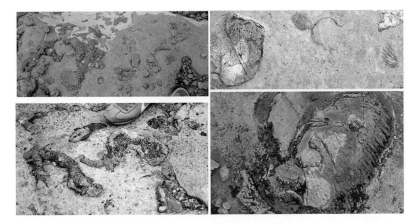

Figure 3.6. Marine fossils at the top of the Grand Canyon (7,200 feet elevation). Upper left: Tracks of marine worms. Upper right: Scallops and clam shells. Lower left: Marine worms in their burrows. Lower right: clam. All of these fossils and tracks are very similar to those that can be found today near seacoasts

they were. The most common interpretation was that they were magical, curative agents deposited in Malta in commemoration of the Apostle Paul's miraculous survival of a snake bite on the island. They were eagerly collected and sold for substantial profit. Although in the previous century Guillaume Rondelet had noted that they resembled shark teeth, this idea did not carry much weight. There was substantial financial and ideological support for not changing the story. It was also considered possible that these and other embedded structures grew spontaneously in the rocks and were an example of the spontaneous generation of life. Unfortunately for this trade, Steno recognized the obvious, that these were the teeth of sharks. Following the dissection of a great white shark that had been captured and the head of which brought to Florence for him to dissect, he published this argument with its implications that the island of Malta had at one time been under the sea and that giant sharks, no longer seen on earth, had lived in that sea (Fig. 3.7).

Several other fallacies were subject to Steno's merciless logic. One was the possibility that the shells grew in the rocks. He pointed out that, if a living organism were to grow while buried deep within a rock, the growth would necessarily split the rock. Anyone who has seen roots push through sidewalks or basement walls will appreciate this argument. Furthermore, Steno pointed out, if mud comes to overlay a rock, the mud will conform to the shape of the rock. He generalized this argument to a principle: if something plastic or semi-solid buries a hard substance such as a rock, or a salt crystallizes around it, the later material will conform to the shape of the earlier. In other words, if mud settles on a rock and, through chemical processes, the mud itself becomes hard or rock-like, the mud will be molded around the rock. He went on to point out that the rock molded around the shells. Therefore,

LAMIAE PISCIS CAPVT

·EIVSDEM LAMIAE DENTES·

Figure 3.7. Fishermen brought the head of a great white shark that had been captured off the coast of Italy to Steno to dissect. Although coastal fisherman had seen the teeth of sharks, they were not familiar with teeth the size of the "tongue stones" that had been found as fossils. Steno realized that the teeth of the great white shark were identical in almost every detail to the tongue stones and argued that these fossils were indeed the remnants of gigantic sharks. This is his illustration of the comparison of the teeth to the tongue stones. Credits: Steno glossopetrae - http://earthobservatory. nasa.gov/Library/Giants/Steno/Images/sharkhead.gif

according to Steno, the shells had been buried in sedimentary mud, and the mud had eventually turned to rock. The shells were not growing in the rock.

Why did he assume that the rocks had been sedimentary mud? He had observed what happens during floods and when streams meet the sea. Floods carry a lot of mud, which settles out on the flood plain when the flow rate decreases and the waters retreat. This is a major means of nourishment of soils such as those of the Nile River. Also, for physico-chemical reasons, much more sediment can be suspended in fresh water than in salt water. When streams enter the sea, because of the slowing of the flow rate and the mixing of the waters, the mud of streams settles into large deltas, as is seen in many parts of the world, such as where the Mississippi meets the Gulf of Mexico (Figure 8.3, page 99). Steno had seen all of this and realized that the hillsides surrounding rivers and bays seemed to be a continuum of the flooding and sedimentation of the valley, even if the hillsides were now rock. This conclusion was not an idle guess. To convince himself, he studied the rocks of the hillsides. Besides the evidence of the sedimentation lines, he looked at the fine structure. They bore the characteristics of floods. First, when mud settles onto a surface, by the law of secondary deposition it takes the form of the uneven surface onto which it settles, but the upper layer, now settling by gravity, will be flat or flatter. The lines of the rocks displayed this characteristic. Second, when material begins to settle out, the heaviest (largest) particles will settle out first, followed by the smaller particles. One can easily see this by suspending any sample of mud or sand in water in a glass, and letting it settle out undisturbed. There will be a distribution of particles, with the largest ones on the bottom (see Fig. 8.4). Steno found this characteristic also to be true of the sedimentation lines of his rocks. Finally, he formulated his (obvious) Law of Superposition: When sediments are laid down, the newer ones will be on top of the older ones. Therefore, even if stones are found at strange angles (Fig. 2.1) it is possible to determine which is the top and which is the bottom.

This logic led Steno to recognize both that the earth was likely to be very old; that much had been formed by sedimentation in water; that the height of the land relative to the sea had changed drastically; and that, from the folds and angles of the sedimentation, the land had undergone violent and tortured existence. In spite of the facts that his logic was impeccable; that he was sufficiently highly considered that he became a bishop and in 1988 was canonized (on October 23, the anniversary according to Bishop Ussher of the formation of the world—see page 51; and he is now considered to be the father of modern geology, his findings and theories were not universally recognized, and it would be two hundred years before the world would come to be comfortable with similar arguments presented by Lyell, as is discussed elsewhere (pages 82 and 168).

Although the ideas were growing, the biggest intellectual limitation was the sense of time. According to Steno, it was evident that the earth had changed. He was aware that during his lifetime he could measure very slow and gradual change, but it was not yet totally acceptable to extrapolate the rate of change. Massive floods such as Noah's flood might leave hundreds of feet of sediment, though one would have to make several assumptions to account for the many layers, since the sediment from one flood should be only one layer. As long as there was no real sense of great age or time, any other interpretation would be pure speculation, creating doubts in the minds of great thinkers, but otherwise not being enormously important or fully impacting the world. Thus the evidence and the logic had not yet fully matured. Evidence that led to the conclusion that the earth was very old will be discussed in Chapter 8, but first we need to address some biological questions. These include the questions of what the word "species" means and whether or not species can change. From there we move to increasing evidence from the New World that confirmed the suspicion that the world was very old, and that not all of its history had been reported in Genesis.

REFERENCES

Cutler, Alan, 2004 (2003), The seashell on the mountaintop. Plume, the Penguin Group, New York.
Sobel, Dava, 1999, Galileo's daughter, Penguin Books, New York.
Mann, Charles G., 2005, 1491, Alfred A. Knopf, New York.
Teresis, Dic, 2002, Lost discoveries. The ancient roots of modern science from the Babylonians to the Maya, Simon and Schuster, New York.

STUDY QUESTIONS

1. At Chinese New Year, crowds go through the streets carrying a model of a giant dragon. From its characteristics, describe its biology. (If you are not familiar with a Chinese dragon, choose any animal from mythology, a comic book, or a video game.)
2. Where roads cut through hills, or where rocks are exposed, try to see if you can recognize how the rocks were formed, and see if you can detect any traces of fossils in the rocks. Hint: Much of the central part of the US is sedimentary; southern Florida is coral; and coal often contains fossils. Granite and other hard stones are soils that have been processed through very high temperatures and pressures as they moved through the earth. You may be able to find more information at a local museum or library; or a website may discuss the geology of the region. Try searching for geology and your local region, rocks and your local region, etc.
3. What were Steno's rules, and why were they important?
4. Observe any small stream or waterway near your residence. Can you identify evidence that the water was at one time higher than it is now? Describe the evidence. Is there any way that you can estimate when the water was higher? If the water of this stream is clear enough, can you make any judgment that the water was at one time lower than it is now?

CHAPTER 4

WERE KANGAROOS ON NOAH'S ARK?

EXPLORATION OF THE WORLD AND ITS EFFECT ON IMAGES OF THE STRUCTURE AND FATE OF THE WORLD

The Age of Exploration, or the Age of Discovery, seriously upset the Western European view of the world. As in other relatively confined societies, the primary theory of creation (the Judaic Genesis, accepted by Christians and Moslems) was reasonably consistent and unchallenged. There were contradictions and inconsistencies as well as earlier pagan legends that were similar enough to be considered ancestral, as is discussed in Chapter 2, page 21, but overall the story of the formation of the earth, night and day, plants, animals, and finally Adam and Eve, coupled with the Garden of Eden, Noah's flood, and sequence of the patriarchs did not seriously defy logic. The existence of marine fossils such as shellfish on mountaintops was known and, although Aristotle had correctly surmised that they indicated the lifting of land from the ocean floor, for the most part fossils were regarded as evidence of the Flood, indication of how life could be generated out of rock, or tricks of the devil. That some fossils were very different from modern animals and plants was not troubling. After all, few Europeans had seen an elephant or a giraffe, and these animals seemed no more-or-less fantastic than basilisks, manticores, or gryphons (Fig. 4.1).

There was one other issue, best explained by discussing a bit of biology from the standpoint of one of the great modern evolutionists. This is the apparently static nature of biology and the earth from a single vantage point.

To a single human living in a specific location, the earth is quite stable. Muddy water might run down a hillside or mountainside, but the hill does not disappear; a river may overflow its banks and cut a new channel, but the river pretty much follows its primary course. Singularities in weather, such as major storms, droughts, floods, or earthquakes soon become legends and even myths. Ernst Mayr, who gave us the basis of our current understanding of the relationship among evolution, species formation, and genetics, emphasized that the same was true for our understanding of animals and plants. A given species might be more abundant in one year than in another, but overall the species was always there. As Jared Diamond has argued, even extinction usually passes unnoticed. The human generation in which a species has become extinct has known the species only as very rare, and has heard of its abundance only from ancestral tales. Thus Diamond, as a young man and expert

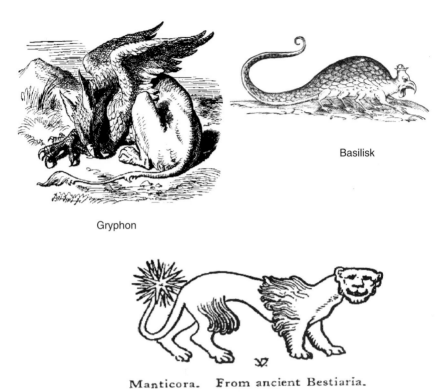

Basilisk

Gryphon

Manticora.   From ancient Bestiaria.

Figure 4.1. Mythical creatures from medieval and early renaissance times. With limited ability to travel and otherwise explore the world, most literate people saw no important differences between creatures such as these and other fantastic animals such as the rhinoceros, the giraffe, or the crocodile. Credits: Gryphon - Gryphon illustration by Sir John Tenniel for Lewis Carroll's Alice in Wonderland (Wikipedia) Basilisk - Source: Ulisse Aldrovandi, "Monstrorum historia", 1642, Austrian National Library, Signature BE.4.G.23 (Wikipedia) Manticore - A manticore from an ancient bestiary (Wikipedia)

ornithologist (one who studies birds), counted all the birds he could identify on one island in New Guinea. Like Ernst Mayr who had preceded him, he asked the pre-literate tribesmen how many types of birds there were, and came up with essentially one-to-one correspondence. (Diamond recognized differences between two extremely similar species of moderate interest to the native population; the native people considered the two to be the same species. In at least one instance the New Guineans were more perceptive in distinguishing species than he was.) Diamond did not conclude that all his education and training had led to no greater sophistication than a pre-literate hunter. He concluded instead that, in a limited territory such as an island, species were quite distinct and easily discriminated. It was only when a zoologist ranged over larger territories and found geographical a zoologist variation—for instance, a frog from western North America might be bigger and fatter, with a slightly different coloring pattern, than a frog from the

East Coast—that it became difficult to tell where, in terms of shape or color, one species ended and another began. Such were the very disturbing observations of the explorers, conquistadors, and missionaries. The European world view was forced to change markedly. But first, let's take a brief look at this European world view.

CLASSIFYING THE SPECIES—IS THE WORLD FULLY KNOWABLE?

Noah took onto the Ark two of every species[3]; and these species came off the ark and repopulated the world. Despite the curiosity, which must have existed in medieval Europe, as to how lions and lambs got along, or plague locusts and wheat, or the jokes that also must have existed about why Noah bothered to take mosquitoes or rats along, it was perfectly within reason to assume that there was a finite number of animals and plants on earth. It would be laudable and even holy to compile a complete list of these organisms. Thus, motivated by theological as well as scientific reasons, Carl Linnaeus (sometimes referred to as Carl von Linné or Carolus Linnaeus), the Swedish father of taxonomy or system of classification, in 1735 (almost 250 years after Columbus' first voyage, and 41 years before the American Revolution) published his first effort to systematize the assorted botanical and zoological information of the period, and to compile a complete compendium of all living creatures. The system of classification was one we still use today, with what one amounts to as a family name and a given name. The equivalent to the family name ("Smith") is the genus name and would include, for instance dogs and wolves (Canis) or leopards and tigers (Panthera) while the equivalent to the given name would define the species itself (the common one, familaris). The name for a dog would therefore be *Canis familaris* (the genus is listed first and is capitalized, like an Asian family name, and the species name is listed second and both are italicized. Following the traditions of Linnaeus' time, all scientists use Latin, and the name simply translates from Latin as the common or familiar dog. We will discuss the problem of classification in greater detail in Chapter 5, page 55.

Linnaeus' self-imposed task was indeed Herculean, but he considered it to be finite—that is, there was an end to the project. It would be possible to identify

---

[3] [Genesis 7, Revised standard Version] And Noah and his sons and his wife and his sons' wives with him went into the ark, to escape the waters of the flood. [8] Of clean animals, and of animals that are not clean, and of birds, and of everything that creeps on the ground, [9] two and two, male and female, went into the ark with Noah, as God had commanded Noah. [10] And after seven days the waters of the flood came upon the earth. [11] In the six hundredth year of Noah's life, in the second month, on the seventeenth day of the month, on that day all the fountains of the great deep burst forth, and the windows of the heavens were opened. [12] And rain fell upon the earth forty days and forty nights. [13] On the very same day Noah and his sons, Shem and Ham and Japheth, and Noah's wife and the three wives of his sons with them entered the ark, [14] they and every beast according to its kind, and all the cattle according to their kinds, and every creeping thing that creeps on the earth according to its kind, and every bird according to its kind, every bird of every sort. [15] They went into the ark with Noah, two and two of all flesh in which there was the breath of life. [16] And they that entered, male and female of all flesh, went in as God had commanded him; and the LORD shut him in. [17]

and classify all living things. Remember that he was working approximately 250 years after Columbus first reached the Americas. It seemed necessary to undertake this classification, because explorers were bringing back plants and animals that had not been known in Europe, and the list of known creatures was beginning to expand. The project, however, still seemed reasonable. However, clouds were beginning to appear on the horizon. We can describe it as the problem of the kangaroo.

The Australian kangaroo is a marsupial, meaning that although it is warm-blooded and fur-bearing, its young are born extremely immature and promptly migrate to a pouch, where they physically attach to a milk-producing gland that is not quite the same as the nipple of a true mammal. There are a few other differences that separate kangaroos, opossums, and their relatives from most other mammals. Using the kangaroo as an example is somewhat misleading, since the first kangaroos were not known to Europeans until 1770, but they illustrate the problem introduced by the raccoons, skunks, and opossums of the new world: How did they get from Noah's Ark to North and South America without being seen, either alive or as fossils, in Europe or the Middle East? One could adjust to the idea that, for instance, lions were seen in northern Africa or the Middle East but not in Europe because, after all, Europe was colder. It was theoretically possible for lions to be in Europe, walking across the land links of the Eastern Mediterranean. Lions simply did not like to be in Europe. However, the climate of North America was not that different from that of Europe, and there was no obvious reason why a raccoon or opossum or skunk could not live in Europe. The same could be said for the true cacti, the spiny flat, branched, or ball-like plants native to the New World deserts. Contrary to old cowboy films and popular images, they did not exist in European, African, or Asian deserts. The world could live without poison ivy (though for a brief period the English considered it to be an attractive houseplant), but creatures of considerable benefit to humans, such as corn, tomatoes, and potatoes, sugar cane, sunflowers, and chocolate, were quite popular among the natives of the New World, as was tobacco, but were unknown in the Old World. Why had God not given Europeans the benefits of tomatoes, potatoes, and corn? Surely the Ark was not a holy Greyhound bus, dropping off passengers on different continents.

Even the explorers were confused. The great explorers were courageous but also extremely knowledgeable people. They had to orient themselves on the ocean so that they would return, for instance, to Spain rather than going too far north and running into England or too far south and running into Africa; they had to be able to locate fresh water and to successfully hunt for food whenever they reached land; they had to locate trees suitable for repairing and waterproofing their boats (pitch pines, named for the waterproof sap they exuded); they had to be able to defend themselves or, preferably, barter and trade with people whose language they had never encountered. The translators, the physicians, the naturalists on these boats were very important members of the crew. Thus it was that Columbus, reaching Hispaniola (Haiti/Dominican Republic) knew that he had landed on an island because there were no large mammals there—an astute observation that would be understood centuries later but nevertheless left the lingering question as to God's logic in not distributing large mammals onto islands. More perturbing, and even specifically noted by Columbus, was the silent or barkless dogs of the Caribbean.

The European view of the world (or, as we might say today, the environment) came very much out of Genesis 1:26.[4] All living things served mankind. Other cultures, in Asia, the Americas, and Africa, had different views, but Europeans understood that, though sometimes the value of something like a flea might be difficult to discern, in one way or another all creatures existed in the reflection of humans at the center of creation. And scholars had set about enumerating the "uses" of all creatures. For instance, the function of a dog was to protect the property of its master, by barking at and if necessary biting an intruder. What then was the "use" of a dog that didn't bark?

Then there was Cuvier. Between 1795 and 1832 Georges Cuvier was professor of animal anatomy at the Musée National d'Histoire Naturelle in Paris. He had recognized the relationship between form and function in an animal, and more importantly had recognized how everything was linked. For instance, a carnivore would most likely have good binocular vision to judge distance, sharp tearing teeth, strong jaws, sharp and strong claws together with strong limbs, and the short digestive tract of a meat-eater. Based on this understanding, Cuvier was considered, probably correctly, to be able to reconstruct a skeleton from a single bone. Since he was so erudite, his opinions were widely respected. His importance in the story of evolution is the following: He could also reconstruct the skeletons of fossils. These buried bones were being found more and more frequently. Cuvier could easily distinguish mammals from reptiles and birds, carnivores from herbivores, and so forth. Fossils were often incomplete, but he could reconstruct from a fragment of the animal its probable size and appearance. And what he found was deeply perturbing. He found that the reconstructions often led to probable animals that could be classified, or grouped into specific categories, but that the animals in these categories were distinctly different from living animals in the same categories. We now recognize this as part of the story of evolution, but in the sense of Noah's Ark, the focus of his argument was a bit different: the species he reconstructed from bones no longer existed. They had become extinct. How did extinction relate to Genesis? Were these creatures from before the Flood (ante-Diluvian)? Had they been carried on the Ark and later been abandoned by God? And how long ago did they disappear? Why would God have put creatures on this earth only to take them away?

Domestic animals were another puzzle. Dogs were dogs, but if an alien arrived on earth, would this alien really consider a Chihuahua or a dachshund to be the same as a St. Bernard or a greyhound or a poodle? Domestication is defined as

---

[4] Then God said, "Let us make man in our image, after our likeness; and let them have dominion over the fish of the sea, and over the birds of the air, and over the cattle, and over all the earth, and over every creeping thing that creeps upon the earth."

human control of breeding, and it was very clear that horses, cattle, goats, sheep, birds, and (in China) fish could be markedly altered by human choice of breeding partners. It was less obvious but at least intuitively understood that domestic crops could be improved and changed markedly from their wild ancestors by selective breeding. So, did Noah take on board a German shepherd or a poodle? By 1809 Jean-Baptiste Lamarck was arguing on this basis as well as that of Cuvier's fossils, that Linnaeus was wrong, that species were not fixed but could change over time. Lamarck proposed that animals and plants changed in response to their environment. He is subject to some ridicule today because we now know that he misinterpreted the causal relationships (see Chapter 12, pages 167–168) but in fact he was a highly intelligent, perceptive scientist who heavily influenced the theory of his time and led to later advances.

Thus the biology of the herbals and zoological books was becoming less and less certain. These concerns were joined with a similar growth of concerns regarding the physical world that had begun to grow in Eastern Europe. The Pole Nikolai Kopernik, better known by the Latin form of his name, Nicolaus Copernicus, in 1514, about 25 years after Columbus' voyage, proposed that the sun, not the earth, was the center of the solar system. Copernicus' ideas were not readily accepted, both for ideological reasons and for reasons having to do with the ELF rule: His evidence was not very good. Copernicus described perfectly circular orbits, but with the calculations of perfectly circular orbits the match to the actual paths of the planets was not exact. The great astronomer Tycho Brahe, who believed in epicycles (wheels spinning on the edges of other spinning wheels, Fig. 4.2) calculated epicycles that came far closer to matching the actual positions of the planets. Copernicus argued on the basis of Logic, similar to that of William of Occam, who argued that the simpler hypothesis was the one to be believed (Occam's Razor), that epicycles were an affectation. However, Brahe's Evidence was stronger. It was not until Johannes Kepler demonstrated that the

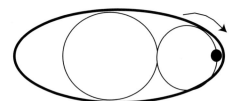

Figure 4.2. Epicycles. If one circle (or sphere) rolls along another, depending on the relative sizes of the two spheres, a single spot on the external sphere will appear to trace an ellipse through space or even go backwards. Since the trajectories of the planets as viewed from the earth follow such patterns, elaborate mathematical schemes based on the position of the spot were devised. Calculations such as those of Tycho Brahe predicted quite well the positions of the planets, but the theory was abandoned when Copernicus and later Kepler argued that there was no physical reason for epicycles and that a simpler model was orbits around the sun. As the concept of gravity developed, it became possible for Kepler to calculate elliptical paths based on the laws of motion and gravity. These proved more accurate than the calculations of Brahe

orbits of the planets were elliptical, that logic and evidence merged. The result of epicyclical movement would be an ellipse, but the hypothesis of an ellipse around the sun was much simpler than the hypothesis of epicycles around the earth. This argument continued to build for almost 100 years, until in 1609 Galileo received a telescope, which had recently been invented, and used it to demonstrate that the heavens were not constructed as had been believed. By 1612 Galileo was convinced that the earth revolved around the sun, leading to the well-known trial of 1616.

The stability of the earth was also less certain, and again the Age of Exploration had some impact. First, mapmakers had been making maps for a few hundred years. Although the outlines of the continents were rather imprecise, coastlines were important to sailors, and especially the locations of small islands and shoals that were hazards to the ships. It was beginning to become apparent that in these details river mouths could change over the years as silt accumulated in some areas and erosion opened others. By 1795, James Hutton from Scotland was suggesting that the features of the earth were not permanent but were gradually changed over time by erosion, sedimentation, and similar processes. His theory was called **gradualism.** And the exploration of the New World was raising other questions. For instance, the Grand Canyon was first reported by Garcia Lopez de Cardenas of Spain in 1540. Though scientists did not really try to understand its construction until 1870, it was clear that the Colorado River had cut it, and any reasonable estimate of how fast a river cuts a channel made one wonder about the age of the earth. In 1650 the Irish Bishop James Ussher had published the first part of a monumental work, in which he had assiduously counted all the dates and ages backward through the Bible, compared some dates with Greek records, and made an assumption or two. Using these calculations, he came to the conclusion that the world had been created on October 23, 4004 B.C. This calculation seemed in line with previous assumptions, based on estimates of the Bible; it was hailed as an achievement, and accepted without excessive circumspection for almost 200 years. However, geological formations like the Grand Canyon made one wonder: was approximately 6000 years enough to cut such a canyon?

Between the 16th and the 19th centuries, many of the apparently solid beliefs on which the interpretation of Genesis was based were increasingly in difficulty. The increasing confusion as to exactly what a species was made it difficult to understand whether Noah would, for instance, have brought on board a pair of eastern bullfrogs and a pair of western bullfrogs, or just one pair of bullfrogs, and it made no sense that the Ark had specific drop-off points or stops on route. God seemed to have made some species only to let them die out. Barkless dogs did not serve humans in the way that Europeans understood. The Bible gave a maximum age for the earth of 6000 years, less if one assumed that the 800+ years of the patriarchs of Genesis were allegorical, but there was indication that some features of the earth would take longer to form. And why were there seashells in the mountains? Several of the changes that came about are described in Fig. 4.3. This figure should be used in reference to the several Chapters 3–8.

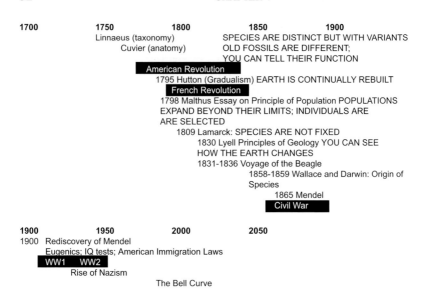

Figure 4.3. The historical context in which the story of evolution was born

REFERENCES

http://nationalacademies.org/evolution/ (Essay on Evolution from the National Academy of Sciences)
http://darwiniana.org/(Darwiniana and Evolution, International Wildlife Museum, Tucson, AZ)
http://www.pbs.org/wgbh/evolution/index.html (Summary of Evolution series from Public Broadcasting System)
Mayr, E (1963) Animal Species and Evolution, Harvard University Press, Cambridge.
Diamond, J.M. (1965) Zoological classification system of a primitive people. Science 151: 1102–1104.

STUDY QUESTIONS

1. Look at any type of organism that you commonly encounter: sparrows, pigeons, dandelions, tropical fish, maple trees. Do you have any difficulty identifying it as a member of a specific species or type? How much variation is there among individuals? How does this variation compare to that of domestic organisms such as cats or dogs?

2. Describe any animal or plant of which you have heard but which you have never seen. In what characteristics does it differ from animals or plants that you know? Do these characteristics match any animal or plant that you consider to be fictitious? How do you know which are real and which are fictitious?

3. Look at a riverbank, a lakeshore, a river delta, a mountain range, a fault that has generated earthquakes, or any geological structure near you that may have changed over the history of the earth. Is there any way that you can estimate the

rate that it is changing or has changed, and from this the time that this feature has been present?

4. What criteria would you use to decide if specific animal or plant species are related to each other, and how closely they might be related? For instance, you might ask what is the closest relative to a raccoon, a skunk, or a bat. Defend the criteria that you choose.

CHAPTER 5

ARISTOTLE'S AND LINNAEUS' CLASSIFICATIONS OF LIVING CREATURES

The human mind seeks to categorize and classify. Thus Aristotle recognized animals as "blooded" (vertebrates) and "without blood" (invertebrates) and described in great detail the characteristics of each (Fig. 5.1). No society has trouble distinguishing in most instances among fish, amphibia, reptiles, birds, and mammals. Some animals do cause trouble: it was widely argued, until decided by the anatomists (Chapter 2, page 28) whether whales, seals, and porpoises were mammals or fish, and there were even some rather amusing accommodations: in Catholic ritual, during Lent and on Fridays, the eating of flesh (meat of mammals) was forbidden. In the American South and in South America, respectively porpoises and capybaras (a pig-size guinea pig-like animal that spends much of its time in rivers) were redefined as "honorary fish" and thus permissible for consumption. However, beyond this crude classification we generally push further into detail as long as our curiosity and economic interest push us. Among the mammals, we distinguish cats and dogs, and among the dogs, hunters, lap dogs, guard dogs, and racing dogs. Whalers knew which whales were either too difficult or too worthless to hunt, and those that were valuable and easy to hunt (the "right whale"). Sailors landing on a foreign shore and needing to recaulk their ships could identify trees that could supply the appropriate sap ("pitch pines"). However, most of this classification remained sporadic, inconsistent (would a penguin be a bird or a fish?), and local. By the 18th C, enough was being learned about the world that such a haphazard structure was clearly unsatisfactory. Carolus Linnaeus changed that by attempting to classify the entire range of known living things. This classification, accomplished in the mid 18th C, accomplished three major feats: binomial nomenclature, non-linear relationships, and stratification. The fourth and most important accomplishment, however, was that the third was so clear and structured that it led to the eventual recognition of its own inadequacy, becoming the basis for the questioning that was an important element for the understanding of evolution.

The first accomplishment attributed to Linnaeus, binomial nomenclature, was, strictly speaking, not his. Many societies have had general and specific means of naming individuals, whether they were by lineage (the Biblical "Isaac the son of Joseph"), by occupation ("William [the black]Smith"), characteristic ("John Short") or origin ("William [of] England"). By the same token, the Latin-versed scholars often referred to animals and plants by two names, but more often by verbose descriptions. The Bauhin brothers in the 16th C had tried to simplify the annotations, much as scientists today do in their writing. A scientist today might abbreviate an unwieldy complete description ("the membrane-bound phospholipid phosphatidyl serine (PS)") and thereafter refer in the text only to "PS". The Bauhin brothers similarly used a general and specific pair of terms to refer to specific organisms. However, Linnaeus was the first to use this shorthand consistently and widely, assigning the equivalent of a family name and given name, as is described in Chapter 1, page 10. Since his classification was widely published and read, he considerably popularized the custom.

Linnaeus' second accomplishment was to show the relationships not as a linear order, in which each organism was necessarily higher or lower in perfection than any other organism (See Table 5.1) but rather alongside each others: rodents were not necessarily higher or lower than horses or dogs. This was the beginning of the branched tree picture that Darwin finally drew (Fig. 5.4) Third, he built the classification into a hierarchy, stratifying it beyond the simplest levels of similarity. Beyond the specific and generic or first name-last name classification ("Dog, the common one" = Canis familaris or "Panther, the spotted one" = Panthera pardus) he grouped organisms into broader and broader categories, each one superseding the previous (dog->carnivorous mammal->mammal->vertebrate-> animal->eukaryotic organism) in a hierarchical tree, as is illustrated in Fig. 5.2 and Table 5.1. Over the years, by this heroic effort, he put into systematic order 4,400 species of animals and 7,700 species of plants. This accomplishment was several-fold. First, although lineage was not understood as a biological phenomenon, it indicated connections, as we might expect levels of increasing similarity in appearance as we move along the familial tree humans->Caucasian humans->European Caucasians->Mediterranean

Table 5.1. Classifications according to Aristotle

God
Humans
Mammalian animals
Flying squirrels
Bats (and birds?)
Fish
Reptiles
Shelled animals
Insects
Sensitive plants
Plants
Short mosses
Mushrooms
Stones
Crystalline salts
Metals
Earth

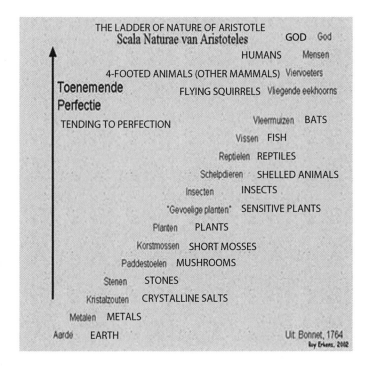

THE LADDER OF NATURE OF ARISTOTLE
Scala Naturae van Aristoteles   GOD   God

HUMANS   Mensen

4-FOOTED ANIMALS (OTHER MAMMALS) Viervoeters

Toenemende   FLYING SQUIRRELS Vliegende eekhoorns

Perfectie

TENDING TO PERFECTION   Vleermuizen BATS

Vissen FISH

Reptielen REPTILES

Schelpdieren   SHELLED ANIMALS

Insecten   INSECTS

'Gevoelige planten' SENSITIVE PLANTS

Planten   PLANTS

Korstmossen   SHORT MOSSES

Paddestoelen   MUSHROOMS

Stenen   STONES

Kristalzouten   CRYSTALLINE SALTS

Metalen   METALS

Aarde   EARTH   Uit Bonnet, 1764
                                       by Erkens, 2002

Figure 5.1. Aristotle's Scala Naturae, or Scale of Life, as described in an 18th C Dutch document. The translation of the terms is in CAPITAL LETTERS. Though Aristotle's classifications improved on others' ideas in that he recognized the difference between the live-bearing seagoing mammals and egg-laying fish, he lumped many types of shelled animals together, and did not clearly recognize that bats and flying squirrels, like humans, were mammals. Credits: http://www.kennislink.nll/upload/78469_962_1020862987092-systhematiek2.jpg

peoples–>Italians–>people from around Naples (as in the last name "Napolitano")–>a individual (Maria Napolitano). The existence of this sequence of groupings could not fail to provoke questions about the significance of the groupings—why do they exist, and why do these groups of animals and plants have the same basic structures?—as well as the inferred limitations of the groupings—why are there only these groupings? Why are there not animals with both four legs and wings (or do such animals exist? See page 45.) Such questions would begin to haunt the 19th C.

Linnaeus' fourth accomplishment was that his third accomplishment, the ordering and classification of all organisms, was so thorough that it sealed its own fate. Linnaeus, in good Protestant style, felt that he was doing God's work and helping to understand God's creation in classifying all organisms, though he understood that he angered local clergy by daring to classify humans in the same general grouping as chimpanzees. In a similar fashion, Jewish scholars were producing tracts on the secular (and therefore forbidden, or at least discouraged) subjects of zoology and anatomy, under the ruse of depicting the animals of the Old Testament. The problem was that, while his project had begun because the exploration of the world had demonstrated that there were too

Figure 5.2. Classification according to Linnaeus, Although he recognized structural similarities among related organisms and vastly improved the systematization of types of animals and plants, he was unable to account for several organisms, which he listed as "paradoxical animals" and many of his classifications by today's interpretation were wrong (see Table 5.2). Nevertheless, by grouping organisms as he did, he created the basis for others to recognize that the similarities meant common origin. Credits: Systema naturae per regna tria naturae, secundum classes, ordines, genera, species, cum characteribus differentiis, synonymis, locis or translated: System of nature through the three kingdoms of nature, according to classes, orders, genera and species, with differences of character, synonyms, places). 1735 (Wikipedia)

many creatures in the world to understand and study unless they were systematized in some fashion, this systemization quickly proved to be a never-ending task, as more and more species were discovered and added to the list. More importantly, by the 19th C, approximately forty years after Linnaeus had published his compendium, arguments were arising as to how to classify new discoveries. It is easy enough to sort coins into pennies, nickels, dimes, quarters, half-dollars, and dollars, but is the steel penny of the Second World War a penny, as its shape indicates, or a nickel, as its color indicates?

In biological terms that are relatively easy to understand, ladybird beetles (ladybugs) are small red beetles with black spots, but they may range from almost entirely red beetles with a tiny black spot to almost entirely black beetles with a trace of red, and the red may vary from crimson through orange to yellow. By today's terms, now that we have worked much of this out, they are the same species if they can successfully breed with each other (see below), but to the scientists doing these classifications, what was the true character of the species, and are all these beetles one species, two species, or several species? This question continues today. One can see in any museum a collection of varieties of butterflies, snails, or other small, easily preserved organisms, demonstrating the variability of animal and plant life; and many passionate collectors collect, photograph, or otherwise document the variation among a single species. In the 19th C, this situation was not a given but a conundrum: What were species? Was there an ideal type for a species, with all variants being imperfect attempts to reach it, as Aristotle argued? Were there boundaries? If so, what were the boundaries? If there was no continuum, what defined and decided the boundaries? If there was no clear boundary— for instance, if the largest green frog (known as *Rana clamitans*) was the same size as the smallest bullfrog (known as *Rana catesbeiana*—did this mean that there was no clear distinction between one species and another? And what did the Bible say about this? Thus the very system that Linnaeus had put in place to structure and classify all of Creation was ultimately to undercut the conviction of unique and specific creation.

Although we now think that some of Linnaeus' classifications were incorrect, the general pattern is essentially unchanged. What he described was a hierarchy of similarities (Table 5.2; a modern version of the table, incorporating current views, is given in Table 5.3). The question is, what is truly biological and what is an artificial construct of human imagination? In other words, does a species really exist? Do genera (the plural of "genus') really exist? The question is more complex than one might imagine. The problem is best illustrated in the phenomena of ring species, discussed further on page 167. If one catches a leopard frog in, say, upstate New York, there is no problem whatsoever in identifying it. Nor, for that matter, do leopard frogs have any trouble identifying each other. Leopard frogs exist from Northern Quebec to Louisiana. Leopard frogs from Quebec look like and easily breed with leopard frogs from Maine; those from Maine with those in Massachusetts; those in Massachusetts with those in New York, etc. The problem arises when one compares a frog from Quebec with one from Louisiana. They look a bit different. The one from Louisiana is a bit fatter, and its nose is pointier than the frog from the north. It has fewer and rounder spots. Its call sounds noticeably different (Fig. 5.3) More troubling, if one tries to get a Quebec frog to breed with a Louisiana frog, the mating does not go well at all, and even if they do mate the eggs rarely if at all hatch successfully. If we define a species as a population that can successfully interbreed, then the Louisiana and Quebec frogs are two species; but if we go province and state by state across its range, frogs in the same region can interbreed, and at no point do we find a point at which *Rana pipiens* from the north is clearly separate from *Rana pipiens* from the south. So by this criterion they are the same species.

Table 5.2. Classification according to Linnaeus (Systema naturae per regna tria naturae, secundum classes, ordines, genera, species, cum characteribus differentiis, synonymis, locis or translated: "System of nature through the three kingdoms of nature, according to classes, orders, genera and species, with differences of character, synonyms, places")

Kingdom	Class	Order	Genera	Species
Animals	I. Quadruped			
	Anthropomorph	Humans	Homo	Europeans (Homo europaeus albesc.(=white))
	Independent hands			Americans (Homo americanus rubesc. (=red))
	Body: hirsute			Asians (Homo asiaticus fulcus (=yellow))
	Feet: 4			
	Female: viviparous, milk-producing			Africans (Homo africanus niger (=black))
		Apes		
		Monkeys & others		
	Wild animals	Bears		
		Cats		
		Porpoises		
		Otters		
		Seals		
		Hyenas		
		Dogs		Dog
				Fox
				Wolf
	Rodents	Bats		
		Squirrels		
		Mice	Rats	
			Mice	

		Shorttailed mouse (Mus brachiurus)	(Mus)	Domestic mouse (Mus domesticus)
		Longtailed mouse (Mus macrourus)		
			Lemurs	
			Marmots	
	Horses	Beavers		
		Rabbits		
		Horses		
		Elephants		
		Hippopotamuses		
		Pigs		
		Camels		
	Split-hooved animals	Bovines (cows)		
		Gazelles		
		Deer		
		Sheep	Ovis	Common sheep (Ovis vulgaris)
				Arabic sheep (Ovis arabicus)
				African sheep (Ovis africanus)
				English sheep (Ovis angolenfis)
II. Birds	Hawks			
	Crows			
	Geese			
	Swans			
	Ducks			
	Chickens			

(Continued)

Table 5.2. (Continued)

Kingdom	Class	Order	Genera	Species
III. Amphibia	Passerines			
	Serpents	Turtles		
		Frogs		
		Alligators		
		Snakes		
IV. Fish No legs; fins; true dorsal fins	Paradoxical animals			
	Whales* (horizontal tail)			
	Sharks (cartilage instead of bone)			
	Short-gilled fish			
	Spiny-finned fish			
	Divided-finned fish			
V. Insects	Hard shelled (beetles)			
	Veined wings	Butterflies		
		Wasps*		
		Flies* (mouth under head)	Houseflies (Musca)	Musca domesticus
				Musca species 1
				Musca species 2
			Horseflies (Tabanus)	

Mosquitoes
(Culex)
Crane
flies
(Tipula)

Dragonflies*
Half-winged   Crickets*
                     Fireflies*
                     Ants*
                     Sucking bugs

Wingless   Lice
                     Fleas
                     Crabs
                     Spiders

VI. Worms   Crawling worms   Earthworms
                             Tapeworms
                             Leeches

            Shelled   Cockles
                             Nautilus
                             Oysters
                             Clams

            Plantlike   Sea cucumbers
                             Starfish
                             Sea urchins
                             Jellyfish*
                             Squid
                             Microshelled

Plants

CHAPTER 5

Table 5.3. Modern hierarchical classification system derived from Linnaeus

Rank	Example	Translation and description	Others at this rank
Domain	Eukaryotes	All organisms with true nuclei (essentially all plants and animals except bacteria and viruses)	Bacteria; viruses
Kingdom	Animals	All organisms that feed rather than live by photosynthesis, not including parasitic plants and fungi	Plants; fungi
Phylum	Chordates	All animals with a notochord (a rigid structure that preceded a true bony backbone: lampreys and some small marine organisms have a notochord but no backbone, and sharks do not have a true skeleton)	Mollusks; arthropods
Class	Mammals	Warm blooded, fur-bearing animals that suckle their young	Amphibians; reptiles; birds
Order	Carnivores	Flesh-eating mammals with teeth that can tear	Rodents; bats; whales and porpoises
Family	Cats	Lions, tigers, panthers, and cats	Dogs; skunks; otters
Genus	Panthers	Large, spotted, or uniformly-colored cats	Lynx; domestic cat; cheetah
Species	Leopard	A specific type of large spotted carnivorous cat ("leopard" literally translates as "spotted lion") that does not voluntarily cross-breed with other large cats.	Lion; tiger; jaguar

There are many similar situations. For instance, the common herring gull of the northern latitudes varies slightly along a geographical pattern, from Greenland westward across North America, continuing into Siberia, and onward to Europe. At no point is there a clear demarcation between one type and another. However, the gulls of Greenland look noticeably different from the ones in England and Ireland, and they do not interbreed. One could argue that the Greenland gulls and the English gulls were different species but, following the variations westward from Greenland, they appear to be one species. It remains theoretically possible for a mutation arising in a gull in Nova Scotia to reach the population in Ireland, as it remains theoretically possible for a mutation arising in a Louisiana leopard frog to reach the population in Quebec.

Situations such as the existence of ring species can teach us much about how new species arise, and was a major argument in Darwin's *Origin of the Species*, and

Figure 5.3. Northern (left) and Southern (right) leopard frogs. To a trained eye, these frogs can be distinguished by the pointier nose of the Southern frog, and the blotchier spotting and more complete leg stripes of the Northern frog. Their calls are very different, as may be heard by listening to http://allaboutfrogs.org/files/sounds/nleaprd.wav and http://allaboutfrogs.org/files/sounds/sleaprd.wav. They do not interbreed, though intermediates do. Credits: Northern leopard frog - http://www. umesc.usgs.gov/terrestrial/amphibians/mknutson_5003869_overview. html (public domain) (Wikipedia), Southern leopard frog - http://cars.er.usgs.gov/Education/sldshw/herpeto -logy /slides.html

will be addressed in other chapters. For our purposes here, it is sufficient to understand that the ability to interbreed is a true biological distinction and is used by biologists to define species. A species is a population that successfully interbreeds, with the progeny (young) being of equal health and fertility as the parents. This rule separates, for instance, the horse and donkey. In captivity, they will interbreed and produce healthy, vigorous mules. However, the mules are sterile, and there will be no grandchild generation. Also, organisms that do not reproduce sexually but simply divide, such as some bacteria, create a problem for this definition but, for most readily visible organisms that people encounter, the definition works.

Beyond species, the definitions become more arbitrary. The branching tree structure, which Linnaeus saw but from which he did not draw any conclusions, became central to Darwin's hypothesis concerning the origin of species (Fig. 5.4) but does not ultimately resolve the definitions of genera, , families, orders, classes, and phyla, in the same sense that it is often difficult to decide, at the outer reaches of a family, who is truly a cousin. To take the Napolitano example above, one can easily identify the Napolitano clan and the Siciliano clan but when a third cousin twice-removed from the Napolitanos marries a grandchild of the Sicilianos, to which clan do they belong? In many countries people vehemently argue that a child born in that country but of foreign parents is not truly a member of that country, and in the US, an individual who represents the third generation born in the US but who bears a surname or appearance that is not from northwest Europe may be considered as not truly American. There is a creature that looks and feels basically like an earthworm, but it has fleshy legs and fleshy antennae like a millipede. If it had a hard shell and hard legs and antennae, it would look like a millipede (Fig. 5.5).

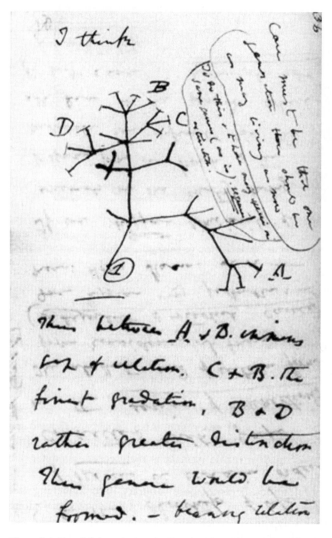

Figure 5.4. Darwin's branched tree. Darwin's original sketch of a potential evolutionary tree (theoretical) found in his notebooks. Note that all organisms start at an original point (1) and that they evolve in ever-increasing diversity, with surviving (extant) families marked by the letters A–D, with others having become extinct. Credits: Tree of Life: the first-known sketch by Charles Darwin of an evolutionary tree describing the relationships among groups of organisms. © Syndics of Cambridge University Library

Does it belong to the phylum of the worms (Annelida) or the phylum of the jointed animals (Arthropoda)? The juvenile forms of sea squirts, marine organisms that basically look like bags with two holes through which they pump water, have a notochord with a brain above the notochord and nerves that run along the back of it, and they have tail muscles in the chevron pattern typical of fish. Do these structures

Figure 5.5. An evolutionary intermediate. Top: the common Annelid (ringed) earthworm. Its body is divided into segments, and the head end is slightly swollen, but there are no limbs, antennae, or obvious other external specializations. Middle, a velvet worm from Australia. It has a soft body like an earthworm, but soft, fleshy feet-like appendages and soft, fleshy antennae. Bottom: a centipede. It has a hard shell and hard, jointed legs, both characteristic of arthropods (lobsters, crabs, insects, and spiders; the name "arthropod" means "jointed-footed animal"). It also has a clearly-defined head, which allows it to centralize information coming in from the senses. Credits: Velvet worm - http://www.ars.usda.gov/is/kids/insects/story5/velvetadventures2.htm

make them card-carrying members of the chordate phylum (which includes the vertebrates) or not? There are egg-laying warm-blooded furry creatures, such as platypuses (Fig. 11.1) and echidnas. Frogs have smooth moist skin and lay their eggs in clumps, while toads have dry warty skins and lay their eggs in strings, but there are amphibia whose skins are intermediate and whose eggs are in what might describe as elongated clumps. The fossil record includes many such intermediates such as feather-bearing dinosaurs that, unlike birds, had teeth and bony tails. Today we can trace by the evidence of DNA when lineages separated from each other, and how far apart they are (Chapters 14 and 15), but ultimately designations above the level of species are human constructs. Most species are easily classifiable, but there are always examples for which the borders are fuzzy. To summarize: The concept of species is as close as we come to a truly biological distinction. The branching tree by which we describe life on earth reflects the manner in which the variety of life appeared on earth, as we will explore in the following chapters but, while many of the classifications appear to be obvious—birds versus reptiles, cats versus dogs, peaches versus apples—we know of many intermediates, and at the borders the distinctions are human decisions.

REFERENCES

http://nationalacademies.org/evolution/ (Essay on Evolution from the National Academy of Sciences)
http://darwiniana.org/ (Darwiniana and Evolution, International Wildlife Museum, Tucson, AZ)
http://www.pbs.org/wgbh/evolution/index.html (Summary of Evolution series from Public Broadcasting System)

STUDY QUESTIONS

1. Compare and contrast the classification schemes of Aristotle, Linnaeus, and Darwin. What are the strong points and weak points of each? What evidence is there to defend each?
2. Taking an example cited in the text, if you found an animal that you could not obviously classify as a frog or a toad, by what criteria would you classify it? Is there a reason to classify it?
3. There are some creatures normally considered to be fish (lungfish) that have scaleless, slimy skins through which they can get oxygen and primitive lungs so that they can breathe air. Their fins are rather fleshy, allowing them to crawl across the land, and they can spend considerable time out of water. Are they amphibia or fish? How can you tell? (This question is discussed further on page 153. Speculating on the topic at this point will help you to understand the issues in this chapter.)
4. You notice that, in a local pond, some frogs sing their mating calls in the early evening, while others, that look the same, sing only in the early morning. Similarly, some of the female frogs seem to listen to songs only in the evening, and others only in the morning. Are they different species? Defend your argument.
5. Argue against the proposition that all classifications above the species level are arbitrary.
6. By what criteria do we classify bats and porpoises among the mammals, and penguins among the birds? Do you agree or disagree with these classifications? Explain.
7. What criteria would you use to distinguish among different groupings of plants? Why?

CHAPTER 6

DARWIN'S WORLD—SPECIES, VARIETIES, AND THE AGE OF THE EARTH. EVIDENCES OF GLACIATION

When Louis Agassiz came to Harvard from Switzerland in 1846, he brought with him not only his considerable expertise in biology, but also a lifetime experience in the Swiss Alps. In the Alps, he had often observed the workings of glaciers and had speculated that the glaciers had once been much more massive. Glaciers form when more snow accumulates each winter than can melt in the spring. The snow continues to pile up until it compresses the snow below into a dense form of ice, so dense that it has a blue color. (The ice is blue for reasons very similar to the reasons why the sky is blue and why the ocean under sunlight is blue. It has to do with the way that water transmits and reflects light. But that is a different story.) As anyone who has ice-skated knows, at appropriate temperatures ice, when compressed, will melt. The ice skate puts the weight of your body on a very narrow surface, compressing the ice and causing it to melt. Thus the skate glides easily along the ice. It does not work if the ice is too cold or the weight is not enough. Try it!

Glaciers do the same thing. With all the weight of the glacier above, the ice at the bottom melts, allowing the glacier to slide down the mountain. Glaciers slip down mountainsides at rates from a few feet to hundreds of feet per year, as has been documented by objects such as abandoned climbers' tents being moved down the mountain. At the upper end, the glacier is renewed by the continuous accumulation of snow. See Fig. 2.1. Underneath the glacier, the movement rolls or pushes rocks and often breaks them; and the glacier often breaks apart small structures such as uneven parts of the earth. The glacier expands in the winter and retreats in the summer, leaving piles of the rubble it produced. The glacier as a whole moves like a river: The glacier is always there, but the water in it changes constantly.

The movement of the glacier produces characteristic marks, very similar to those that would be produced if you scoured a dirty pot with cleanser or a soft stone such as a pumice stone. The uneven surface (the remnant food) would be ground away, and the pot, if it were soft metal, would be scratched by the cleanser. The remnant food would accumulate at the edge, where the scouring stopped. Glacial valleys look very much like those shown in Figs. 6.1 and 6.2, whether the glacier is still there or not. The walls are steep and give the appearance of having been gouged

Figure 6.1. Termination of a glacier in Alaska. The glacier extends much farther into the sea in the wintertime and previously was much larger. Note the characteristics of the land through which it has come: steep-walled carved and scored valleys (arrows) and piles of rubble, mostly stones, boulders, and pebbles, along the sides of the glacier. The rubble along the side is called lateral moraine. At the front of the farthest extension of the glacier is the terminal moraine

Figure 6.2. Edge of a fjord, or valley carved by a glacier (Norway). Note that the physical characteristics are the same, allowing its identification long after the glacier has disappeared

out; the stones at the base are scratched or polished; and there is considerable rubble piled up at the sides and at the front of the glacier. These piles of rubble are called moraines, from the French dialect word meaning referring to the types of hills formed by rubble.

You can imagine Agassiz' bewilderment, astonishment, and finally sense of wonder as he toured the US and began to realize that the characteristics of the landscape that he was viewing—the rocky terrain of New England and farther north, compared to the deep soil of land south of New England; the Great Lakes and Finger Lakes; the scratched appearance of rocks and odd placement of huge boulders—were similar to the characteristics of the Swiss glacial valleys that he knew, but on a vastly larger scale, rather like Jack and the Beanstalk, where Jack encounters a world of giants on a much greater scale of measurement than he is. The glacial valleys of the Alps are not more than a few miles long, and much less than a mile across. (Glacial valleys are typically much longer and deeper than they are wide. This also is a characteristic of the fjords of Norway, and is how a fjord is defined (Figs. 6.3 and 6.4). Fjords were formed by glaciers, though this association

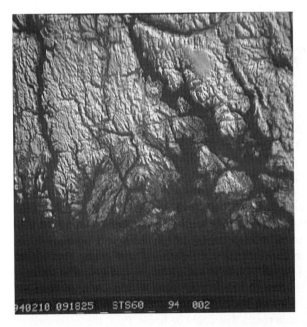

Figure 6.3. Satellite view of fjords in Norway Note the long, narrow, straight channels. Part of the definition of a fjord is that it is deeper than it is wide, characteristic of valleys carved by glacial tongues. Credits: Image Science and Analysis Laboratory, NASA-Johnson Space Center. 25 Sep. 2006. "Astronaut Photography of Earth - Quick View." <http://eol.jsc.nasa.gov/scripts/sseop/QuickView.pl?directory=ISD&ID=STS060-111-3> (6 Dec. 2006). Image Science and Analysis Laboratory, NASA-Johnson Space Center

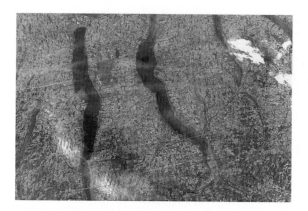

Figure 6.4. Satellite view of Finger Lakes District, New York. The long, narrow, deep lakes are characteristic of fjords. Though less visible by satellite, the entire Lake Champlain-Hudson River valley system is also a fjord. Credits: Image Science and Analysis Laboratory, NASA-Johnson Space Center. 10 Jul. 2006. "Astronaut Photography of Earth - Display Record." <http://eol.jsc.nasa.gov/scripts/sseop/photo.pl?mission=ISS010&roll=E&frame = 23284&QueryResultsFile = 116542430891802. tsv> (6 Dec. 2006). Image Science and Analysis Laboratory, NASA-Johnson Space Center

was not realized until Agassiz recognized it. What Agassiz saw, though, suggested glaciers the width of continents, and tens of miles thick!

There were many features that Agassiz identified, most of which can be readily observed throughout the northern parts of the US and in Canada. First, there were fjord-like lakes, most notably the Finger Lakes of upstate New York. Then there were the scooped-out areas resembling the water-beds at the front of a receding glacier, but much more vast: the basins of the Great Lakes. The mountains of New England were often smooth and polished on their north sides, but rough and steep on their south sides. And then there were the scratch marks. On embedded rocks throughout the north, there were scratch marks—glacial striations—mostly in a north-to-south direction. They are very prominent in Central Park (Figs. 6.5 and 6.6). Only something very massive and universal could have done all this. Finally, there were huge boulders, much larger than could be moved by humans, animals, or floods, sitting randomly on mountain tops, in valleys, and in various locations. They had two characteristics in common: They showed no relationship to the land or the stones in their neighborhood, but they did resemble the rocks of land hundreds of miles to the north. These are called erratics, more correctly glacial erratics, and we now know that they were, indeed, carried to their current location by riding piggy-back on glaciers (Fig. 6.7).

There were three other characteristics that could be noted. First, the vegetation of New England and upstate New York is very different from that of areas immediately to the south, and not just because of climate. North of Long Island Sound, in Connecticut, there is very little topsoil, and the trees are consequently shallow-rooted, light trees like birch and aspen. Ten miles away, on Long Island, there is much more soil, and deep-rooted heavy trees like maple and oak are prominent. This

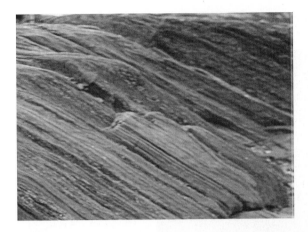

Figure 6.5. Glacial striations visible on rocks in New York City. Many rocks are scored in this fashion, as is this boulder in Central Park. Most of the gouges are in a north-south direction, and some are much deeper. Such marks are found wherever the glaciers scraped the soil to bedrock. There is a Glacial Striations State Park in a small island in Lake Erie off the Ohio shore. Striations are not found south of a line running roughly from Long Island, New York through the flat area surrounding the Great Lakes

is very visible in the amount of light that reaches the floor of the forest (Fig, 6.9). Second, the soil of the north shore of Long Island is rather peculiar. It is rather sandy, filled with small pebbles of various sizes. Third, the north shore of Long Island is very hilly, with the hills separated by deep north-to-south ravines (Fig. 6.8).

All of these different observations could be accommodated by one overall hypothesis, shocking for the time but perhaps the only reasonable interpretation. Long

Figure 6.6. Smaller scale glacial gouging visible on rocks in New York City

Figure 6.7. Shelter Rock, a glacial erratic, Long Island, New York

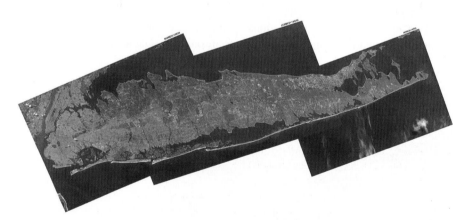

Figure 6.8. The terminal moraine of Long Island, seen in a satellite view. The end of the moraine is marked by the transition between the hilly region to the north (darker color) with deeply carved north-south valleys ending in bays in Long Island Sound. These valleys were the last fingers of the glacier. North of the Sound, the land is heavily scarred in a north-to-south direction. To the south of the hills is flat sandy soil, the outwash from the glacier. (At the western end of the island, neighborhoods in Brooklyn have names such as "Flatlands" and "Flatbush".) Credits: http://eol.jsc.nasa.gov/sseop/clickmap/ image ESC_large_ISS011_ISS011-E-8036 and 8037

Figure 6.9. Left: The Long Island forest, which is typical for this latitude and elevation. Deep-rooted large-leaf trees such as maple and oak predominate, leaving the undersurface quite dark and with little vegetation. Right: Forest in Westchester County, New York, less than 30 miles north of the forest pictured on the left, but north of the terminal moraine and Long Island Sound. This forest consists of slender, shallow-rooted trees such as birch, poplar, and aspen, and light penetrates to the floor, allowing substantial undergrowth. Contrary to the moraine and areas south of the moraine, which have topsoils one to several feet deep, north of the moraine the topsoil is only two or three inches deep before yielding to bedrock, and the forest reflects this difference. The pictures were taken during the same week

Island was a giant terminal moraine, and many of the features of northern US and Canada could be best explained by the existence at a previous time of huge glaciers that covered most of the territory, tens of miles thick. Of course, the possibility of the existence of massive glaciers carried several implications, none of which were compatible with the Biblical description of the origin of the earth. The first concern, of course, was that Genesis neither described a period of ice nor a situation in which the climate was substantially different from the current climate. Second, it was evident that ice of this order of magnitude would take an extremely long time to form and to melt, and the state of the remnants, including sediment on top of them, erosion such as river cuts into the remnants, and estimates of the age of the remnants, suggested a time span well beyond the calculated six thousand years. To estimate the age of the remnants, one had merely to look at the land. The area immediately to the south of the Great Lakes, perhaps three times the surface area of the lakes themselves, is very flat and sandy, and aquatic fossils such as fish can be found throughout the region. In Indiana, starting immediately south of Lake Michigan, there are sand dunes that clearly seem to be related to the lake bottom. The farther one is from the lake, the more settled and mature the forest is: the topsoil is thicker, and the trees are more like the rest of the region. In these and other areas, if one can count tree rings on current and fossil trees, one can get a lineage that goes back into the thousands of years (Fig. 6.10; see also Chapter 8, page 95).

Even the plants of the New World supported the argument of glaciation. Glaciers covered Europe as well as North America, but in Europe the glaciers drove the flowering plants off the continent. To return to Europe after the glaciers melted, they would have to cross the rapidly drying Sahara desert and the Mediterranean

A

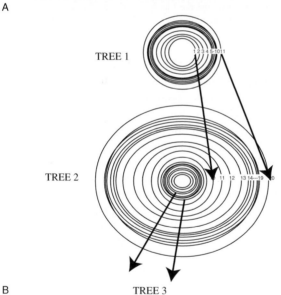

TREE 1

TREE 2

B   TREE 3

Figure 6.10. A: Tree rings seen on a felled relatively young sequoia tree. B: Principle of counting tree rings. Where there are many fossils of trees, it is possible to recognize similar patterns depicting years of abundant growth (presumably with adequate water) and years of very poor growth (see inset on photograph). These can be compared on different trees and a sequential record built. Sequences have been covered for a few thousand years

Sea. In North America the plants were driven to the southern part of the United States and Mexico, and as the glaciers melted they spread northward again. Ponce de Leon's choice of a name for what is now Florida ("Land of flowers") was not simply a publicist's gimmick. To the Europeans, the Americas had far more varieties and numbers of flowers, including far showier flowers, than Europe did. As is described in Chapter 8, many of these were eagerly collected for cultivation in Europe.

All of these observations suggest that the Great Lakes were once much more massive, as if they were a giant reservoir for melting ice, and that they have retreated considerably.

There is also biological evidence for an Ice Age (actually several successive Ice Ages). Most animals and plants live in a preferred characteristic climate. We do not expect to see palm trees in Maine or spruce trees (one of the common Christmas trees) in Florida. Coral reefs are characteristic of warm water, while moose and caribou are common in colder climates. There are many indications from the fossils of the types of animals and plants that during the period in which these geological formations were created, the living organisms were characteristic of climates colder than the current climate in the region. This conclusion can also be reached by penetrating into more subtle analysis: pollen grains, for instance, are preserved in many locations, and the type of plant that gave rise to them identified. Also, many chemical and biochemical reactions proceed differently at different temperatures, and all of the remnants of these reactions indicate a colder climate.

It was not only astonishing but, because of the many converging sources of evidence it soon became incontrovertible that many thousands of years ago much of the earth had been much colder than present. At the very least it suggested that Genesis was incomplete and that efforts to calculate the age of the earth from Genesis were most likely wrong. In more general terms, as is explained in Chapter 8, these results were congruent with the emerging analyses of geologists such as Lyell in that they indicated a much greater age for the earth. When other sciences developed, the several lines of evidence led to the conclusion that the universe was roughly one million times older, and that even the continents were at least 100 times older, than Bishop Ussher's calculation.

There were several other biological considerations as well. The creatures that live in the northern part of North America are very different from those that live in the southern part. The moose-spruce forests of Canada and Alaska are very different from the oak-maple forests of Virginia and Tennessee, the alligator-infested cypress swamps of Florida, the live oaks and pitch pine of southern United States, or the palms of Mexico. If, for instance, Kentucky once had a climate like northern Ontario, many animals and plants would not have been able to live there, so the distribution of living organisms would have been quite different. Even those that managed to hang on through the climate changes would have either to be very resourceful or be modified to deal with the severely changing climate. As the slogan goes, "adapt or die". Over the time scale of generations, the changing climate must be very stressful for all organisms.

An obvious example is how humans adapt to different climates. As we will discuss in Chapter 29, all fossil evidence suggests an origin for humans in east central Africa, and our physiology also reflects this origin. Unlike many mammals, we sweat through our skins to get rid of excess heat, and have become almost hairless, presumably to allow the sweat to evaporate and cool us. All animals have what is called a "neutral temperature": one at which they are comfortable. The neutral temperature can be recognized as the temperature at which the animal's energy consumption is lowest, because it neither has to shiver or move to keep warm or sweat or pant to keep cool. The neutral temperature for cold-weather animals such as polar bears and wolves is much lower than that for tropical animals such as monkeys. The neutral temperature for humans is identical to that of a tropical animal. Furthermore, we adapt to heat by sweating and altering our metabolism much more effectively than we adapt to cold. So how is it that humans live in all continents except Antarctica? We simply move our tropical climate wherever we go. We build houses that we can keep warm by using fire, and we use the skins of cold-weather mammals to keep our bodies at tropical temperatures. An unprotected human in a blizzard can lose enough heat in 20 minutes to die. Without these abilities, humans could never have left the tropics and, if for any reason, the climate had turned colder, we would have perished. Our closest relatives, the great apes, which have a physiology extremely similar to ours, do not know how to kindle or control fire, and they do not clothe themselves. They are therefore confined to central Africa and Asia, while humans, while humans, cover the earth.

We can assume that the same demands are placed on all animals and plants. Thus the fact that the climate of the earth has fluctuated strongly means that species have been pushed from one location to another and that they have frequently been under severe stress, which perhaps forced them to change. Also, interpreting the evidence for glaciation in a consistent manner implies an extended history of the earth, which is a further issue to address. This understanding is growing in the early 19th C. For European and American science, it is almost an awakening: If the world is not what we thought it was, what is this world in which we live? And how did it come to be? It was this atmosphere that Darwin encountered in college, and of which he was thinking when he joined the Beagle for its trip around the world.

REFERENCES

http://www.amnh.org/exhibitions/expeditions/treasure_fossil/Treasures/Giant_Sequoia/sequoia.
   html?acts (Site from American Museum of Natural History, New York, image of giant sequoia
   tree rings)

STUDY QUESTIONS

1. Is there evidence in the region in which you live for a previous ice age? What is the evidence? What other interpretations can you give for what is claimed to be your evidence? What other hypotheses might there be?

2. Is there any way to assess the age of the evidence that you can identify?
3. What might you expect would happen to the animal and plant life in your region if the climate got noticeably colder or warmer? Can they all migrate, or can the species spread elsewhere?
4. What would happen if the climate in your region got noticeably colder or warmer and a species of plant or animal could not move? For instance, fish in a lake might not have a means of moving; or animals and plants within a mountain valley might not be able to cross the barrier; or there might not be suitable climate at a different elevation on the mountain.
5. Would the amount of rainfall be likely to change if the climate got colder or warmer?

CHAPTER 7

THE VOYAGE OF THE BEAGLE

"After having been twice driven back by heavy southwestern gales, Her Majesty's sip Beagle, a ten-gun brig, under the command of Captain Fitz Roy, R.N., sailed from Devon-port on the 27th of December, 1831. The object of the expedition was to complete the survey of Patagonia and Tierra del Fuego, commenced under Captain King in 1826 to 1830—to survey the shores of Chile, Peru, and of some islands in the Pacific—and to carry a chain of chronometrical measurements round the World"

Thus begins one of the most remarkable books of all time, The Voyage of the Beagle, by Charles Darwin. This was a most fortuitous voyage. The young Charles was rather lucky to be on board. Born on the same day as Abraham Lincoln and of a distinguished family—his father was a well-known physician and son of the distinguished natural philosopher, Erasmus Darwin, and his mother was a Wedgwood of the Wedgwood China family—Charles had not been too promising a student. He had started medical school but, nauseated by surgery at a time when there was no anesthesia, he abandoned this career. He then tried theology but, to all appearances, seemed to spend his time collecting beetles and hunting, and not passionate about a career in the clergy. Although he formed a close attachment to a minister who was also a naturalist, it was easy to picture him as a ne'er-do-well. Thus, when he proposed following up the lead from his mentor John Stevens Henslow to apply for the position of naturalist and companion to the captain of the Beagle, his father—who considered that the young Charles really should be settling down into a career—offered only the most grudging tolerance: "If you can find any man of common-sense who advises you to go I will give my consent." Luckily, Charles' uncle, Josiah Wedgwood, whom his father highly respected, recommended that he go. One might guess that Mr. Wedgwood was prescient or merely despaired that Charles was going nowhere and, with the time for reflection on a long trip, might yet find a goal in life. There is however evidence that young Charles had shown some spark of talent, for Wedgwood described him as a man of "enlarged curiosity" and Henslow had recommended him by saying, "I consider you to be the best qualified person that is likely to undertake such a situation, amply qualified for collecting, observing, and noting anything worthy to be noted in natural history." Once on board, Darwin earned the nickname "The Philospher" because of his propensity for noticing, questioning, and analyzing everything that went on.

As for Captain Fitz Roy, Darwin was his fourth choice. The captain wanted a traveling companion on the very long trip, a good, intelligent, conversationalist who could be of service on the boat. Given the conventions of the time, it was a given that the companion should be of good family and well bred. This aspect was most likely what most recommended Darwin to Fitz Roy. In fact, the Beagle already had a naturalist, the ship's senior surgeon Robert McCormick, on board. Charles' easy conversation and culture rather quickly endeared him to the captain, and within four months, McCormick returned home, convinced that Charles was very much the favorite of the captain.

In any case he was an ideal choice for the trip. He had been eagerly reading Lyell's latest findings in geology, which had prepared him both for the argument of gradual change in the surface of the earth and the possibility that these changes took extensive time to accomplish. Furthermore, as he himself concludes, it was now possible to explore the world as never before: *"The short space of sixty years has made an astonishing difference in the facility of distant navigation. Even in the time of Cook, a man who left his fireside for such expeditions underwent severe privations. A yacht now, with every luxury of life, can circumnavigate the globe."* This was indeed an improvement that made this sort of exploration possible, but we should not underestimate what it would take to spend five years at sea. By today's standards it was still very demanding. Over water, Darwin was most of the time seasick. Whenever the boat docked, he went ashore to explore. Although he frequently stayed with English or other families to whom he had an introduction, many of his explorations were by horseback, one to two hundred miles at a time, camping, or climbing to the tops of mountains or through forests so dense that *"Here we were more like fishes struggling in a net than any other animal."* see Figures 7.1 and 7.2

The book itself is remarkable for many reasons. First, it is essentially a diary of a trip by a young man, and mostly it describes the geology and animal and plant life of strange lands. Nevertheless it sold very well. The 19th C was an

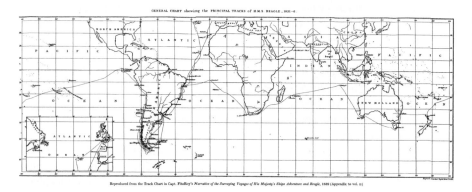

Figure 7.1. The journey of the Beagle as recorded by Captain FitzRoy. This picture has sufficient resolution to be examined using a magnifying glass. Note how long the Beagle stayed at the various ports, and the inland excursions that Darwin undertook. Credits: Charles Darwin's Diary of the voyage of the H.M.S. Beagle edited from the MS by Nora Barlow, Cambridge, University Press, 1933; Kraus Reprint Co. New York, 1969

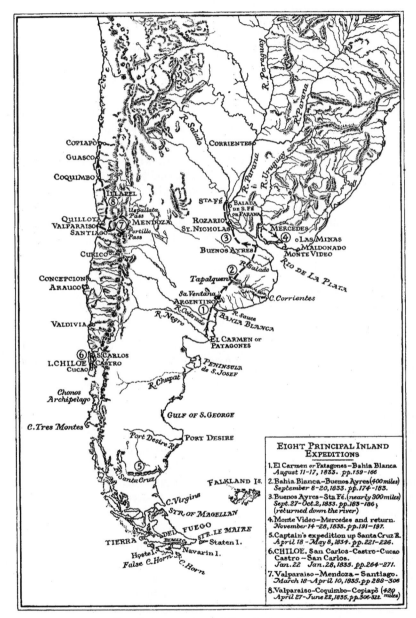

Southern portion of South America

Figure 7.2. Darwin's inland trips according to his notes. He spent a considerable part of his trip riding inland on horseback, rejoining the boat at a port further along. His diary for these excursions reflects the thoughts that occurred to him, especially in the Pampas of Argentina, at Tierra del Fuego, and in his excursions into Chile. The earthquake was at Concepcion. Credits: Charles Darwin's Diary of the voyage of the H.M.S. Beagle edited from the MS by Nora Barlow, Cambridge, University Press, 1933; Kraus Reprint Co. New York, 1969

exciting time for Europe, with explorers going to examine foreign countries, mostly looking for business opportunities or opportunities to exploit other countries—it was, after all, the period of the building of the British Empire—and many who never left home were eager to learn all they could about these lands that they could not imagine and would never see. Second, written as it is with the ingenuousness of a young man, it provides a wonderful example of his growth and maturation, including his struggles with, and indeed his abhorrence of, the concept and practice of slavery, indenture, and corrupt societies. Third, it most masterfully reveals the fundamentals of science and the mind of a scientist. For Darwin never notices an anomaly but that he questions how it came to be: why some animals and plants are found in one location and not another; why similar islands in different oceans have different flora and fauna; how mountains are raised and valleys are formed; how coral reefs and atolls are formed—in short, how the world works. Finally, the book is tantalizing in that each time Darwin notes something and questions its origin, one sees the roots of what will become The Origin of the Species. As he touches on the edge of these great ideas, he becomes most poetical: *"It is not possible for the mind to comprehend, except by a slow process, any effect which is produced by a cause repeated so often, that the multiplier itself conveys an idea, not more definite than the savage implies when he points to the hairs of his head. As often as I have seen beds of mud, sand, and shingle, accumulated to the thickness of many thousand feet, I have felt inclined to exclaim that causes, such as the present rivers and the present beaches, could never have ground down and produced such masses. But, on the other hand, when listening to the rattling noise of these torrents, and calling to mind that whole races of animals have passed away from the face of the earth, and that during this whole period, night and day, these stones have gone rattling onwards in their course, I have thought to myself, can any mountains, any continent, withstand such waste?.... Daily it is forced home on the mind of the geologist, that nothing, not even the wind that blows, is so unstable as the level of the crust of this earth."* Here he expresses the first inklings of his realization of the great age of the earth, a necessary understanding if there is to be time for evolution to occur. The book is eminently readable, even to a non-scientist, and is highly recommended, for its importance in the culture of the 19th C; for Darwin's vignettes of the social structures of the societies that he visited; and for its image of the growth of a remarkable young man.

When Darwin signed on as a naturalist for the Beagle, he had a large but relatively simple task in front of him. Bolstered by reliable and relatively safe ships and means of identifying both longitude and latitude, as well as the potential of finding new lands and resources for the economy, many nations were exploring the earth. They were interested in the gold, silver, copper, and lumber of the New World, as well as the possibility of converting to Christianity (and most likely subjugating) the inhabitants thereof. Their curiosity was piqued by the strange animals and plants described by the explorers. Given the sense that all living things were made to serve humans, Europeans wanted to know what was out there and, since the comfortable and certain world of Linnaeus was confused by the new findings, the scientific community was trying to relate the new materials to the ordered structure of the universe. As a further incentive, valuable new animals and plants had been brought back to Europe by earlier explorers. These creatures included American cotton, tomatoes, potatoes, corn, chili peppers, chocolate, avocado, sugar cane, and tobacco, as well as fur-bearing animals such as raccoons and edible animals such as turkeys (misnamed because explorers confused it with the helmeted gamecock or turkeybird, an African bird known in Europe) as well as plants potentially useful for medicinal purposes, many flowers, and other less valuable but nonetheless unusual animals and plants such as armadillos. (Cotton and sugar cane existed in the Middle East and Asia but had not attracted the attention or interest of Europeans.) There was always the hope that more would be found. Thus Darwin's job was to collect, catalog, and classify any plants or animals that he could.

Darwin, however, was also a thoughtful man, considering the relationship of the provocative ideas of his grandfather, Erasmus Darwin to his training in Theology, as well as the exciting ideas of geologists such as Lyell. There were at least two issues that were quite difficult to resolve. First, unless each species was quite constant and discrete, it would be quite difficult to understand how Noah could have accommodated on the Ark the whole gamut of life. Second, if species were not constant but could change, as domesticated animals surely could under selective breeding by humans, could that change have produced everything that exists on earth? In 6000 years one might get different varieties of dogs, but could one generate all types of animals and plants? Many thoughtful "naturalists," including Charles Darwin's grandfather Erasmus Darwin, had formulated suggestions that species could change, but in all cases either their arguments were forced and easily undercut, or no mechanism was suggested and therefore there was no compelling reason to believe these naturalists. What one required was (a) evidence that species were not stable; (b) sufficient time to allow for the change of species; and (c) a mechanism by which evolution could occur. During the voyage of the Beagle, Darwin accumulated the evidence for (a) and observed enough to convince him of (b). Since he continued to ask questions of his questions and doggedly pushed until he had solutions, over the next twenty years he worked out the mechanisms by which the evidence that he had accumulated could be explained. By the time that he published his explanation, in 1859, it was so clear and convincing that, like all great ideas and poetry, one could only ask, "why did I not think of it?" However, it was quite painful, since the argument was that the creation of the species did not need a Creator but could come about through natural mechanisms. To Darwin, admitting this possibility was "like confessing to a murder". Nevertheless, Origin of the Species today is rather boring to read. What makes it boring is that Darwin makes a point, then assiduously documents it with many examples. By the third example, we are convinced. This is because we are today so thoroughly imbued with the idea that we do not need the convincing that was necessary in the middle of the 19th C.

THE VOYAGE OF THE BEAGLE—DISCOVERIES AND PHENOMENA THAT CAUSED DARWIN TO DOUBT

The first issue that Darwin confronted was the sense of time. Most of the previous efforts to argue the evolution of organisms had foundered on the ridiculousness of the idea that everything on earth could have been produced by any non-Divine mechanism in 6000 years. However, as is discussed in Chapter 12, page 168, Lyell had argued that the great features of the earth could be explained by gradual processes that were known in the present earth: sedimentation, erosion, and earthquakes but that reading the structure of hills and geological formations in this way would suggest far more vast extensions of time. Darwin was reading and thinking about Lyell's arguments while at sea—he dedicated the book to Lyell—and what he saw stunned him. The great plains of Patagonia appeared to be outwash from the far distant Andes Cordillera, itself a magnificent range of mountains reaching in places 23,000 feet. It was this realization that provoked the expostulation quoted above. While he was still mulling these thoughts, the boat reached Concepción, Chile, where there was a great earthquake. When it was over, subtidal flats that had been underwater were now above the sea, having been lifted three to eight feet. Darwin, who had been in the mountains marveling at fossils of seashells 1300 feet above sea level and even up to 14,000 feet, realized that earthquakes such as the one he had witnessed could explain their elevation. Later (though not in *The Voyage of the Beagle*) he would estimate, from Spanish records, the frequency of earthquakes of that magnitude in the Andes and calculate the time it would take to lift the mountains 20,000 feet. His calculation had too many assumptions and was quite inaccurate, but it was well beyond any biblical calculation. By the time that he wrote *Origin of the Species* he was calculating from the rates of erosion of cliffs and the rate of accumulation of sediment the ages of various tracts of land in England. His figures, over 360 million years, were still not correct, but not terribly wrong, and 6,000 times longer than biblical time. It would be sufficient to allow the evolution that he described.

The second issue that became important during the voyage was the apparently idiosyncratic manner in which animals were distributed throughout the world. The farther he goes, the more these issues bother him: the Cabo Verde Islands and the Galapagos Islands are very similar in physical structure and proximity to the equator, but life on these two islands is very different. Why do the living organisms of the Galapagos look similar to those of South America, while those of the Cabo Verde Islands resemble those of Africa? Why is the fauna of islands so limited, in particular, lacking large mammals and frogs? Why does one find the fossils of giant armadillos only in the lands where one now finds small armadillos, and the fossils of giant sloths only in the lands where one now finds smaller sloths? Why are there camels in Africa, but llamas in South America? Why do similar species not share territories? For instance, there is a large, flightless bird called a rhea in South America (Fig. 7.3); but, in fact, there are two species, as Darwin realized and pointed out, such that one is now named after him. The ranges of the two species abut but do not overlap. Darwin wondered why. If species were created by direction of the Creator, why did they distinctly differ by location, so that some systematists would insist that the variants were different species? Darwin wondered about all of these things. In fact, he wondered why people did not wonder: *"My geological examination of the country generally created a good deal of surprise amongst the Chilenos: it was long before they could be convinced that I was not hunting for mines. This was sometimes troublesome: I found the most ready way of explaining my employment, was to ask them how it was that they themselves were not curious concerning earthquakes and volcanos?—why some springs were hot and others cold? – why there were mountains in Chile and not a hill in La Plata? These bare questions at once satisfied and silenced the greater number; some, however (like a few in England who are a century behindhand), thought that all such inquiries were useless and impious; and that it was quite sufficient that God had thus made the mountains."* It is a matter of some note that, by informal survey, a large number of today's practicing scientists had a nickname "questions" or the equivalent. All children are curious. Many of those who do not lose that curiosity become scientists.

Finally, the Beagle reached the Galapagos Islands (Turtle Islands), a group of volcanic islands approximately 500 miles to the west of Ecuador. As he understood, the islands were of relatively recent origin and had never been connected to land. What he saw on the Galapagos greatly troubled him and, though he did not understand what he saw and even missed one of the most important points, the difference between birds from different islands, so that he lumped them all together. Nevertheless, he sensed that it was terribly important: *"Considering the small size of the islands, we feel the more astonished at the number of their aboriginal beings, and at their confined range. Seeing every height crowned with its crater, and the boundaries of most of the lava-streams still distinct, we are led to believe that within a period geologically recent the unbroken ocean was here spread out. Hence, both in space and time, we seem to be brought somewhat near to that great fact—that mystery of mysteries—the first appearance of new beings on this earth."*

The islands were named because of their large population of giant turtles and lizards. Consistent with the lack of large, predatory mammals, these reptiles show no fear of humans and, when frightened, run from the sea to shore rather than to the sea. As Darwin learned from residents and observed for himself, the turtles from each island could be readily distinguished. Why should they exist here, of all places, and then vary from island to island? Even more curious were the birds. They were all unique, but relatively similar to each other and there were several varieties, distinguishable by size of their beaks. The beak size was important, for the different birds ate different kinds of seeds. There was even one bird that acted like a woodpecker. Though it did not have a woodpecker's very hard beak and

Figure 7.3. Top Left. Ostrich Africa Top Right. Rhea (South America); Bottom: Emu (Australia); Originally these birds were considered to have evolved by convergent evolution, or selection for the same traits
among unrelated animals owing to similar circumstances. Today it is recognized that they are descendents of the same ancestor. See Chapter 22. Credits: Ostrich - http://www.dreamstime.com/Ostrich-
rimage329921-resi191750 © Photographer: Steffen Foerster | Agency: Dreamstime.com. Emu -
(http://www.dreamstime.com/Australian Emu-rimage864548-resi191750 © Photographer: Martina Berg
| Agency: Dreamstime.com)

strong neck, it tore spines from cacti and used the spines as probes to dig insects from within the cacti. Even more curious than this, all the birds appeared to be finches, closely related to, but distinct from, a finch found in Ecuador. Why should these several unique species be found only in a small archipelago, 100 miles across, 500 miles from the coast of South America; and why should these birds be similar to a continental species? If God created these species at Creation, why should they not be the same as those found elsewhere in the world, for instance in the Cabo Verde Islands?

A final marvel of *Voyage of the Beagle* is a passage that only indirectly relates to the story of evolution but represents a scientific triumph of itself and for which, among geologists, he is justly famous. For he observed the atolls—curious rings of coral, many miles across, surrounding a shallow lagoon and, sometimes, a central island—of the South Pacific, and he gave the first clear and convincing explanation of their origin. You may have some image of what they may be, for these are the famous and romantic South Sea islands, such as Bikini, the Coral Sea islands, the Caicos Islands, and the Marshall Islands, that are well known idyllic hideaways. What is interesting about Darwin's analysis, though, is his perfect use of ELF logic. Falsification by experiment is not possible, but what he does is to collect as much evidence as possible—the slopes of the sea bottom within and outside of the lagoon; the slope and configuration of the central islands, including rivers and valleys; the biology of the coral-forming organisms, which survive only from the sea surface to a depth of 20–30 feet; the texture and composition of the underlying soils; the similarity of atoll corals to fringing and barrier corals; and many other features. He then applies logic to the assembly of this information, relating the patterns that he sees to known physical forces such as the destructiveness of waves at the top of the coral and what determines a specific angle of slope, and in this manner rules out (falsifies) most of the competing hypotheses, leaving him with the one surviving hypothesis. The surviving hypothesis might be surprising but, in the face of his logic and evidence, is the inevitable conclusion of his argument: that an atoll started life as a volcano. The volcano ultimately became extinct and was ringed with coral. Then, over the course of millennia, the volcano slowly sank in the sea. As it did, new coral grew on top of old coral, maintaining the ring even as the mountain disappeared beneath the waters (Fig. 7.4).

If this argument does not convince you, as it should not, given that no evidence has been presented to support it, you should certainly read the 30-some pages in which he develops his argument. It is a brilliant exposé, and a masterful demonstration of the power of ELF logic. It remains the accepted interpretation of the origin of atolls.

By the time that Darwin returned to England, he had seen a great deal, and he, like any scientist, wanted to know how it worked–how the distributions of species came about. He was deeply troubled by the strange and seemingly idiosyncratic distribution of animals and plants throughout the world; he appreciated the evidence for great age of the earth; he had seen how populations could expand, as had the wild horses in Argentina, which had escaped from the Spanish and now numbered

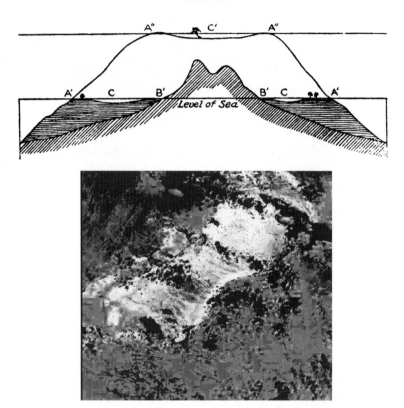

Figure 7.4. Upper: Configuration of an atoll correctly surmised, from the evidence of the slopes of the land beneath the coral and the correspondence of the reef to any island within the atoll, that an atoll started with a volcano projecting above the ocean. Coral reefs grow on the shores of the volcano. Over time, the volcano ceases to be active and, for several reasons, gradually sinks beneath the ocean. However, coral can survive only near the surface of the ocean, where it can get light. As the land sinks, the coral grows upward, building a wall up to the surface. Ultimately the volcano disappears, living only its ring of upwardly-grown coral and a shallow lagoon where the volcano once stood. What is illustrated is his diagram, with the following text:

A'A'. Outer edges of the barrier-reef at the level of the sea, with islets on it. B'B'. The shores of the included island. CC. The lagoon-channel.

A"A". Outer edges of the reef, now converted into an atoll. C'. The lagoon of the new atoll.

N.B.-According to the true scale, the depths of the lagoon-channel and lagoon are much exaggerated.

Lower: A Caribbean atoll, Los Roques. Credits: Configuration of atoll - Charles Darwin's Diary of the voyage of the H.M.S. "Beagle" edited from the MS by Nora Barlow, Cambridge, University Press, 1933; Kraus Reprint Co. New York, 1969. Los Roques - http://eol.jsc.nasa.gov/sseop/clickmap/ image ISS010-E-14222

in the thousands; he had seen species merge into one another, and very similar species, like the greater and lesser rhea, abut territories, but not commingle, while no others existed in the world; and he had seen fossils of unique animals, such as giant armadillos and sloths, in lands where smaller versions existed, but nowhere else. He had even noted in passing a peculiar land crab on the Cocos Islands that subsisted by eating coconuts. It later would become a major point in *Origin of the Species,* for the crab was adapted, both in behavior and in shape, to eat the coconut. It would first tear away the husk at the end where the eyes of the coconut were found. Then, with one strong pincer it would hammer at one of the eyes until it broke through, and finally it would turn around and with a smaller hind leg it would reach into the coconut to extract the flesh. In the sense of "how does it work?" he would ask himself how it was that one species could so perfectly adapt itself to another species, as did as well hermit crabs that perfectly fit the shells that they had borrowed or how a hawkmoth had a tongue that would just fit into a trumpet-shaped flower that seemed to be designed for the hawkmoth (Fig. 7.5).There was such a flower, an orchid, in which the nectar lay 30 cm (12 in) from the opening. In 1862, Darwin predicted that a hawkmoth with a 12 in tongue would be found. Twenty-one years later, the insect was identified and given the subspecies name *praedicta,* meaning "predicted". Thus, by the time that Darwin returned, he had the evidence. He sought the logic of "how does it work?" which would allow him to construct intellectual arguments to test by attempting to falsify predictions.

Darwin mulled over the implications for many years. A few years after returning to England, he read Malthus' essay (chapter 10), and he realized that the argument that Malthus presented for cities applied also to the animal world. He did not invent the terms—"survival of the fittest" was first used by Herbert Spencer, while "nature red in tooth and claw" was from Tennyson—but he made the connection between variation, the value of some variation, and the culling of a species. Darwin was

Figure 7.5. Left: Adaptation of a hawkmoth to a flower. The flower is tobacco (*Nicotiana*). The inset at the upper right is the head of a hawkmoth that drinks nectar from the flower, at approximately equivalent magnification. The hawkmoth's tongue, here coiled, is long enough so that it can hover over the flower while reaching the nectar at the base of the flower. Right: Darwin's Star Orchid. The nector is at the bottom of long spurs, and reached by a tongue entering in the center of the flower. See text. Credits: Robert Raguso, printed in Int. J. Plant Sci. 2003, 164 (6): 877–892 (reprinted with permission)

to some extent afraid of the type of scandal that greeted Robert Chambers (who was quite viciously attached by scientists, preachers, and newspapers) and to some extent he desired to address every implication of his theory, as he worked toward a grand encyclopedic presentation of his idea, when Alfred Russel Wallace wrote to him. Wallace, likewise a naturalist who had visited Brazil and then worked his way to Malaysia, had observed the geographic and individual variation of species. Then, while recovering from a bout of malaria, he also read Malthus, and he recognized the same connection. In 1858, he drafted an essay to describe the connection between Malthusian logic and the evolution of species, and he sent it to the by-now-renowned Darwin in the hope that Darwin could comment on it and perhaps get it presented. Darwin, an honorable man, got in contact with Lyell, saying that he, Darwin, was obliged to present Wallace's paper as the first publication of the idea. Lyell however knew that Darwin had written notes outlining the theory much before and had published some articles in which the theory was hinted at or quietly mentioned. Therefore Lyell arranged to have both papers presented simultaneously in 1858. Wallace's argument, though less detailed, is very clear and easy reading (9 typewritten pages), and is available on a website (see references). It differs from Darwin's book mainly that it is a précis (a summary of ideas) based on the logic of animal overbreeding and the normal absence of population explosions. He also places more emphasis on the selection of varieties rather than of individuals and considers controlled breeding of domestic animals to be so different from the wild as to be non-instructive, whereas Darwin argued that they were different manifestations of the same process (human selection of traits desired by humans, as opposed to natural selection of traits suited for survival.) Nevertheless the similarity of Wallace's thesis to Darwin's is remarkable, and Wallace's article is highly recommended.

We talk today of Darwin but rarely mention Wallace for a few reasons. First, Darwin was in England, was already known and respected for his analyses of beetles and mollusks, and for his explanation of atolls; and he was the descendent of a distinguished family. Wallace remained in the East, was far less known, and did not have the backing that a more prestigious family would have given. Much of this sounds as if the differences were entirely social, but Darwin followed up by publishing 18 months later what he considered to be an abridgement of his full theory, which was what we consider today the rich and voluminous *Origin of the Species*, while Wallace published far less, was reticent, and ultimately stayed in the background. *Origin of the Species* sold out immediately, was reprinted many times, and was hotly debated. Although Darwin was not in good health and did not do much public speaking, he remained in close contact with his strongest supporters, such as initially Lyell and later Thomas (T.H.) Huxley, who argued his case publicly. Wallace did continue a distinguished career and today is also known for his recognition of what is now known as Wallace's line—an imaginary line running more-or-less east to west through the Malaysian islands. On the north side of the line, the flora and fauna are essentially all Asian, while on the south side of the line, they are predominantly Australian. Today we know that very strong currents and winds prevented the various species from expanding from their continental origins beyond those islands. The geographical boundaries that Wallace observed helped him to formulate his hypothesis.

The last question is how the theory of natural selection occured to two individuals almost simultaneously. The simultaneity provides an excellent example of the ELF rule. By the mid-19th century, the evidence based on exploration had accumulated and the initial step of the logic was presented by Malthus. The time was now ripe to ask where species came from, and it was a "hot" question at the time, much as there was a race to understand the structure of DNA (Chapter 14, page 191) and to elucidate the genetic code (Chapter 16, page 227). The final component was the falsification, which in this case allowed both Darwin and Wallace to test their hypothesis against several new situations, and to conclude that all other theories fell by the wayside. We shall encounter a similar apparent coincidence in Chapter 13, page 175, in the situation of the simultaneous rediscovery of Mendel's experiments. Again, as here, the intellectual attitude of the moment is very important in driving science.

REFERENCES

Darwin, Charles, Diary of the Voyage of the H.M.S. Beagle (edited from the MS by Nora Barlow, Cambridge, University Press, 1933; Kraus Reprint Co., New York, 1969, 440 pp.

Darwin, Charles, 2004, The voyage of the Beagle, Introduction by Catherine A. Henze, Barnes and Noble, New York (notebooks first published 1909).

Larson, Edward J. 2001. Evolution's workshop. God and science on the Galapagos Islands. Basic Books (Perseus Books Group), New York.

http://www.wku.edu/~smithch/index1.htm (Essay on Wallace from Western Kentucky University).

http://www.clfs.umd.edu/emeritus/reveal/pbio/darwin/darwindex.html (Darwin-Wallace 1958 paper on Evolution, from University of Maryland).

STUDY QUESTIONS

1. What was the strongest evidence from the biology that Darwin encountered that led him to his hypothesis?
2. What was the strongest evidence from the geology and geography that led Darwin to his hypothesis?
3. What was the strongest evidence from the fossil record that led Darwin to his hypothesis?
4. What characteristics of Darwin's personality and style were important to assure that he would draw the theories from his experiences?
5. Argue for or against the hypothesis that it is the time or the historical moment, not the individual, that determines when great advances in science are made.

CHAPTER 8

IS THE EARTH OLD ENOUGH
FOR EVOLUTION?

THE PROBLEM

One of the most important passages in Voyage of the Beagle is Darwin's description of his explorations in the Andes Mountains. He had climbed the mountains in Chile, observing marine fossils at high elevation and noting that the higher he went, the more they differed from those at the shore. He was pondering Lyell's remarks about them and debating their origin. When he returned to Santiago, he observed the effects of an earthquake that had occurred while he was in the mountains. The land had lifted approximately three feet, and shelves of land that had formerly been in shallow water were now lifted above the water, and the shellfish that lived there were dead and drying. Seeing this, Darwin wondered if the Andes had been lifted by such incremental shifts as Lyell had proposed. The frequency of earthquakes and their effects had been noted since the Spanish had occupied the land 500 years before. Using these figures: height of Andes, feet lifted per episode, and number of episodes in 500 years, Darwin was able to calculate an approximate age of the Andes. The figure that he came up with, 100,000 years, proved to be quite inaccurate, but it served one major purpose. The figure was way beyond the theological age of 4000 years, and suggested that the earth was much older than that.

HOW DO YOU MEASURE THE AGE OF THE EARTH?

How old is the earth? And how could we possibly measure something like the age of the earth? To measure something one needs a ruler, a scale, or a clock. The measuring device must be calibrated in some fashion, even if the calibration is crude. For instance, an inch was the length of the end segment of the thumb (it still retains the name "thumb" in French) and a foot was the length of the king's foot. It can be more precise, such as a specific fraction of the earth's diameter (kilometer) or the weight of a specific volume of water at a specific temperature (gram), but one aspect of calibration and measurement is that you must be able to establish both ends of the measurement. We can get very accurate clocks, but you cannot just read a clock backwards to find a beginning of time. It is a problem like that of a digital clock. If the power goes off, the clock stops. It will restart, showing midnight, when power is restored, and will continue from that point on. If you come home, for instance, at 6 PM, to find that the clock has stopped and restarted, and that it now reads 2 AM, you can conclude that power was restored two hours earlier, at 4 PM. However, you cannot determine at what point power was lost.

Scientists in several disciplines asked if there were any way to judge the age of the earth. During the 18th and 19th centuries, a few techniques became accessible, and we have several far more complex means today, and they all converge on a common number. The presentation of this story emphasizes two major issues of scientific inquiry: the convergence of evidence from multiple, independent means of evaluation, and the concept of falsification of a hypothesis. Dating by tree rings is discussed in Chapter 6, page 75–76. Others are described below.

In the time since Galileo had first seen craters on the moon, astronomers had recognized that the craters resembled those made by the impact of meteorites on the earth, and the realization grew that they were in fact meteorite impact craters. On the moon, there is no wind and no water, so that a crater, once formed, does not deteriorate, erode, or fill up with silt or dust. There are two features about such craters that allow one to make an estimate of the age of the moon. First, meteorites still strike the moon, so that at least the current rate of formation can be calculated. Second, one can determine the order in which the impacts occurred. For instance, in the photograph in Figure 8.1, the small, heavily shadowed crater just below and

Figure 8.1. When one impact follows another, the second will obliterate the first. A. drawing of sequence. The impacts occurred in the succession 1, 2, 3. B. Photograph of sequential impacts in craters on the moon. The sequence of several craters can be seen. From the Apollo 11 flight. Credits: http://www.hq.nasa.gov/office/pao/History/alsj/a11/AS11-44-6609.jpg

to the right of the center sits in the wall of a heavily eroded crater in the lower half of the picture. Therefore, the bottom crater must have formed first and was partially buried or destroyed by the second impact.

Such observations allow the construction of a crude calendar. If one assumes that the rain of meteorites is constant—it is not, but the simultaneous impacts can be identified—then by counting the number of impacts, one can establish an estimate of how long ago the first impact occurred, in the same manner that one can determine the age of a tree by counting rings and knowing that one ring is formed each year.

LORD KELVIN AND THERMAL COOLING

You are certainly familiar with the fact that when a hot object is removed from the source of the heat, it cools slowly, with the outside becoming cool before the inside. If a large object and a small object are heated to the same temperature, for instance by being placed in boiling water, the smaller object will cool faster than the larger object. The total amount of heat in the object is called heat capacity. The rate at which an object cools depends on its size, how much surface it has, its heat capacity, the difference between its temperature and that of its surroundings, and the ability of the surroundings to absorb the heat. Each of these factors is known or can be measured. Thus by 1841 Lord Kelvin (for whom the Kelvin temperature scale is named) had calculated the temperature of the sun, based on the fact that the color emitted by an object changes with temperature (red hot steel is about 1800° K (2800° F), white hot steel is 5500° (9500° F), and a hot blue flame is 16,000° K (28,000° F)). Knowing from miners and from volcanoes that the center of the Earth was hot, he made the theoretical assumption (hypothesis) that the earth had broken free from the sun, starting at the same temperature, and from that time was a VERY large spherical object cooling in space. Using the same laws of heat transfer that were used for common objects, and correcting for the warming effect of the sun, he calculated how long it would take for the Earth to cool to its present temperature, testing his hypothesis by seeing if he could come up with a reasonable figure. He came up with a value that the world was hundreds of millions of years old. He later revised his calculations, finally settling on somewhat less than 20 million years, but it was still at least 5,000 times longer than the biblical age. This calculation proved to be very inaccurate, since radioactivity, of which he knew nothing, contributes substantial heat to the earth. Nevertheless, it was a further argument that the Earth was quite old. Today's calculations, corrected for radioactivity, give a figure in the low billions of years.

EROSION

Erosion can be measured in several ways. A fairly easy one, if one has a historical record, is to follow the change in shape of land over time, and to recognize the extent that the earth reflects a continuous process. For instance, barrier islands usually shift or recede with time, and the changes can be followed by comparing

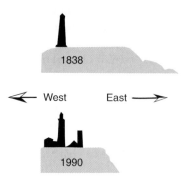

Figure 8.2. Erosion on a barrier island. Judging by the position of the lighthouse, the eastern end of Long Island, New York is disappearing at approximately one foot per year. Credits: Montauk - Redrawn from http://www.montauklighthouse.com/erosion.htm

documents. The eastern end of Long Island, New York, is receding at approximately one foot per year, as can easily be determined by comparing 19th C measurements to those of today (Fig. 8.2). Similarly, Niagara Falls has worked its way back from the original cliff face over approximately twelve thousand years, and the Mississippi

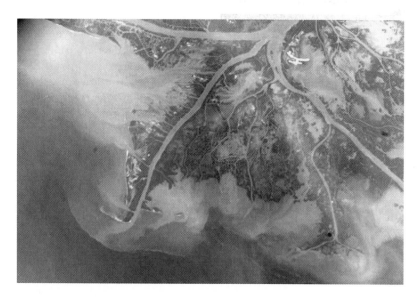

Figure 8.3. Deposition of sediment. Salt water can hold much less sediment than fresh water. Thus mud flowing from the Mississippi is deposited in the Gulf of Mexico in an alluvial plain. Modern deposition is quite prominent in this satellite image, and the entire region south of New Orleans has been built in this fashion. Credits: Mississippi delta - http://eol.jsc.nasa.gov/sseop/clickmap/ cite as above image Miss_deltaISS011-E-5949

River has continually deposited mud where it meets the Gulf of Mexico, forming a large delta (Fig. 8.3).

SEDIMENTATION

Mud and stones settle out of water onto the bottom, and the character of what is being washed into the water changes the nature of the sediment, forming layers. From the patterns of sedimentation, one can easily interpret the order of events. For instance, the later sediment will be on top of the earlier sediment (Fig. 8.4) and, in a single incident, larger stones will settle out faster than pebbles and sand. This argument was well and forcefully made by Nicholas Steno at the end of the 17th C (page 40). Thus one can distinguish between sediment brought by a flood and sediment building slowly as a muddy river settles out. The limitation of this

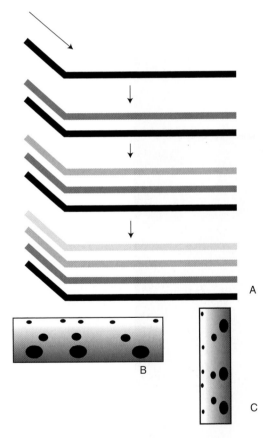

Figure 8.4. Reading sedimentation lines. A: Owing to gravity, later sediments are on top of earlier sediments. B: in one period of rapid deposition, heavier stones will settle out faster than smaller stones or pebbles, so that the original orientation can be determined even if the piece is tilted (C)

analysis is that, although the order of events is clear, it is difficult to establish a time scale. A river might silt out at a rate that accumulates an inch of soil every ten years, but a major flood might bring enough sediment to add ten inches of sediment in one event. What made this question important was the realization that there were massive depths of soil that appeared to be sedimentary. Some of the best examples are found in the American West (Figs. 8.5 and 8.6). The striping indicated three aspects of considerable interest. First, conditions must have alternated during the formation of these soils. For instance, the red stripes were red because they contained a lot of iron (rust), which indicated that they contained sea salts and were formed in sea water. The alternating white stripes represented more freshwater

Figure 8.5a.

Figure 8.5b. Extensive sedimentation lines, Bryce Canyon, Arizona. In this and the following figures, these lands are 5–7,000 feet above sea level

conditions. Black stripes usually contained much organic matter and evidence of marshland plants, but the white stripes with which they alternated were white because of the remnants of shellfish, indicating deeper water. Thus the level of the sea relative to the land must have fluctuated. Second, allowing any reasonable rate of accumulation of sediment comparable to what we can observe today, it must have taken many thousands of years, perhaps even millions, to build sediment in this

Figure 8.6. Extensive sedimentation lines, unnamed hill, Arizona

fashion. This would be much longer than the Biblical 6000 years for the age of the earth. If one argued for the continuity of processes, or gradual formation of soils, as Lyell argued (page 168) as opposed to a catastrophic, sudden formation of these soils, then there was a conflict to be resolved. Finally, these sedimentary rocks are found at considerable elevation (over 5000 feet) and far inland in the United States. Why should one find marine sediments a mile in the air, and over one thousand miles from the sea? Minimally, it was plausible to argue that the earth might be older than 6000 years. This argument would be particularly forceful in the instance of the Grand Canyon (Fig. 8.7), which, though first seen by Europeans earlier, was studied with considerable interest in the early 19th C. Not only were the depths of sediment enormous, but by making the supposition that the Colorado River had cut the canyon—which seemed reasonable based on its appearance and the obvious structure of small canyons and creek beds in the West (Fig. 8.8)—then one was faced with the possibility that it took hundreds of thousands of years for the river to cut the canyon. Furthermore, there are marine fossils at the top of the canyon, at 7,000 feet (see Fig. 3.6, page 41). Making even reasonable estimates for rates of accumulation of sediment and rates of erosion, we come up with numbers that, prior to the exploration of the world, would have been inconceivable. Current dating indicates that the sediment was laid down over a period of 1.6 billion years and that the Colorado River cut through 4,600 feet of this accumulation in approximately 5 million years. Although for the reasons suggested above the true numbers were

Figure 8.7. Sedimentation lines extending thousands of feet as exposed in the Grand Canyon

Figure 8.8. Incipient formation of a canyon, seen as the creek cuts through the land. The hillside in the background was probably formed by the same creek in an earlier era

not known in the 19th C, the possibility that the numbers might be very large was shocking.

MOUNTAIN BUILDING

Lyell had proposed that lands could be lifted from the seas by unknown processes that we would describe today as mountain building. Although the mechanisms were not known, the existence of marine fossils on the slopes again argued that such processes did occur and must be painfully slow. Lyell and others tried to estimate the speed at which they occurred, again arguing that the process was gradual rather than catastrophic. Darwin took Lyell's latest book with him on the Beagle (Chapter 7, page 81) and, encountering an earthquake on the coast of Chile, used it to examine Lyell's hypothesis. In brief, he did a simple calculation. The earthquake lifted the shoreline approximately 3 feet. According to records kept since the Spanish had been in Chile, earthquakes of this magnitude occurred approximately every twenty years. The Andes reach heights of 14,000 feet. How old are the Andes, if this hypothesis is correct?

$$3 \text{ feet}/20 \text{ years} = 14,000 \text{ feet}/X \text{ years}$$

$$X = \frac{14,000 \text{ feet} * 20 \text{ years}}{3 \text{ feet}} = 94,000 \text{ years}$$

This figure is far from what we calculate today, but its importance is that it is very far from the estimate of the Bible, and it began to address the biggest conundrum for the acceptance of the hypothesis of evolution, that there was not enough time for it to occur. That humans could breed dogs or fish or corn or peaches to their liking was undisputed, but the issue was that, in 6000 years, there was no way to create the variety of living things seen on earth. To believe in evolution, one would have to believe that the earth was very old.

All of these measurements are highly suggestive and provocative, but they are relative and are based on assumptions. It is rather like the old joke in which the watchmaker tells the church bell-ringer that he is so impressed with the precision of the bell ringer that he sets the clock he has on display by the noon bells. The bell ringer answers, "That's interesting, because I ring the bells according to the display clock." In other words, there was no means of calibrating clocks based on sedimentation or erosion.

RADIODECAY

One type of clock that can be read backward is that of radioactivity. Although it was unknown to Darwin, once radioactivity had been discovered by Pierre Jolie and the Curies toward the end of the 19th C, physicists quickly determined its primary properties. The one that is most relevant to our argument is that radioactive decay occurs among individual atoms without reference to other atoms. This property, called zero-order kinetics, means that the rate of decay is constant and depends neither on the concentration of the radioisotope nor on temperature. First-order reactions, which depend on the encounter of two molecules such as an acid and a base, increase in rate as the temperature increases (speeding up the molecules) or the concentration increases (increasing the likelihood of a collision between the molecules). For zero-order reactions, the rate of change is constant regardless of temperature or concentration. For instance, if there are one million atoms of tritium, 500,000 will decay in 11 years whether the tritium is found in Antarctica or Brazil, and whether the tritium is found as an extremely low percentage of water or as a concentrated pure laboratory preparation. Radioactive decay can also be detected in vanishingly small amounts of chemicals. Tritium can be easily detected if there is 5×10^{-15} g of radioactive water present. To put this in perspective, if you take the smallest amount of a fine powder such as powdered sugar that you can see, dissolve it in a bathtub, and then remove one drop of this solution, you can easily detect the tritium. There is also one other property of interest: A radioactive molecule decays to an identifiable molecule. Thus ^{238}uranium (the radioactive form of uranium) decays to ^{208}lead. As is explained below, these properties can be used to calculate the age of a rock and thus, potentially, the earth.

There is one slight complication, but one that is easily resolved by mathematics. This is that a specific fraction of radioactive atoms will decay in a given time. What this means is that if one has 1000 atoms of tritium, 500 of them (half of them) will decay during the first 11 years. Of the remaining 500 atoms, another half, or 250, will decay during the next 11 years. We define this property as the half-life of the isotope. It is the period during which half of the remaining radioisotopic

Table 8.1. Decay of radioisotopes as a function of half life

Starting time	Radioactive atoms	Decayed atoms
0	1,000,000	0
11 years	500,000	500,000
22 years	250,000	750,000
33 years	125,000	875,000
44 years	67,500	942,500

atoms decay. The curve of radioactive decay will not be linear but will gradually decline in a mathematical form called an asymptotic decline (Table 8.1, Fig. 8.9). During the first half life, half of the atoms will decay. During the next half life, half of the remaining half, or one quarter of the total, will decay. During the next half life, half of the remaining quarter, or one eighth of the total, will decay. This will continue until the last atom decays. Different types of isotopes have different half-lives, ranging, for common isotopes, from hours to billions of years (Table 8.2).

How does all this help us to measure the age of the earth? Suppose, as Lord Kelvin suggested, that the Earth was molten at first. In liquids, heavier materials such as metals will sink (toward the center of the earth), while lighter materials will float toward the top. As the earth cooled, the materials would solidify. Some materials would contain substantial amounts of metals. Now suppose that in one rock there is a measurable amount of ^{238}uranium, which can decay at a known rate, independent of the temperature or the concentration of uranium, to ^{205}lead. The amount of lead will increase, and the amount of uranium will decrease, as a function of time. In 4.5 billion years, half of the original uranium will have turned

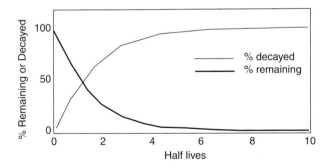

Figure 8.9. Radioactive decay. Over time, a defined percentage of the remaining radioactive atoms will decay, leading to an exponential decrease in the total radioactivity and an exponential increase in the product of the decay. Half lives for different elements range from fractions of seconds to hundreds of thousands of years. Although these are curves rather than straight lines, it is easy mathematically to determine the age of the material by comparing the ratio of precursor radioactive atom to product atom, or the ratio of precursor radioactive atoms to non-radioactive atoms. Not all atoms are radioactive. For instance, approximately one atom of carbon in one trillion is radioactive

Table 8.2. Half lives of elements

Uranium 238 (used in bombs)	4.5 billion years
Carbon 14 (used to measure age of earth)	5,730 years
Strontium 90 (can get into bone)	29 years
Tritium (Hydrogen 3, used in watch dials)	11 years
Phosphorus 32 (used in medicine)	14.3 days
Iodine 131 (used in medicine)	8 days
Most material from Three Mile Island	Less than one second

to lead. Using a similar logic, the time that the uranium had been in the rock could be calculated from the ratio of lead to uranium. Using this type of logic, scientists looked all around the Earth for rocks containing uranium. In several locations, most notably in western Canada and Australia, some rocks gave calculated ages of 2.5 billion years or more.

RED SHIFT

You have probably noticed that as an airplane, a race car, or a train passes, the sound of its motors changes. This is called the Doppler Effect, after the physicist who first described and explained it. It works as follows: Imagine the steady hum of an electric motor, which generates sound according to the 60 cycles/second of the electric current, and the sound comes toward you at the speed of sound, approximately 600 mph. If a jet airplane is heading toward you at near the speed of sound, each peak of the cycle will start closer to you than the last, arriving sooner and effectively increasing the frequency to approximately 120 cycles/second, increasing the tone or pitch of the sound. As the plane passes by you, the frequency will drop to 60 cycles/second. As it goes away from you, each peak will start farther from you and reach you later, effectively delaying the cycles to 30 cycles/second—a distinctly lower pitch (Fig. 8.10). The equivalent to this change in pitch is seen also with light. Because the mechanism is the same for light and sound is the same, the phenomenon is called "red shift". Long wavelength light is red, and short wavelength light is blue (see Fig. 19.2, page 273). (A rainbow is generated because when light is reflected through glass (a prism) or water (a raindrop), the shorter wavelength rays are bent more than the longer ones—see Fig. 8.11.) The equivalent of a lower tone when an object moves away is a change in color toward red; of a higher tone as an object approaches; a change in color toward blue. Fortunately, there are absolute values for light: When sodium is heated, it emits yellow light at a very precise wavelength, termed a "line" on a spectrum of colors. The light from sodium and other elements can be detected in the emissions of stars. If the star is moving toward the Earth, the lines from sodium and other elements will be compressed, or shifted toward blue. If the star is moving away from the Earth, the lines will be shifted toward red. Almost all stars appear to be moving away from the Earth. From the magnitude of the red shift, one can tell how rapidly the star is receding from the Earth. By using figures such as these and other measurements

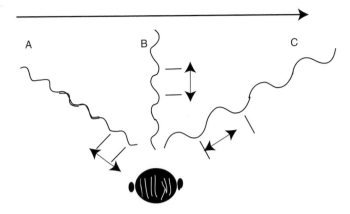

Figure 8.10. Doppler effect or red shift. As an object moves (arrow at top) and at first comes toward the observer (A), since each wave travels at the same speed but the later waves start from a position closer to the observer, the waves are perceived as occurring at higher frequency (higher pitch or shift to the blue end of the spectrum). As it passes directly in front of the observer (B) the pitch is heard correctly. As the object moves away (C), waves are perceived as occurring less frequently (lower pitch; shift toward red). Since the light given off by heated atoms is always at a specific wavelength, the shift of the wavelength to the red or blue indicates the relative motion of the object

that give estimates of the current distances of the stars, one can calculate how long it would have taken the stars to expand from a common starting point, assuming that they continue to move along their original paths (the Big Bang theory). The calculation indicates an approximate age for the beginning of the universe of about 6 billion years ago.

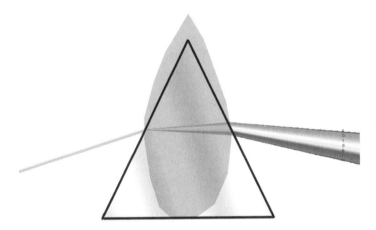

Figure 8.11. When light enters a prism or a raindrop at an angle, it is refracted (bent) during its passage through the medium and back into the air. This is why a spoon in a glass of water can appear to be broken. Short wavelength light (violet) is refracted more than long wavelength light (red), breaking the light into a spectrum (from top to bottom: red, orange, yellow, green blue, indigo, violet)

Branching and Interconnection of Sciences

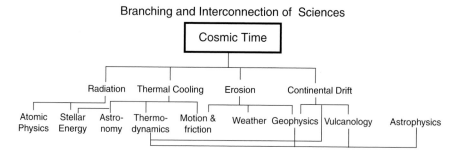

Figure 8.12. The age of the earth is identified through many sciences, each using different methods. Some of the methods serve more than one function, but all converge on a common understanding of the age of the earth

Of the many lessons to be learned here, one more general principle is quite important. The **convergence of multiple independent means of assessment** provides very solid argument in support of a hypothesis. This topic is the subject of Chapter 9, immediately following, but it is worth noting that the mechanisms of erosion, of earthquakes, of thermal cooling, of red shift, and of radioactive decay, in no way depend on each other. Nevertheless, when all factors are taken into account (for instance, that radioactivity adds to the heat of the earth), they all give dates for the age of the earth that are consistent with each other. We can conclude that the value that they give does not derive from our misunderstanding of our means of measurement, such as using an inaccurate thermometer or mistaking a centimeter ruler for a yardstick. All these sciences, and more, give the same result (Fig. 8.12). To claim that one interpretation is false would require us to disprove the findings of all the other sciences as well. There is essentially no doubt that the world is very old.

REFERENCES

Dalrymple, G. Brent, 1991, *The Age of the Earth*: Stanford, Calif., Stanford University Press, 474 p.

Darwin, C., The Origin of the Species

Harland, W.B.; Armstrong, R. L.; Cox, A.V.; Craig, L.E.; Smith, A.G.; Smith, D.G., 1990. A Geologic Time Scale, 1989 edition. Cambridge University Press: Cambridge, p.1–263

Powell, James Lawrence, 2001, *Mysteries of Terra Firma: the Age and Evolution of the Earth,* Simon & Schuster

Wilde S. A., Valley J.W., Peck W.H. and Graham C.M. (2001) *Evidence from detrital zircons for the existence of continental crust and oceans on the Earth 4.4 Gyr ago.*

Nature, v. 409, pp. 175–178 http://nationalacademies.org/evolution/ (free downloads) http://pubs. usgs.gov/gip/geotime/age.html

STUDY QUESTIONS

1. A hospital that uses [131]iodine spills a little. Assuming that a safe level is 1/1000 of the amount that is spilled, how long will it take to get to that level if it is not cleaned up?

2. The next time that a jet airplane passes overhead or a high-speed trans-continental freight train passes, notice the change of tone as it passes. Diagram how this works.

3. Why do we not notice the red shift in the sound of a passing car or a color change when a baseball is thrown past us?

4. Take two similarly shaped containers, for instance two jars, of very different sizes. Fill both with hot water from the same source, and cover the jars. Without otherwise disturbing them, measure the temperature of the water inside the jars and the temperature of the surfaces of the jars for several hours after you begin the experiment.

5. If you *really* want to try something challenging, try to extrapolate this measurement to the size of the earth. Suppose that your small jar has a diameter of 5 cm (about 22½ inches) and your large jar has a diameter of 10 cm. The surface temperature of the small jar dropped from 60° C (about 140° F) to 30° C in 30 minutes, or 1° C/minute. The surface temperature of the large jar dropped from 60° to 45° in the same time, or 0.5°/minute. You can plot these results with diameter on the X axis and rate of cooling on the Y axis. Now assume that the jar is as wide as the earth—12,700 km, or 12,700,000 m, or 1,270,000,000 cm. How long would it take for the jar to cool so that the surface was a comfortable 25° C? (Obviously the calculation is much more complex, but this is essentially what Kelvin did.)

6. If you live along the west coast of the US, identify a suitable website or library source that can give you an estimate of the rate at which earthquakes lift the land. Now look up the height of the mountains in the region. How long would it take for earthquakes to create those mountains?

7. How far has Niagara Falls moved since the last ice age?

8. What evidence in your area is there for the age of the earth?

PART 3

**ORIGIN OF THE THEORY OF EVOLUTION:
SOCIAL ASPECTS**

CHAPTER 9

EVALUATING DATA

THE TESTABLE HYPOTHESIS AND MULTIPLE INDEPENDENT MEANS OF CONFIRMATION

Science deals with the mechanisms of how the world operates, and one of its basic tenets is that we determine how things operate by building hypotheses and then designing experiments to attempt to falsify or disprove the hypothesis. It is therefore extremely important to design the experiment and to describe it in such a manner that other scientists can repeat it to convince themselves of the validity of the results. Although it is not impossible, in some fields such as astronomy and evolution, for obvious reasons it is difficult to conduct experiments. In these cases we use other, less direct, means of validating our hypotheses. For instance, we might predict what we should find in a situation we have not yet investigated and then investigate it. Such an approach was used to test Einstein's Theory of Relativity by realizing that the theory predicted that light rays could be bent by gravity. The next full eclipse of the sun provided the opportunity to determine if the light coming from stars almost behind the sun was bent by the gravity of the sun (Fig. 9.1, discussed in more detail on page 114).

The experiment was conducted and proved to be one of the first convincing arguments in favor of the theory. This is called a thought experiment or, since this type of experiment was first elaborated in Germany, a Gedanken experiment. Most of us do such experiments on a regular basis. For instance, as a child and left-hander, in baseball I routinely hit balls to right field. Fielders knew this and positioned themselves to catch my hits. I hypothesized that I started my swing too early. If I could resist starting my swing for a fraction of a second, I could get the ball away from right field. I tried this approach and achieved at least a partial success.

Another approach, especially useful when one has little option to modify the situation, is to accumulate many independent lines of evidence that point to the same conclusion. This is a difficult concept to understand, even though it too is fairly commonly used in everyday life. For instance, if you see a flash of light, hear a boom, and smell smoke, you are quickly convinced that an explosion has taken place. Your seeing the flash of light depended on light waves, which you detected with your eyes. What you heard depended on sound waves, or vibrations of air (and sometimes more solid material, in which case you would feel the shaking) and were

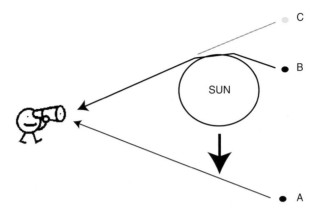

Figure 9.1. The Einstein Gravity Experiment. Since even if light is bent the observer interprets it to have moved in a straight line, if it is bent the star appears to shift position. As the sun moves across the heavens, during an eclipse the position of stars distant from the sun can be correctly assessed (A). When the star is behind the sun, the light is bent or deflected by the gravity of the sun (B) so to the observer the star appears to have jumped ahead (gray line, C). An object the size of the sun is large enough to produce a measurable shift, but the effect can only be seen when the light of the sun is blocked by the moon in a solar eclipse

detected with your ears. The odor is chemical and is detected by odor-sensing cells in your nose and sinus. You could see a flash, like a camera's flash, without hearing a boom; you could hear a boom, if an object fell, without light; and you could smell an odor independent of either. Not one of these sensations (inputs of data) depends on the other. Yet each one supports the theory that an explosion has taken place. Even more convincing, the direction from which each comes can be ascertained. If they all appear to come from the same place, the direction of each also supports the theory of an explosion. Each by itself could suggest an explosion or something else. An acrid odor alone might suggest a fire but not an explosion; a boom without a flash of light or an odor might suggest that something has fallen. But all three together indicate an explosion in a particular direction. A charred hole remaining in the region would be another bit of evidence, as would the information that a tank of acetylene gas (welding gas) had been stored in the area. This is what is meant by *multiple independent sources of evidence*. It is different, for instance, from the argument that you heard a boom and a sound-recording device indicated a loud noise. Both of these sources of data derive from the same source, the generation of sound waves, and are therefore not independent. They are independent evidence that there was a sound, but not independent evidence of an explosion. A worse example would be if, in an earthquake in a rainstorm, a building wall cracked and water got into the building, the entrance of water would not be further proof of the tremor, because it would depend on the first proof, the crack in the wall. It would be evidence dependent on other evidence. (Table 9.1).

Such documentation is routine in daily investigations, now so beloved of forensic police stories on television. For instance, one can deduce a car's speed from the

Table 9.1. Dependent evidence and Independent evidence

Independent evidence: The world is old

Evidence 1	Evidence 2	Evidence 3	Evidence 4	Evidence 5
Isotope concentrations	Red shift	Thermal calculations	Uplift of mountains	Erosion
Theory of decay of radioisotopes	Wave theory (properties of waves in water as well as in sound and light)	Theory of energy, radiation, and specific heat of objects	Theory of plate tectonics (movement of continents)	Geophysics; friction and integrity of matter

Independent evidence: The world is round

Evidence 1	Evidence 2	Evidence 3	Evidence 4	Evidence 5
Measurement of angle of sun at noon	Shadow of earth on moon	Observation that other heavenly bodies are round	Circumnavigation of globe (Magellan)	View of earth from moon
Geometry (Greek)	Light propagation (light travels in straight lines)	Inference	Direct demonstration	Direct visualization

Dependent evidence: The earth was created 6000 years ago

Evidence 1	Evidence 2	Evidence 3	Evidence 4	Evidence 5
Calculations from the Old and New Testaments give this figure	The Catholic Church states this as belief	Jewish Fundamentalists and older philosophers state this as belief	Islamic Fundamentalists state this as belief	Fundamentalist (Protestant) Christians state this as belief
Written document	Ultimate source is same as first column	Ultimate source is Old Testament as in first column	Ultimate source is Old Testament as in first column	Ultimate source is same as first column

(*continued*)

Table 9.1. (continued)

Evidence 1	Evidence 2	Evidence 3	Evidence 4	Evidence 5
Dependent evidence: Little old ladies shouldn't drive				
I saw a little old lady driving dangerously	My wife saw her too	The driver coming the other way honked	My child says so too	
One observation is not a generalization	Confirms the fact that she was not driving well, but does not extend the generalization	Confirms the fact that she was not driving well, but does not extend the generalization	My child got the information about the incident from me	
Dependent evidence: News item is true (in several papers & book)				
A bomb was found on a plane Reported on CNN	A bomb was found on a plane NY Times reports that CNN reported it	A bomb was found on a plane Time magazine reports that CNN reported it	A bomb was found on a plane NY Post reports that CNN reported it	
Dependent evidence: Large cells are more likely to be cancerous				
There are more large cells in cancers than in normal tissues	If one separates large cells from small cells, one gets more large cells from cancers	If one isolates large cells, they are frequently cancerous	QUESTION: WHAT WOULD CONSTITUTE INDEPENDENT EXPERIMENTAL EVIDENCE TO TEST THIS HYPOTHESIS?	
Microscopic or other measuring observation	This is essentially the same observation as the first	This is essentially the same observation as the first		

length of skid marks, because the faster an object travels, the longer it takes to stop it. This is called momentum. One can also judge the speed by the amount of damage in an accident. This is also a matter of momentum, but is a measure of the amount of force needed to stop the vehicle, while the skid marks are a calculation of the amount of time and distance needed to stop the vehicle, making a reasonable assumption about the force applied by the brakes. Another means of judging speed would be the amount of time elapsed to cover a specific distance, for instance if a traffic surveillance camera recorded the car as it moved down the road. Or a bartender could remember that the driver, leaving the bar, walked in front of the television just as a home run was hit at 10:07, and the accident occurred 20 minutes later but 30 miles away, allowing a calculation of a minimum average speed of 90 miles/hour. Typically, if all three calculations gave approximately the same result, the conclusion would be reasonably certain. However, if one of the calculations gave results noticeably different from the other two, any good lawyer would be able to convince a jury that the speed was uncertain.

This type of argument holds even in much simpler cases brought to trial. I could claim that you insulted or threatened me and try to press charges against you on that basis. You of course would deny that you had, and a police officer or judge would simply say that it was a case of "he said, she said," and that there was no way to tell who was lying. In other words, there was no corroborating evidence. However, if other witnesses, not known to you, me, or the others, voluntarily came forward and each reported a version of the incident very similar to either your or my claim, we could describe this corroborating evidence as independent confirmation of the claim. If furthermore a video surveillance camera captured us gesticulating in a fashion that was consistent with that claim, and both of our subsequent behaviors were consistent with that claim, this would also be independent confirmation that a judge would accept as evidence.

This understanding of the concept of *multiple independent means of verification* is very important to the study of evolution because, in the narrowest sense, we cannot experiment to determine the age of the earth. We therefore must deduce it from any sources of information that we have. The two major types of source are creation legends and extrapolations of physical data. For instance, the Abrahamic creation legend, Genesis, as followed by Judaism, Christianity, and Islam, suggests an approximate age of the earth of 6000 years. Other legends, such as those of Asia, the natives of the Americas, and Africa, generally vary in the range of 6 to 10,000 years. On the other hand, several physical measurements suggest ages over one million times longer: 14.5 billion years for the universe, 4.5 billion years for the earth, 2 billion years since the origin of life, 400 million years since the rapid expansion of life, and 65 million years since the disappearance of the dinosaurs. For those accepting physical evidence, the 6–10,000 year figure more appropriately reflects the development of civilization, following the appearance of true modern humans approximately 50,000 years ago.

So which figures are we supposed to believe, and why? Scientists believe that in all cases, evidence and logic rule. But what is evidence? In many cultures and times, holy writings were (are) considered to hold greater validity than what is perceived or calculated. Did the Red Sea divide? Did Joshua command the sun to stand still? Did Jesus walk on water? Most of us are familiar with the situation that different witnesses to the same event will later remember the event differently. Psychologists and sociologists can readily demonstrate that people's memories can be changed by social or psychological pressures, and that suggestion can alter memories. My children have vivid memories, from their formative years, of events that did not happen (I think) but which appear to be an amalgam of different incidents and suggestions.

This is the primary argument of Chapter 8, page 95. To a scientist, several branches of science produce data that has not been refuted, and each source of data points to the same conclusion of an age for the earth of approximately 4.5 billion years. The ratio of lead to uranium, and other ratios, in rocks are based on our understanding of the mechanisms of radioactive decay, with each calculation based on our measurement of the rate of decay of that element. Calculations based on the current temperature of the earth take into account various laws of physics relating to the dispersal of heat, radiational warming of the earth by the sun, generation of heat by friction and by radioactive decay. Calculations of the age and size of the universe rely on the properties of waves (of light) as one speeds toward or away form the source, as well as on laws of gravity and momentum. The age of the continents, and the age of fossils, are estimated from rates of sedimentation and erosion as well as the principles of how layering occurs and variations in magnetic fields. Today by direct measurement from satellites we can measure the movements of continents, previously inferred from magnetic fields and the types of fossils found on the continents. Thus to a scientist the hypothesis that the earth is billions of years old is supported by many independent lines of evidence, while the hypothesis that the earth is 6,000 years old is supported by one primary document. Whether or not one considers this document to be evidence or simply a hypothesis depends on one's faith, but to a scientist it does not constitute evidence in the sense that it can be tested and subjected to falsification. On the other hand, to refute the hypothesis that the earth is a few billion years old, one would have to deny the logic, experimentation, and evidence of several huge branches of scientific analysis and experimentation (geology, several branches of physics, astrophysics, biochemistry, molecular biology, among others) or at least explain the exceptions to theory that a much younger earth would pose. The data from these different lines of evidence converge in the same sense that a group of people in a circle, trying to locate an owl in the woods, each individually point in the direction that they hear the owl. Where the indicated lines intersect, the owl is likely to be (Fig. 9.2).

BIOLOGICAL DNA CLOCKS

Why biochemistry and molecular biology? We will return to this question in Chapter 15, page 243, but for the moment a simple explanation will suffice. Assume that DNA is one long string of chemicals that carries the information to construct an

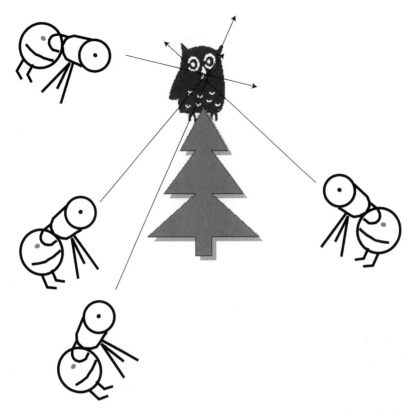

Figure 9.2. If several observers can each identify the direction of the sound of an owl, then the intersection (convergence) of the directions will give the position of the owl. If one considers all possible interpretations of a phenomenon as representing all possible directions in a circle, and a hypothesis as one of these directions, then the fact that many independent hypotheses converge on one central point adds considerable strength to the confidence that the hypothesis is correct

individual. You can consider it to be a very long string of natural pearls, each of which is distinct and identifiable. We will add one further consideration: rather than being loose on a string, each pearl has a snap-together end like children's toys.

Now let us make a further assumption: the string falls apart, or breaks, on a fairly regular basis. The actual date of breaking may vary, but if it breaks apart 36 times in a year, it averages 10 days between breaks. This is the same type of calculation that indicates the expected lifespan of light bulbs. We will also assume that the string is put back together each time it breaks, but the pieces are not necessarily rejoined where they broke. Thus the strand will look a little different each time it is repaired. If, for instance, the pearls were originally arranged in the order of size or color, the distribution of sizes and colors will become more random. For this argument, most importantly, the length remaining of the original ordered strand will become shorter with time. If we know how often it breaks, we can calculate

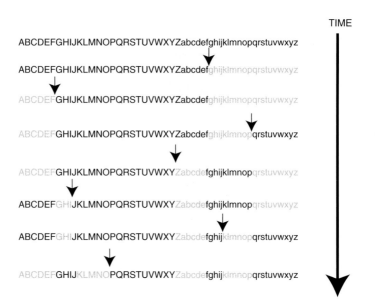

Figure 9.3. DNA breaks as a clock. Since each break is random, the probability of preserving the intact piece A...z decreases with time. This would be true for any arbitrary segment in the chain. Thus, the longer the time has passed since the chain was intact, the shorter the fragment will be

how long it has been since the first strand existed. We can also tell what that strand looked like by studying the different strands and determining the order of breaks.

All we have to do is substitute "DNA" for "pearls". DNA strands tend to break at a predictable average rate. We also know of short stretches of DNA that are found in almost all animals and plants. These must be very ancient and represent something similar to the original DNA. The shortness of these pieces gives us some estimate of the time that the DNA has been changing.

We can also use these comparisons to assemble an order of relationship. For instance, there are many more long pieces in common between apes and humans, or between lions and tigers, than between humans and tigers. We can conclude that there were stretches of DNA common to the ancestors of apes and humans long after the last time that the DNA was common to apes, humans, lions, and tigers. In other words, we can establish an evolutionary tree. See Fig. 9.3.

EVALUATING POPULATION MEASUREMENTS: BELL CURVES, STATISTICS, AND PROBABILITY

A major difference between a research laboratory and "the rest of us" is the level of control that can be exercised in a laboratory setting. In a laboratory if we wish to determine, for instance, that a given chemical can cause cancer, we might use mice that are so inbred that they are effectively identical twins or even clones. We would have records of how long they normally lived and the causes of their deaths. We would give them all identical food, and maintain them with equal numbers of animals in each cage, at standard temperatures and lighting conditions. To test our drug, we would administer it in various doses at known times during the day, during a known point of the ovulatory cycle if we are using female mice, to mice of a specific age. We would treat the control group in exactly the same fashion, including giving them an injection of the solution in which our drug was dissolved if it were given by injection since the stress and pain of the injection might affect even a normal animal. We would be concerned that all environmental conditions should be the same. We would establish a standard protocol that would determine how long we would wait to examine the outcome and the criteria by which we would evaluate the outcome. We might carry the study further to examine the effect of the chemical on cells in culture, in this case controlling all parameters (measurable variants) to the point of artificiality. For instance, we might use cells that were altered so that they could not degrade the chemical, or cells that were rigged so that they were far more sensitive to potential carcinogens (cancer-causing agents) than most cells. Our goal would be to assure that every conceivable parameter was identical in the control and experimental situation, so that the only difference between the control and the experimental situation was the chemical.

Of course nobody is perfect, and there is always likely to be something that we cannot control, as scientists learned when they realized that animals responded differently to drugs depending on the time of day at which the drugs were administered. Other unanticipated factors have proved to be increasing levels of fear if, for instance, mice hear the squeals of the control group being injected before the experimental group is injected; the sound of a watchman doing his rounds at night being sufficient to synchronize certain aspects of physiological rhythms; the fact that females caged together will synchronize their ovulations; a slight difference in weight between muscles on the left and right sides of the body; or the fact that, when one opens a mouse cage, there is a difference between the mice that come up to see what the activity is and those that flee to the back of the cage. Even the most highly inbred animals do not all die at the same time or of the same causes.

Thus the issue of control is an extremely important one in science, and at meetings many arguments ensue concerning the adequacy of the control experiments. In fact, if one reviews Nobel Prize-winning research and asks how it differs from most other research, very frequently a major difference turns out to be that the future laureate was suspicious that the presumptive control for his or her experiment was not adequate and decided to verify that control, thus stumbling upon a surprising new interpretation of the data. This issue is discussed further under the heading of controls (page 135).

Try as we might, we never have perfect control over our experiment and, particularly in the life sciences, we can never guarantee that an exact copy of an experiment done yesterday will come out exactly the same today. This is why we must repeat experiments. Perhaps yesterday something distracted me, and I inadvertently incorrectly diluted the drug so that it was too concentrated. Perhaps one of the vials or test tubes that I used was not clean, and it contributed color to the reaction. Perhaps the reagents had thawed and refrozen during shipping and were no longer good. Perhaps the mouse that I used for the experiment was already sick and I hadn't noticed. Perhaps the machine that I used to measure the results was not calibrated correctly. (The level of sensitivity that we rely on today is outstanding. We can easily measure a femtogram, or 0.000000000000001 (10^{-15}) g. In closer to real terms, if we dissolved a cube of sugar in a volume of water equivalent to 200 Olympic-size swimming pools, we would still be able to measure the sugar in one milliliter, or 1/16th of an ounce. As I tell my students, we can never actually see our results. We rely entirely on what our machines tell us. Therefore we have an urgent need to understand what the machine is measuring and to know that the machine is working properly.)

For these reasons a single repetition of an experiment is not acceptable and would not be accepted for publication. You are familiar with this logic. You know that a vaccination will not provide a 100% guarantee that you will not get the flu: you might have the flu already; the vaccination might not "take"; or you might get a different kind of flu. The same is true for our experiments. Not all of the mice exposed to the carcinogen will die of cancer. How then can we interpret our results?

EVALUATING DATA: CORRELATION AND STATISTICS

All of the above is an inferential or deductive logic, as is illustrated in Table 9.2. However, in most other instances, scientists attempt to test hypotheses. They try to devise experiments that establish evidence and can generate a logical hypothesis of

Table 9.2. Formal logic

Proposition (Hypothesis)	Prediction	Result	Comment
True	True	True	
The earth rotates toward the East	The sun will rise in the East	True	If the hypothesis is true then the prediction will also be true
False	True	True	
The earth is still; the sun traverses the Heavens from East to West	The sun will rise in the East	True	However, a result can be true even if the hypothesis is false
False	True	False	
The earth is still; the sun traverses the Heavens from East to West	If this is true, a point in the Heavens should be fixed relative to the earth, and should not move	False	With a suitable prediction, a hypothesis can be proved false

causality. Furthermore, this hypothesis is constructed in such a manner that it can be tested by trying to disprove or falsify it. These are the three tenets of the method presented in this book: **Evidence, Logic, Falsification** (ELF).

In many instances different phenomena can be correlated even if they are not related. A baby born in December gets larger as the weather grows warmer. We cannot conclude however that the temperature causes the baby to grow, and in fact the correlation will deteriorate as autumn comes. However, for a particular type of caterpillar that always hatches in early May and spins its cocoon in early July, we could not dismiss the correlation. A well-known example is the correlation between poplar trees and polio, seen in the United States in the 1950's. Poliomyelitis was a frightening disease in the US during the 1940's and 1950's. It killed 6,000 people and left 27,000 paralyzed during an expansion of the epidemic in 1916. Just before the appearance of the first polio vaccine, in 1952, there were 20,000 cases. Public parks, swimming pools, and cinemas were closed to attempt to limit exposure. During this period, some researchers noted that there was a pronounced correlation between the number of poplar trees in a neighborhood and the likelihood of a case of polio in the neighborhood (Fig. 9.4).

Do we conclude that poplar trees have something to do with the spread of polio? Well, it is certainly worth checking out, but in this case it proves to be a false correlation. What happened here was the intersection of two variables that were both correlated with income. First, in the Midwest and elsewhere, following the 2nd World War, there was considerable construction of housing for returning soldiers and to accommodate the growing economy. Of this housing, the cheapest housing tended to be apartment blocks with few trees. The most expensive housing was individually designed, with carefully selected landscaping and individual choice of trees or sparing of the original trees during construction. The middle-class housing tended to be in large tracts, where the developers used inexpensive fast-growing trees such as poplars. Similarly, in the face of the polio epidemics that expanded each summer, wealthier parents sent their children to isolated summer camps that the disease, spread from person to person, did not reach. In the poorest, crowded, neighborhoods, children were exposed to polio as infants. A peculiarity of the disease is that it can be a very mild disease in the youngest children. Some children die, but for many it seems to be a brief flu, not diagnosed as polio, and they recover with no nerve damage, immune to further attacks. So, many of the poorest children proved to be immune to polio. In the middle classes, however, attention to cleanliness and protection of infants was such that they were not exposed as infants. When these children were more sociable and went to movies or public swimming pools, they were exposed and came down with polio. Thus polio was more common in the middle classes, who tended to live in neighborhoods where poplar trees were planted. There was a reason for the association of poplar trees and polio, but the association did not establish that poplar trees caused polio (or for that matter, that polio caused poplar trees). There was some evidence for the argument, but there was no logic behind it, and indeed it was relatively easy to falsify it.

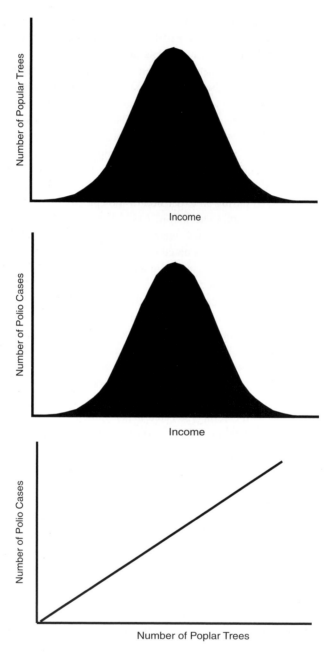

Figure 9.4. False correlations. Poplar trees were most commonly planted in developments intended for lower middle income families (A). For other reasons, frequency of polio was highest in this group as well (B). Thus there is a correlation between income and polio (C), meaning that the numbers are similar, but the correlation in no way implies that poplar trees cause or otherwise directly affect the chance of getting polio

This kind of false syllogism, that correlation proves causality, is extremely common and can be seen almost every day. In the 1990's there was considerable publicity over the possibility of transmission of AIDS by mosquitoes, based on a cluster of AIDS in Florida. In this instance, the ability of mosquitoes to transmit AIDS was examined from several angles and ruled out, and the cluster was later related to a truck stop near a swamp that was frequented by prostitutes. In another instance, there was a lawsuit that received considerable attention, in which a man claimed that his wife used a cell phone constantly and ultimately developed a rare brain tumor. He argued that the cell phone must have caused the tumor.

The point is that correlation may be used to support an argument, but it cannot be used to prove an argument. A cigarette smoker may get lung cancer. In terms of a large population, the correlation between smoking and lung cancer argues that it is extremely unlikely that they are not related, and many other potential explanations for the correlation have been disproved. However, consider it in terms of a lawsuit. Suppose the smoker changed brands of cigarette several times during his or her lifetime. Since cancers take several years to develop, the smoker cannot sue brand B claiming that brand B caused the cancer, since there is no way to determine if brand A, B, or C, or some combination of the brands, caused the cancer, or even if indeed the cigarettes caused the cancer. The problem in brief is this: you cannot determine which cigarette, on which day, caused the cancer, or even if a random cosmic ray caused the cancer. In other words, a single instance—as is often presented in commercials for non-prescription health supplements—is anecdotal; it is not proof. The fact that a single cell phone user gets cancer does not create a connection of cause and effect. The fact that a dog barked at my car on the same day that I found a $100 bill on the street does not prove that barking dogs cause money to fall in my path (or that money in my path causes dogs to bark). This is why we need the criterion of falsifiablility. In a correlation, you can never prove that one action caused another. However, you can prove that a hypothesis is false. If I hypothesized that eating lettuce caused lung cancer, you could set up an experiment in which lots of humans or guinea pigs ate lots of lettuce and did not develop lung cancer, proving my hypothesis false. Note the issue of the experiment. We can try to eliminate every variable except the eating of lettuce. We can have lots of guinea pigs of the same genetic background, control the quality of the air they breathe, and give them diets alike in every way except for the lettuce. In humans, we would have to consider all other types of variables: age, weight, sex, occupation (Do they work in environments in which the air is polluted in any way?) history of smoking, diet, genetic background, family history of lung cancers, etc. If we have a large number of extremely well-matched people who differ only in that one group smokes and the other does not, and it turns out that the smokers develop lung cancer, then we have very good correlation, and also a logic, since we can demonstrate in mice that cigarettes contain products that cause cancer. However, it always remains possible that we have missed a factor that produces the correlation. For instance, why do some people smoke and others not? Is it a difference in nervousness, and this difference can in some manner lead to greater susceptibility to cancer? By and large, smokers drink more alcohol than non-smokers; is it the alcohol? Was there some difference in the childhood of the smoker compared to the non-smoker that led the former to smoke? Would this be associated with the difference in rate of cancer? In brief, correlation is evidence, but it does not provide the logic, and we need the falsifiability because we can disprove one hypothesis but it is never possible to rule out a universe of alternatives and therefore actually prove a second hypothesis.

EVALUATING DATA: STATISTICS

Since in many circumstances we cannot directly prove causality, we are often left with correlation. In the best of all possible worlds, we can construct an experiment that will create a situation so unlikely or restricted that it is nearly impossible to imagine any alternative to the primary hypothesis. This is discussed below. However, particularly in dealing with human data, we ethically cannot conduct an experiment such as injecting people with bacteria to demonstrate that the bacteria cause a disease, or the situation is simply too complicated to allow us to rule out other causes, such as diet, climate, ethnicity, etc. In this case we necessarily fall back onto the use of statistics. But what is statistics? What does it mean, for instance, that a "poll has a margin of error of 5 percentage points"? What is "statistically significant"? Let's start by considering a fairly common statistical issue, seen in every pediatrician's office, a growth chart (Fig. 9.5).

This chart gives heights and weights for growing boys. What does it mean? First, look at the heavy center line. This line marks the average, or mean for each age. It represents the number at which 50% of the boys are above the line and 50% are below the line. For instance, at age 10, 50% of boys are 54 ½" (4'6 ½") or taller, and 50% are shorter than that. Other lines could be drawn to mark the 25th and 75th percentile–the heights at which 25% of boys would be taller and 25% would be shorter, and 50% in between; or between the 10th percentile and 90th percentile, which would include the 80% of boys who are neither in the top 10% nor the lower 10% in height. At the 5% level (tallest 5% and shortest 5%) we consider the heights to be statistically significant. This does not mean that there is a problem, but that we might consider investigating further.

SIGNIFICANCE

The interest of this curve is to answer the question, when should we worry if a child is too small or too tall? The answer is that each decision will be individual, but we use statistics in a specific way to give us a hint. What we do is to define 'significant' in a highly technical fashion. To a scientist, 'significant' does NOT carry the popular sense of "having meaning" as in "a significant glance"; "having influence or effect" as in "a significant piece of legislation"; or "a substantial amount" as in "a significant number of votes" or even "much more than casual" as in "a significant other". To a scientist, the word "significant" has one meaning only: "unlikely to occur by chance alone more than five times out of 100". For instance,

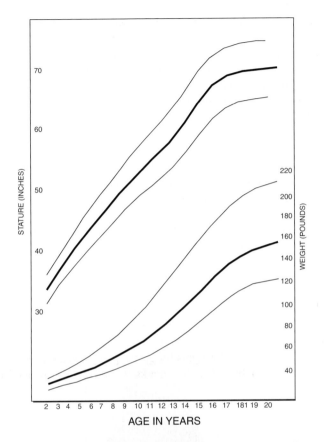

Figure 9.5. A growth chart. The middle line indicates the height at which, in a very large population, the boy with exactly average height for a specific age would fall. The upper line indicates the height above which 5% of the boys would be, and the lower line the height below which 5% of the boys would be. Since there will always be 5% in the upper or lower 5th percentile, no abnormality is implied. However, if a boy's height is at either extreme, a pediatrician might choose to verify that there is not a problem

in the example of the growth curves, we would perk up our ears if a boy was above the 95th percentile or below the 5th percentile—this would be significantly far from the mean—but we would not conclude that the boy had a problem. After all, 5% of the boys will be above the 95th percentile and 5% below the 5th percentile. We would simply say, "This boy is sufficiently far from the mean that we should investigate whether or not there is a problem." Is his height consistent with that of his parents? Is it clear that he is eating properly and digesting and absorbing his food properly? Are his hormone levels normal? Is his bone development normal?

Statistics comes from the property of all data to cluster around a mean. When the causes are random or numerous, the clustering takes on a particular shape, called a bell curve. If we were simply to measure the heights of a large number of boys (for instance, in a city) of a particular age, the numbers would distribute as is illustrated in Fig. 9.6.

This shape of curve is so predictable that it has been analyzed mathematically. It is used to handle numbers in a particular sense. We state that something is statistically significant if it falls in the upper or lower 5% of this curve. Again, this is proof of nothing; it merely means that the difference from the mean is worth investigating to see if there is a logical or other explanation that would cause us concern. It suggests that we can place bets, but does not answer our questions.

A bell curve is used in two distinctly different ways. First, this information can tell us whether an individual is in the outlying regions of the normal distribution, as in the case of the growth curves. In this case we use the term *standard deviation* to describe the variability. A little less than 2/3 of any population falls within the range of one standard deviation from the mean. (The source of the number comes from the mathematics and is not important here.) The size of the standard deviation can vary. For instance, if the color of one variety of flower ranges from white to deep

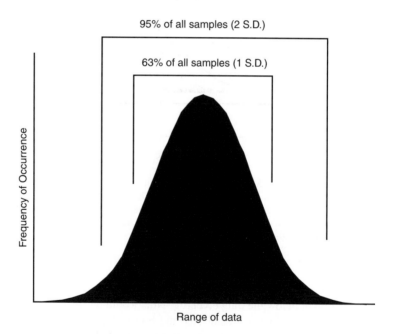

Figure 9.6. The bell curve or Poisson distribution. Fig. 9.5 is based on this distribution. All that it says is that, for any normal variable (height, weight, amount eaten at a meal, temperature in Chicago on June 20, number of points a given basketball player scores in a game) the most frequent numbers are those closest to the mean, with the probability of the more extreme numbers being much less. Thus, if you flipped 1000 coins, you would not be surprised if you got 500 heads and 500 tails, 499 heads and 501 tails, etc; but you would be very surprised if you got 300 heads and 700 tails. The Poisson distribution predicts how often this might occur. For a statistician, if the number falls in the upper or lower 5%, the result is considered significant and the situation should be explored for an explanation. As in Fig. 9.5, this result can occur simply by chance

red, with all possible shades in between, the standard deviation of the color intensity
will be broader than for a variety of flower that ranges from pink to middling red
(Fig. 9.7).

Also, because of the way that the number is calculated, the higher the number
of samples, the more accurately the curve can be calculated. For instance, if you
took a middle-school class that had 17 boys in it, the mean height (the total of all

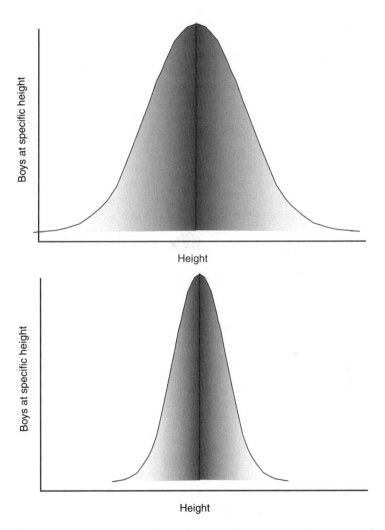

Figure 9.7. The range of variation can change from situation to situation. For instance, the range
of weights of new pencils is much smaller than the range of weights of strawberries. However, in
both situations a Poisson distribution exists, and it is possible to find the abnormally large and small
individuals. Here, the range of heights of boys in the upper figure, perhaps from a highly heterogeneous
community like a big city, is much greater than in the lower figure

heights divided by the number of boys) might be different from the median height (the height of the 9th boy), and the range between the tallest and the shortest might be very broad or very narrow. This is illustrated in Table 9.3 Let's compare those figures:

In the first school, the heights are very evenly distributed. The mean height of the boys is 54 inches (4'6") and the standard deviation is 5, meaning that about 2/3 of the boys (11 boys) will be between 49 and 59 inches. In the second school, the mean height is the same, but there is a greater range and variation in heights. From this school, we would expect that about 2/3 of the boys would be between 46 and 62 inches. In the third school, the mean height is 3" more than in the first school. Should we worry about the 46" child or the 62" child in school 1? These are falling out of the range of expected. On the other hand, they are within the range of the children in school 2, while the 49" child is abnormally short in school 3 but not in schools 1 or 2, and the 65" child is much more out of the range of school 1 than for schools 2 and 3.

Obviously, if we counted only three boys (choose randomly any three from the table) we would get a wide variation in numbers. Perhaps not so obviously but just as true, if we counted all the boys in one city, with the intent of comparing them to the boys of another city, we would expect the means and the standard deviations to be much more reliable (predictable) and useful for us. In brief: if I compared the height of my children to the height of my brother's children, it is very likely that there would be a difference, and we would not make much of an issue of it.

Table 9.3. Variation of heights of boys in different schools

Boy number	School 1	School 2	School 3
1	46	42	49
2	47	43.5	50
3	48	45	51
4	49	46.5	52
5	50	48	53
6	51	49.5	54
7	52	51	55
8	53	52.5	56
9	54	54	57
10	55	55.5	58
11	56	57	59
12	57	58.5	60
13	58	60	61
14	59	61.5	62
15	60	63	63
16	61	64.5	64
17	62	66	65
MEAN	54	54	57
STANDARD DEVIATION	5.0	7.6	5.0

If, however, my child's height was way out of the range of the whole city, I might well have reason to be concerned.

The second use of the bell curve is to compare populations. For instance, one might compare the heights of 11-year-old boys in Sweden, the United States, Thailand, and Liberia. In this case we would get individual population distributions as above, but the curves for each population might be shifted to the left or the right, and we could state that the mean height of Swedish boys is greater than the mean height of Thai boys. We can also plot the means of each population. If we have enough populations to compare, these means of populations would also distribute on a bell curve. We could use that information to ask if a population is on an outlier from the normal range. For instance, if the bell curve illustrated in Fig. 9.7 represented not individuals but different populations, and we found that the mean for a group of boys on a small island was in the lower 5%, we could ask if nutritional, genetic, or other factors were responsible for their short stature. This is the standard arrangement for an experimental set-up; we try to structure the experiment so that everything that we can think of is equivalent, except the one factor that we hypothesize is important, such as cigarette smoke. We then compare two populations, one in the presence of smoke and the other without smoke. This procedure is described in the next section.

Pollsters use the population statistics (the second group). When they say, "candidate A is ahead 54% to 46%, with a margin of error of 7 percentage points," what they mean is, "If we repeated this same survey, with equivalent numbers of respondents, in 90% of the surveys, candidate A would win by 8% or less. However, in 5% of the surveys candidate A would lose, and in 5% she would win by a larger margin."

EVALUATING DATA: EXPERIMENTATION

Since it is so difficult to interpret causality from correlation, we attempt to do experiments. Experiments have several properties. They are founded on a hypothesis that itself derives from certain observations. Then the hypothesis is stated in such a way that a means of testing the hypothesis is suggested. This test is structured in such a way that other interpretations are ruled out, and one remains as both logical and supported by the results. The hypothesis may at some point be proven wrong, when new evidence comes to light, but at present it is the best interpretation available. An example of the latter point would be Lord Kelvin's calculation of the age of the earth. Knowing the amount of heat absorbed by the earth from the sun, and the physics of cooling of a sphere, he made the assumption that the earth had split from the sun and since that time had been cooling. Based on the temperature at the surface of the earth and deep within it, he came up with a figure of between 20 and 400 million years. He later settled on the lower figure. However, he did not know about decay of radioactive materials, which are naturally common within the earth. This decay, like an atomic bomb or a nuclear reactor, generates substantial amounts of heat. When radioactive decay is factored into the calculation, we come up with a much more reasonable few billion years.

Two relatively striking experiments can illustrate the meaning of testing a hypothesis: Einstein's theory of relativity, and the dietary origin of pellagra. In the first instance, it was well understood since Newton's time that light traveled in straight lines. Mirrors, prisms, and lenses depend on this property, as does the "peek-a-boo" game so beloved by infants of all cultures. Einstein's theory, however, predicted that light could be bent by gravity. How could one test the idea, since the force and the bending would be miniscule in any laboratory experiment? It became evident that the force and the bending would be measurable if the gravity was produced by a relatively large heavenly body. The sun would be adequate, but it is so bright that one cannot see the stars behind it. However, an eclipse provides a different situation. Since the trajectories of the stars were well known, one could calculate very precisely their positions at any moment. At the height of a solar eclipse, if gravity could bend light, a beam of light that should have shot past the earth might be bent to hit the earth, and we ought to be able to see a star that should be behind the sun but that seemed to "pop ahead of itself" in its trajectory (Fig. 9.1). After its angle moved past the sun, it would resume its normal trajectory. Thus the solar eclipse of May 29, 1919 was eagerly awaited. If a star "speeded up" then slowed its trajectory, there was no logical explanation of such a bizarre event other than Einstein's hypothesis. Thus the confirmation of the hypothesis was heralded as a major event.

There are many other examples, including some of the most famous of the Renaissance, but in the 19th and 20th Centuries, two of the most dramatic were those of Pasteur demonstrating the bacterial origin of fermentation and Goldberger's demonstration that pellagra was a nutritional disease rather than a contagious or genetic disease. Pellagra is an ugly disease caused by the lack of the vitamin niacin. This was not known when a physician from the Surgeon General's office, Joseph Goldberger, went to the U.S. South in 1912 to investigate the spread of the disease. At the time, two theories were popular: first, that it was an infectious disease, as Pasteur and Koch had demonstrated during the previous century for other diseases; and, second, propounded by Eugenicists (see Chapter 31, page 405) that it resulted from inferior, less resistant, genetic background.

What he first did was to travel through the South, taking notes on everything that he could. (See the discussion on control experiments directly below, page 134). The first common factor that he noted was that poor people, including also prisoners and orphans, had a very monotonous diet, consisting of cornbread, molasses, and pork fat. The poorer the people were, the more likely they were to get pellagra. There was another peculiarity: in institutions such as prisons, the prisoners had pellagra, while the guards did not. To Goldberger, this meant that pellagra was not infectious but rather was caused by something such as a problem with the diet. (Remember that at this time there was no concept of such things as vitamins, and the idea that food could vary in quality was generally scoffed at.) Finally, in 1915, he went to one prison that had its own farm and fed prisoners a varied diet. He arranged that prisoners would be pardoned if they tried the diet common to other prisons. During the experiment, these prisoners lived with the others inmates and no effort was made to prevent infection. Nevertheless, after a couple of months, they began to show signs of pellagra. Meanwhile, Goldberger and his team tried to infect themselves with pellagra. As Walter Gratzer describes, conducting "filth parties," the eight researchers "injected themselves with blood from severely affected victims ... rubbed secretions from their mucous sores into their nose and mouth, and after three days swallowed pellets consisting of the urine, feces and skin scabs from several diseased subjects." They did not contract pellagra. He finally convinced prison wardens and heads of orphanages to try to feed their wards more varied diets and, where he succeeded in convincing them, the pellagra disappeared.

Goldberger had first made a critical observation: though pellagra was common in prisons, it affected only the prisoners, not the guards. Thus he had a means of falsifying the first hypothesis. Since prisoners and guards were in daily contact, though they ate separately, contagion by personal contact seemed unlikely. Since the guards ate balanced meals while the prisoners were given only fatback (dried and salted fat from the back of a hog) and cornbread, he hypothesized that the limited diet was the problem. He therefore conducted his first experiment by feeding a balanced diet to children in two orphanages, who were on a similar diet and likewise suffered from pellagra. Their pellagra disappeared within weeks. Since the curative material in the diet was not known and in any case it would have been completely unethical to conduct the converse and more definitive experiment, to produce pellagra by restricting the diet, he based his argument primarily on the first half of the experimental protocol. However, his results did not convince his opponents. The Eugenicists persisted in their belief that weaker constitutions were involved, and the proponents of infection refused to relinquish their preferred hypothesis. Therefore Goldberger undertook his colorful and unusual efforts to falsify the hypothesis that infectious agents caused pellagra. This left standing the alternative hypothesis that it was the result of poor nutrition. Nevertheless, it took 20 years before the mood in the country came round to accepting the idea and actively promoting better nutrition. Part of the issue was a weakness in the perceived logic: the idea that foods contained materials of magical properties in vanishingly small amounts was not readily absorbed, but also most societies find it difficult to abandon old ideas when new evidence contradicts the ideas. As the sociologist Leon Festinger noted, when the world fails to end on the day predicted by a cult, the response of the cult members is not to abandon their prediction but to proclaim that the failure of the world to end is proof that their prayers were heard and that their belief is valid.

Other factors played a role, as we have seen in other circumstances. Here, for instance, the world had finally accepted the idea of infectious disease, and all focus was on infectious diseases. Second, the general perception of the social order, abetted by growing awareness of the implications of evolutionary theory, made the idea that poverty could cause disease, as opposed to disease being a natural aspect of poverty, relatively unpopular. (To a politician, the suggestion that poverty could cause disease would lead to the conclusion that the state should invest effort in overcoming poverty, which could be politically and financially costly.) Finally, Goldberger was a Yankee and perhaps even worse, an immigrant Jewish physician from New York, working in a South still hurting from the Civil War and deeply suspicious of ulterior motives behind Yankee science. Thus Evidence, Logic, and Falsifiability, all brilliantly achieved by Goldberger, do not necessarily and obviously win. To many, the logic had not truly been established.

THE CONTROL EXPERIMENT

The purpose of a laboratory experiment is to restrict the range of possible interfering factors. For instance, if you wished to interview 100 people to determine the potential outcome of an election, you would get very different results if you stood in front of a school in a prosperous neighborhood, a store that sold computer parts, a courthouse, a dress shop, an ethnic restaurant, or a fast food restaurant; and you would likewise get different results if you stood in front of a supermarket at 8 AM, 10 AM, noon, or 8 PM. You would have to determine how each variation of location or time affected the distribution—sex, age, income level, race or ethnicity, etc.—of people who came by and thereby biased the numbers that you collected. Are young, well-off mothers likely to vote the same way as computer jocks? It becomes very complicated. There are many means by which statisticians try to address such issues, but there is another, more subtle type of bias, deriving from the experimenter's interest in the results. For instance, Mendel (see Chapter 13, page 175) classified peas as yellow or green. What did he do about yellowish-green peas? When several researchers tried to replicate his exact experiments, they got results that were similar, but never as close as Mendel's results to the ideal 3:1 ratio predicted by Mendel's Law. We suspect that, once Mendel realized the principle that he later espoused, he unconsciously classified the yellow-green peas in such a way that it tended to improve his numbers. We know this problem well. Almost any educational reform seems to work, because the greater involvement of the teachers in the new project of itself improves the learning environment, no matter what the change is. A physician or patient who feels that a medicine should have an impact will notice details of the condition that signal improvement even though by more physical criteria there is no improvement. This is so common that it is called the "placebo effect," meaning that patients will claim improvement, or physicians will see improvement, even if the "pill" is only sugar.

One way to defeat this bias of self-interest is to do a double-blind experiment, in which neither the experimenter nor the patient—if it is a medical experiment—knows which subject has received the experimental treatment and which has not. The way that it works is as follows: The size and shape of the nucleus of a cell can be an indicator of its health. The nucleus might be large and elliptical or small and rounded. I know how I conducted my experiment and therefore know which preparations were exposed to which chemicals. If I expect the nuclei to shrink and become rounded, I am very likely to classify intermediates as small in the samples exposed to the chemical. I therefore number my samples with jumbled or meaningless numbers and ask a graduate student or a technician, who is experienced enough to make the measurements but who has not participated in the experiment, to examine the preparations, do the classifications, and give me the results. I may ask more than one person to do this. The technician comes back to me with the results, "This is what I got for the percent of nuclei that were small and round: In sample A, 80%; in sample B, 30%; in sample C, 5%; and in sample D, 50%." I go back to my notes and come back to say, "That's great! Sample C did not receive any of the drug, sample B had a low dose, sample D, an intermediate dose, and sample A, the highest dose!" (And I may give the technician a hug.)

Note the description of sample C. We call it a control experiment. In this preparation, we try to do everything that we do to the other preparations, except for one crucial step or component. Expressed differently, we attempt to assure that, between our test situation and our control situation, every variable that we can think of is the same except for the variable that we wish to test. We need to do this to convince ourselves that the mere act of conducting the experiment has not produced the results. For instance, surgical experiments involving removal of organs always include "sham operations" as controls, since anesthesia and the wounding of an animal will always produce dramatic responses by themselves, whether or not the organ has been removed. In other situations (in most experiments in biology, chemistry, or physics today), we rely on the readings of instruments to tell us things that we cannot see. Thus, a common means of measuring whether something has been taken up by an animal is to give the animal a radioactive form of the material and then to look for the radioactivity in a process called scintillation counting. In scintillation counting, the radioactivity collides with a material that will fluoresce (glow) when hit by radioactivity, and we can measure by machines the light that is emitted. You have seen this kind of fluorescence in wrist watches that glow in the dark. The machine gives us a count of the number of flashes of light given off in a second or a minute and therefore, ideally, the amount of radioactivity taken up. However, it gets more complicated: if our solution is too acid, the fluorescent material will be destroyed, and we will get no glow. If the solution is too alkaline, it will glow spontaneously without radioactivity. If we have too much water present, the radioactivity will be absorbed by the water and will not hit the fluorescent material. If the solution is cloudy, the light may be emitted but blocked or reflected away before the sensors detect it. Sometimes, especially if we are working with extremely small levels of radioactivity, variations in the natural background radiation, from cosmic rays and other sources, may be a substantial portion of the total radioactivity. If the radioactivity is too high, two emissions may occur simultaneously and be counted by the machine as only one count. Or the sensors may not be functioning properly and may produce too few or too many counts. Therefore, since we cannot actually see the differences, we need to have a sample in which everything is the same—the same amount of sample and the same reagents, processed in exactly the same way, except that there is no radioactivity present. This is one of our controls, and we call it the negative control, meaning that it should give us the lowest possible value, which is not necessarily zero. We will also prepare a positive control, in which we will add a known amount of radioactivity to the mixture, to be certain that our counter detects it as it is supposed to. This positive control will allow us to establish that the fluorescent material has not been degraded by acid or other means, that the solution is not cloudy when it is counted (in a light-tight chamber), and that the counting machine is functioning properly. All good experiments include appropriate controls.

In fact, editors of journals look closely at the reported controls to verify that they are the best controls for the experiment and that they have been done correctly. When one looks at the speeches of former Nobel laureates, it is striking how often their prize-winning research started by their wondering if the control to the experiment was really adequate and then turning to verification of the control. One example was the work of Joseph L. Goldstein and Michael S. Brown. They discovered the proteins that bind cholesterol in the blood, which are now commonly known as HDLs and LDLs. They received the Nobel Prize in 1985. They had been attempting to determine what effect cholesterol had on cells in culture, and they added the cholesterol to the culture medium the same way everyone else did. They found that the cholesterol had no effect. Unlike other researchers, however, they reasoned as follows: cholesterol is very insoluble in water. How much actually dissolved in the water and reached the cells, as opposed to, for instance, sticking on the walls of the pipettes and flasks? Therefore they attempted to locate the cholesterol, in essence asking if the control (no cholesterol) was an effective control for the presumed experiment (cholesterol added). They found that the cholesterol was neither on the flasks nor in the cells but rather on proteins in the solution in which the cells were held. These proteins were what are now known as HDL and LDL.[5] The next year, the Nobel Prize was awarded to Rita Levi-Montalcini and Stanley Cohen, who similarly doubted their controls. They had been attempting to isolate and identify vanishingly small amounts of a factor important for the development and maintenance of nerves, called nerve growth factor (NGF). As the name indicates, it causes nerves to grow and survive. Since they could not get enough to analyze, they attempted to determine if it was protein or nucleic acid by digesting it with enzymes that specifically attacked proteins or nucleic acids. If the enzyme worked, the biological activity of NGF would disappear. In the experiment in which they attempted to destroy the nucleic acids, the activity did not disappear.

---

[5] Incidentally, it is not difficult to remember which is the "good cholesterol" and which is the "bad cholesterol". Some proteins are designed to bind one and only one type of molecule, such as cholesterol. They typically bind the molecule tightly and cannot hold much of it. Other proteins sop up all sorts of fats rather loosely, like a paper towel. They can hold a lot of cholesterol, but it can come off very easily, again like a dripping paper towel. Also, fats are lighter than water, and float. LDLs (bad cholesterol) are "low density lipoproteins," since they contain a lot of loosely-bound fat (cholesterol) that can easily be released to cause problems. HDLs (good cholesterol) are "high density lipoproteins," in essence heavier, because they do not carry much fat. However, they really hang on to the cholesterol, and keep it out of harm's way. It is much easier and more reliable to understand than to memorize!.

However, as a control, they tested whether the enzymes themselves would affect the growth of nerves. To their astonishment, the nucleic acid-digesting enzyme also stimulated nerve growth! In fact, the enzyme preparation was contaminated by nerve growth factor, revealing a new, rich source of NGF. With the new source of NGF, they were able to purify and characterize NGF, leading to the prize.

CAUSALITY

The purpose of experimentation is to establish a relationship that, by logic, timing, or sequence indicates causality. The relationship is "when the sun shines, snow melts". What we mean by timing or sequence is that, everything else (such as temperature) being equal—that is, we have a type of control experiment—the snow begins to melt after the sun begins to shine. Therefore, the melting of the snow could not have caused the sun to shine, and it is more reasonable to hypothesize that the shining sun causes the snow to melt.

The essence of setting up the experiment is to phrase the question in an appropriate manner. A well-phrased question will suggest an experiment. Here, "Why does the snow melt?" does not suggest an obvious means of finding an answer, but "The sun provides heat. Is this heat enough to melt the snow?" suggests that one could compare the amount of heat produced by the sun in various circumstances, including different heights of the sun in the sky, different levels of cloudiness, and different starting temperatures, to the rate of snow melt. It also suggests an experiment: On a bright sunny day, if one blocked the sun's rays by shading a patch of snow, would this affect the rate of melting? A well-designed experiment will address both ends of the statement, "if and only if": the result (snow melting) will occur if the sun shines and will not occur if the sun does not shine (only if the sun shines). We will see this result, of course, only if all other variables are equal and accounted for (the control experiment)—that is, the temperature is constant and slightly below freezing, the humidity is the same for both preparations, the wind is the same or absent, and there are no other sources of heat or cold. Even the weight (or more exactly, pressure, which is weight per square inch) on the snow should be the same. Ice skates work because the weight of your body on the narrow blade creates sufficient pressure to melt the ice, and glaciers move because their weight causes the bottom of the ice to melt. But the important element of the well-designed experiment is that we can test the "if" portion ("If this condition occurs, then we will see this result,") and the "only if" portion ("If this condition does not occur, then we will not see this result.") The "if" constitutes the Evidence and the "only if" constitutes the Falsification of our "ELF" rule. We also need the Logic, an explanation of the mechanism that led us to hypothesize that the result would follow from the condition.

An example of why the principle of falsification is so important is an incident involving the closure of a hazardous waste site. The community where the site was located had access to a scientifically-trained consultant for advice on the closure, and the community members were concerned that the approximately three-year closure process would create further hazards to their health. They wanted the city and state to provide cancer screening during the process, and asked the consultant for his opinion. The consultant recommended against the screening for the following reasons:

- The development of cancer is a very slow process, often taking twenty years. If the closure process caused any cancer, it would never be seen during the monitoring process.
- The criterion would not be that a specific person in the community developed cancer but that the frequency of cancer was unexpectedly high in the community.
- The neighborhood near the landfill differed by ethnicity, diet, smoking habits, recreation, age distribution, and drinking habits from nearby neighborhoods. All of these could affect the frequency of cancer. It would be very difficult to establish a baseline or comparison to this community.
- The neighborhood had a relatively high turnover of population. Even if it were possible to recognize an increased frequency of cancer, one would have to know at what age the people were exposed to the landfill, and where they lived before and after they were exposed to the landfill. Perhaps a specific age would prove more susceptible; perhaps there was a problem in the community in which many of them were born and spent their childhood before moving to this community.
- The community was bounded by a major highway, which produced a lot of car fumes, all of which could affect cancer rates.
- The community was downwind of a major airport, so that fumes from planes landing and taking off wafted into the community. This also could affect cancer rates.

As you can see, the problem that the community faced included both the fact that it would be exceedingly difficult to define a suitable control by which to evaluate the results, and the inability to eliminate alternative hypotheses, such as the hypotheses that increased cancer was caused by fumes from the highway, by favored foods of a specific ethnic community, by heavier smoking by community members, or many others. This again turns to the issue that one cannot prove a hypothesis true; one can only eliminate competing hypotheses. The community finally agreed that a combination of monitoring air quality (looking for cancer- and asthma-causing chemicals or particles) and local hospital admissions for asthma attacks, with the ability to stop operations if any value was too high) provided a more immediate and direct response to the hypothesis that the closure of the landfill would produce disease-causing conditions.

THE CHARACTERISTICS OF GOOD SCIENCE: ELF

Although the scientists that we cite today as having performed classical experiments, a careful description of what scientists do awaited the writings of Carl Popper in the early 20th C and many authors since then. To summarize the idea, science studies mechanisms that determine how the world functions, and they do so by collecting Evidence, using Logic to generate a hypothesis of how one element or process affects another, and then designing experiments to attempt to Falsify the hypothesis that they have made. The essence of good science is the ability to structure a question so that a good, definitive experiment can be performed. The answer cannot be wishy-washy, but must definitively rule out an opposing hypothesis. This culture is embedded in every function that scientists perform. All the following quotes are taken from conversations with scientists: professor to student, grant reviewers, manuscript reviewers. "Never do an experiment unless you have a table or a figure in mind." [This comment refers to the fact that an experiment must test a hypothesis. The table or figure makes the comparison of the control to the experiment and demonstrates the falsification. The comment also emphasizes the importance that scientists give to figures and tables.] "Is this application hypothesis-driven?" "These are shotgun experiments." "This is just a fishing expedition." [These last two comments refer to a paper or a grant application that is not based on an underlying hypothesis but, rather, is simply trying chemicals or processes that are known in the hope that something finally falls out of the study. Contrary to the interpretation given by Francis Bacon, this random collection of data is highly disfavored by working scientists.] "Does this contribution have a sufficiently biochemical or mechanistic focus to justify publication?" [A presentation that merely presents new data but does not have experimental justification to argue mechanism will be dismissed as "purely descriptive" and not accepted for publication.]

Ultimately, as is discussed on page 123, the experiment and the control must satisfy what we might describe as a "truth table," for which it becomes apparent that there is no absolute proof for the truth of a proposition, but one can prove the falsity of another one: Table 9.2 is in essence a truth table.

Thus the criterion becomes the true meaning of the expression "if and only if". One can declare a relationship IF, when A occurs, B always occurs AND IF, when A does not occur, B never occurs. If the tide rises when the moon is aligned with the sun and falls when the moon is opposite the sun, we can declare a relationship between the position of the moon and the tide. There may be exceptions, but we should be able to explain them without violating our original proposition. For instance, a fierce storm may push water into a bay, creating what appears to be a high tide at a different time, but we can determine that the wind and decreased atmospheric pressure from the storm are sufficient to account for the extra water. When we correct our measurements for the weather, we find that the movement of the tides has not really changed.

CLASSICAL EXPERIMENTS

Antonie von Leeuwenhoek

We will describe below a few classical experiments for which, in addition to the ones above, you should identify the elements of evidence, logic, and falsification. Do not underestimate the importance of logic or the social and intellectual situation of the time! As is argued in Chapter 14, page 191, DNA was not considered a likely repository for genetic information until experimental evidence had generated the logic that almost required it to be the genetic material. An even more spectacular example was Antonie van Leeuwenhoek's development of the first microscope and his observation of micro-organisms in materials such as water, the scum that he scraped off of teeth, and other media. He was very excited by his findings, and described them in terms that he knew:

(Plaque).... I then most always saw, with great wonder, that in the said matter there were many very little living animalcules, very prettily a-moving. The biggest sort had a very strong and swift motion, and shot through the water like a pike does through the water; mostly these were of small numbers."
"In structure these little animals were fashioned like a bell, and at the round opening they made such a stir, that the particles in the water thereabout were set in motion thereby...And though I must have seen quite 20 of these little animals on their long tails alongside one another very gently moving, with outstretched bodies and straightened-out tails; yet in an instant, as it were, they pulled their bodies and their tails together, and no sooner had they contracted their bodies and tails, than they began to stick their tails out again very leisurely, and stayed thus some time continuing their gentle motion: which sight I found mightily diverting."

Remember that, before this time, 1675, no one had ever seen an organism smaller than the eye could resolve. Although it was clear that diseases could propagate, bad air ("malaria") or vapors from water were considered likely causes. Also, Leeuwenhoek was an extremely skilled craftsman, and the lenses that he made were far superior to those of anyone else. Thus, when he attempted to publish his findings in the proceedings of the prestigious London-based Royal Academy of Sciences, he received what may have been the worst rejection letter ever written:

"When I observed for the first time in the year 1675 very tiny and numerous little animals in the water, and I announced this in a letter to the Royal Society in London, nor in England nor in France one could accept my discovery, and so one still does in Germany, as I have been informed."
In a letter, Hendrik Oldenburg, the Secretary of the Royal Society, London, wrote the following to Antoni Van Leeuwenhoek, Delft, Holland, 20th of October, 1676:
"Dear Mr. thony van Leeuwenhoek, Your letter of October 10th has been received here with amusement. Your account of myriad 'little animals' seen swimming in rainwater, with the aid of your so-called 'microscope,' caused the members of the society considerable merriment when read at our most recent meeting. Your novel descriptions of the sundry anatomies and occupations of these invisible creatures led one member to imagine that your 'rainwater' might have contained an ample portion of distilled spirits–imbibed by the investigator. Another member raised a glass of clear water and exclaimed, 'Behold, the Africk of Leeuwenhoek.' For myself, I withhold judgment as to the sobriety of your observations and the veracity of your instrument. However, a vote having been taken among the members–accompanied I regret to inform you, by considerable giggling–it has been decided not to publish your communication in the Proceedings of this esteemed society. However, all here wish your 'little animals' health, prodigality and good husbandry by their ingenious 'discoverer'.

There was little concept of the ability of a lens to magnify, and no concept of microscopic life; a group of prestigious scientists simply could not accept the idea that an unseen world existed. Of course it did, and improvements in microscope manufacture and many confirmations of van Leeuwenhoek's findings finally won out. This rejection of not-obvious new findings has often been repeated. Similar disbelief greeted August Semelweiss' demonstration that sterilizing a delivery room with carbolic acid eliminated childbed fever, of which many women died after giving birth. Of course his demonstration carried the baggage of suggesting that other obstetricians were responsible for contaminating their patients. In another example, the Nobel Laureate Rosalind Yalow is one of many scientists who has been known to open a speech with a slide showing the letter rejecting her first submission of her findings. That letter is particularly revealing. The editor noted, "The experts in this field have been particularly emphatic in rejecting your positive statement..." because it contradicted then current theory.

Here, as in the case of Vesalius (Chapter 30, page 406), it is always unacceptable to defer to the wisdom of sages, but we do. As the father of modern physiology, Claude Bernard, noted in the mid 19th C, "When the fact that one encounters opposes the reigning theory, one must accept the fact and abandon the theory, even though the latter, supported by impressive names, is generally adopted." To be totally fair, however, the rejection letter received by Yalow did emphasize the conviction of the reviewers that the conclusions were not sufficiently justified— meaning that the reviewers wanted to see more definitive and unequivocal experiments. Thus, they may have been pig-headed, but they still relied on the triumvirate Evidence, Logic, Falsification. The scientists of the Royal Academy, however, did not seriously consider the possibility that the evidence was real, and they certainly did not propose any attempt to falsify it by attempting, for instance, to document the distortions that a piece of glass might produce.

Let us therefore look at four classical experiments that are in one sense related, in that they form a sequence documenting that germs are living creatures and that they can cause disease. The experiments are as follows: Redi's demonstration that maggots do not generate spontaneously; Pasteur's demonstration that spoiling of broths was caused by bacteria; Koch's establishment of rules for identifying disease caused by bacteria; and Snow's tracing of cholera to a living, water-borne organism.

In the 17th C, about the time that van Leeuwenhoek was building his microscope, the world below the resolution of the human eye was still totally unknown. There was not even a magnifying glass. Humans can recognize two dots or lines as being separate down to a limit of approximately 0.2 mm (about 3/64"—look at a good tape measure). Fly eggs are approximately 0.1 mm in diameter. Thus when maggots appeared on meat that had been left hanging in open-air markets, the presumption was that the maggots spontaneously generated on the meat. This was in line with what seemed to be obvious at the time. Although trees and grain crops obviously grew from seeds, molds and many other plants seemed to spring up out of nothing, frogs would appear suddenly after a rain, and small worms would appear in standing water. Life must arise spontaneously from inanimate objects and materials.

Francisco Redi

Francisco Redi was unconvinced, and he knew that the maggots, while seeming to appear magically, eventually turned into flies, and flies were always flying around hanging meat. He therefore conducted the following experiment.

In his book "Experiences around the generation of the bugs", Francisco Redi wrote: " I put in four flasks with wide mouths one sneak [snake], some fish of river, four small eels of Arno river and a piece of calf and I locked very well the mouths of the flasks with paper and string. Afterward I placed in other four flasks the same things and left the mouths of flasks open. Short time later the meat and the fishes inside the open flasks became verminous, and after three weeks I saw many flies around these flasks, but in the locked ones I never seen a worm ". http://utenti.quipo.it/colettisb/ipertesto-redi/redi/redi-exp.htm

Later, facing the argument that the air inside the flasks would go stale, he improved the experiment. He took many different kinds of meat and covered them with a cloth fine enough to allow air to circulate but too fine to allow flies to pass. He exposed the covered meat alongside meat that was uncovered at different times and temperatures, and waited to see what happened. As you might expect, no maggots appeared on the covered meat, whereas they did on the uncovered meat. He then uncovered the original covered meat and demonstrated that, once uncovered, it too would generate maggots. He therefore concluded that maggots did not generate spontaneously but instead were produced by the flies that landed on the meat.

There are many elements to this experiment that are worth noting. The most obvious is that he used several kinds of meat and several conditions. This setup permitted him to make a general statement rather than a specific one, such as, "Maggots do not generate when a dead eel is covered with cloth and hung outside on a rainy day in June." He seeks a more general principle, that maggots do not spontaneously generate under any condition. Thus he varies the conditions so that he might argue, "Maggots do not generate under any of the fifteen conditions that I have tested. Therefore I can extrapolate my findings to any other situation that others might wish to test." In other words, he has made a hypothesis that can be tested by others.

Second, though not obvious in the paragraph quoted, he repeated the experiment several times to confirm that he always got the same result. In other words, the result was not a quirk of something that had happened that day. For instance, the wind might have been too high for flies to land, or the flies might have liked to lay eggs on all meats other than that of a snake.

The third and most critical issue is that he has established a control experiment, a preparation of meat that was, as far as he could tell, identical to the experimental meat with the single exception that flies could reach it. Thus his conclusion was, "all other things being equal, the ability of flies to land on the meat makes the difference between maggots and no maggots". The "all other things being equal" phrase is important, since very slight differences can change the outcome of an experiment. The very act of injecting a drug into an animal may frighten it enough to cause its behavior, growth, ovulatory pattern, or other feature to change, notwithstanding the effect of the drug. Thus it is necessary to inject a saline solution or other harmless solution into a control animal, so that the controls have experienced equivalent stress. You can find many more examples yourself. A good exercise would be to imagine what other variables could have influenced Redi's experiment.

Pasteur and Pouchet

Redi's results were sufficiently convincing that by 1651 the English physician William Harvey, himself an elegant experimentalist who demonstrated by both logic and evidence the circulation of the blood, could declare "Ex ova omnia" ("Everything [comes] from eggs"). However, the connection between insects and bacteria, and between bacteria and disease, was not yet established. By the mid-19th C, this argument was still open, and there were several practical consequences. These included questions as to the origin of diseases such as cholera, to be discussed below, and problems in France concerning spoilage of foods and disease among the grape vines of the wine industry. The issue of what caused spoilage of food was of such practical and theoretical importance that the French Academy of Sciences offered the Alhumbert Prize for the best proof of whether or not bacteria, or putrefaction, would generate spontaneously. Pasteur had demonstrated by this time that boiling milk or food would delay putrefaction, but many believed that the process of boiling had damaged either the foodstuff so that bacteria would no longer thrive on it, or had damaged the air so that bacteria could not survive. According to this argument, bacteria could generate spontaneously but needed an appropriate environment to grow. Thus prizes were offered for the proof of whether or not life could spontaneously generate. The leading contenders were Louis Pasteur and Félix Archimède Pouchet. Although in retrospect there were some elements that Pasteur did not understand, such as the ability of some bacteria to sporulate (go into a sort of hibernation, during which they resist heat and other killing agents) and in fact he was very lucky in the choice of his preparation, the experiment that he designed was elegant. The hypotheses were the following:

1. Bacteria arose spontaneously but required undamaged (uncooked) food to grow.
2. Bacteria arose spontaneously but required something from the air to grow, and whatever was in the air could be destroyed by heat.
3. Bacteria arose from other bacteria that could easily contaminate even a clean preparation.

Pasteur extended the third hypothesis by assuming that bacteria could be airborne and could drift in on breezes. However, they were heavier than air and would settle out in still air. He had already established that if a meat broth was boiled and the flask sealed, then it would remain uncontaminated. This experiment was very similar to one Lazzaro Spallanzani had done in 1767, in which he had demonstrated that small animals could not generate in boiled flasks unless and until the flasks were opened to the air. Others, however, protested that either the broth or the air had been damaged by the boiling and could no longer support the generation of life (hypothesis 2). Pasteur therefore constructed an elaborate flask that had an S-shaped loop (Fig. 9.8). The flask was open to the air which, it was already known, could diffuse even without a breeze. However, the narrow neck of the flask blocked breezes, and the air would penetrate only by diffusion. At this slow pace, bacteria would settle into the lower part of the loop. He then took some boiled meat in its juice, basically a bouillon, and let some cool in an ordinary beaker and in his flask. The bouillon in the beaker quickly became infected, proving that

Figure 9.8. The flask that Pasteur used for his famous experiment. See text for explanation

the boiling had not destroyed its ability to support bacteria. However, the bouillon in the special flask did not become contaminated. One could still argue that there was some problem with the interaction of the boiled bouillon and the air above it. Pasteur therefore tipped one of his flasks so that the bouillon reached the low point in the neck, where he hypothesized that the bacteria had settled, and then sloshed it back into the main part of the flask. Within a couple of days, it was apparent that the tipped flask was contaminated, whereas the one that had not been tipped was still clean. Pasteur was awarded the prize, and one of the flasks that he prepared over 150 years ago is still on exhibit, still open to the air, and still uncontaminated. Today thousands of laboratories studying bacteria and cells in culture use a modification of this experiment, called a Petri dish after the designer, Julius Richard Petri, to grow their cells. The Petri dish works on the same principle as Pasteur's flask, allowing potential contaminant bacteria and fungi to settle out. Although air can freely circulate in the dish, it remains uncontaminated because potential contaminants settle out by gravity (Fig. 9.9). Note that what most people tend to assume is the correct position of the Petri dish is upside-down.)

The logic and structure of the experiment is best illustrated by Table 9.4 It is a matter of some curiosity that the subject was sufficiently interesting to scholars that Pasteur's experiments were carried out in the context of a contest to prove or disprove the existence of spontaneous generation.

Koch's postulates

There were many further demonstrations that Pasteur was correct, and scientists and physicians turned to seeking the causes of infectious disease. In 1890 Robert Koch published a list now called Koch's Postulates of what would be required to argue that a microorganism caused a disease (note what is Evidence, Logic, and Falsification, and what constitutes the "if and only if" criteria):

1. The organism must be found in all animals suffering from the disease, but not in healthy animals.
2. The organism must be isolated from a diseased animal and grown in pure culture.
3. The cultured organism should cause disease when introduced into a healthy animal.

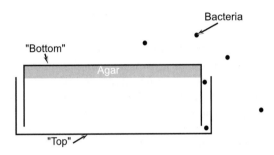

Figure 9.9. Upper A modern Petri dish. Note that its proper position is "upside down". (Lower) The mechanism by which a Petri dish functions. Note that this is the same mechanism as the Pasteur flask

Table 9.4. The logic of the Pasteur experiment

Aspect	Demonstration	Explanation
Hypothesis	Bacteria arise from other bacteria, as opposed to hypothesis of spontaneous generation	Broths in which bacteria can grow are contaminated only when bacteria have access to them; boiling kills them and sealing the flask prevents new ones from entering.
Evidence	Bacteria do not grow in flask that has been boiled and sealed.	According to hypothesis of spontaneous generation, bacteria should be able to generate (falsifies or nullifies this particular hypothesis, except in special case of damage to medium (broth))
Logic (if)	Bacteria do not grow in flask even if air can be exchanged, but without air currents	Addresses qualification of damage to air; nullifies hypothesis that quality of air is essential
Falsification (only if)	If flask is tipped to wash in dust that is fallen, bacteria will grow	Addresses qualification of damage to medium, since medium has not been changed; nullifies hypothesis that quality of medium has changed
Conclusion	Bacteria do not arise by spontaneous generation	Started with two hypotheses, of which one has been completely nullified; the hypothesis that bacteria arise only from other bacteria stands until a new hypothesis can be tested and shown to nullify it.

4. The organism must be reisolated from the experimentally infected animal.

Not all of these are fully attainable. Some organisms can be found in healthy animals and cause disease only rarely; other organisms cannot be easily grown in culture. Nevertheless the rules have validity.

Sir John Snow and Cholera

A final classic experiment using the logic of science was Sir John Snow's demonstration that cholera was an infectious disease. Cholera was a devastating disease. Essentially a severe diarrhea, but one that could drain so much fluid from a person that it could kill a person by dehydration in a few hours, it would break out in cities and spread rapidly, killing hundreds or even thousands in the space of a few weeks. There were two major hypotheses as to what caused it: "Effluvia," by which was meant odors or gases escaping from infected patients, who thus poisoned the air for healthy individuals; or biological or chemical agents in the bodies of the victims. Snow therefore looked at the logic of what the evidence was telling him:

1. Cholera traveled from city to city at the same rate that people traveled. Thus, if cholera broke out in Rome or Paris, it would not reach London faster than the time that it took stage coaches or boats to reach London.
2. If cholera came from another country, it would be seen first at a seaport. It would not appear suddenly in the Midlands of England.
3. It would break out on ships, but only if the ships came from cholera-infected countries. If cholera had broken out in Rome, a ship coming from Rome might develop cholera, but cholera would not appear on a ship coming from Stockholm.

All of this evidence suggested that cholera was transmitted from person to person, but it still did not resolve the two hypotheses. But then Snow encountered a new patient who had no personal contact with any cholera victim. However Snow, an astute observer, learned that the patient had received clothes from a recent cholera victim. This was not unusual; if someone died young, the clothes were often recycled. Snow then refined his hypotheses:

4. Hypothesis: If cholera is passed by effluvia, then all persons in contact with patient should get cholera, and those in contact with only the clothes should not.
5. Hypothesis: If cholera is passed by liquids, then those in contact with the liquids should get cholera, whether or not the patient is present. Since in a medical situation one does not usually have recourse to a lab experiment, Snow re-examined his evidence to see if the evidence supported one or the other of these hypotheses.
6. It might be disgusting, but it was an issue that physicians could note. In the later stages of cholera, patients have vomited and lost through diarrhea so much that their digestive tracts are empty, and anything further that is lost is clear and watery, and may not even be noticed. In such conditions, Snow considered the possibility that the last clothes the patients wore had not been washed after their death.

7. Those who washed more frequently, such as workers who handled mud and clay and other materials that they would want to get off their hands, did not get cholera.
8. Nurses and doctors, who washed frequently, did not get cholera even though they worked with cholera patients. This evidence suggested to Snow that the disease was spread not by the air but by liquid excretions from the body. Since these excretions normally went into the sewers, Snow then turned his attention to the distribution of disease and the distribution of water in the cities. The disease tended to be clustered, with some exceptions that caught Snow's attention:
9. In the city of Manchester, those getting water from a well near a leaky sewage pipe got cholera.
10. In Essex, there was an outbreak in one district served by a single well. A washerwoman living in that district was the only one who did not get cholera, but she used water from another well.
11. In Locksbrook, a landlord who lived elsewhere was accused, during an outbreak, of providing poor water to his tenants. To prove that it was safe, he drank water being delivered to those buildings. He subsequently died of cholera.

With this information in hand he looked at a new epidemic in London. Ultimately 300 people died during the outbreak. Snow plotted, on a map, the residences of all the victims, and saw that they all clustered around one source of water, known as the Broad Street pump. A brewery nearby was not involved in the epidemic, but the brewers had their own source of water for the beer and did not use the pump. Six cases were in a different neighborhood but, when Snow got a map of the water pipes, he realized that the people in that neighborhood also got their water from Broad Street.

From this evidence Snow argued that the source of the epidemic was the pump. Furthermore, he argued, it was not a chemical contamination, since a chemical would be expected to dilute out with time and thus cause less disease; but the severity of the epidemic was continuing, suggesting that the cause could reproduce. Therefore the cause was likely to be biological, in other words, a germ. With that information, he finally did his experiment. He removed the handle from the Broad Street pump, rendering it inoperable. People in the neighborhood had to go to the pumps in the surrounding neighborhoods to get their water. Within a few days the epidemic was over.

Note what he proved and what he did not prove. He demonstrated that it was likely that the contamination came from one pump, that it was carried by the water, and that it was biological in origin. He did not identify the organism. In fact, the germ that causes cholera is extremely difficult to grow in the laboratory and, though widespread, does not often cause cholera. But its existence is one of the reasons we chlorinate water. Snow did not prove anything, in the sense that he had no true experiment to falsify his hypothesis. He falsified competing hypotheses, leaving his hypothesis standing. It was enough to signal him to intervene, and the success of his intervention convinced everyone of at least the pragmatic value of sterilizing water.

In this digression from the subject of evolution, we have looked at the issues of what constitutes evidence, what we mean by multiple independent means of verification, what constitutes an adequate control, and the complexity of interpreting data that must be assessed by statistical comparisons. We have considered the relationship between evidence and the logic of the experiment, and have seen that the "if and only if" basis of experimental logic is the same as the ELF logic emphasized throughout this book. However, the necessity of the logic may not be apparent either to scientists or to the public until other information becomes available. Often, the interest in a question is driven either by new findings or by new social concerns.

REFERENCES

Best, Joel (2001). *Damned Lies and Statistics: Untangling Numbers from the Media, Politicians, and Activists.* University of California Press.

Desrosières, Alain (2004). *The Politics of Large Numbers: A History of Statistical Reasoning*, Trans. Camille Naish, Harvard University Press.

Huff, Darrell and Geis, Irving, (1993) How to Lie With Statistics (Paperback reissue) W. H. Norton, New York

Tijms, Henk (2004). *Understanding Probability: Chance Rules in Everyday life.* Cambridge University Press.

Dobell, Clifford, ed. 1960. Antony van Leeuwenhoek and his "Little Animals", Dover Publications, N.Y.

Festinger, Leon, 1964. When Prophecy Fails: A Social and Psychological Study of a Modern Group That Predicted the Destruction of the World, Harper Collins Publishers, New York.

http://www.ucmp.berkeley.edu/history/leeuwenhoek.html (biography of Leeuwenhoek from U. of CA, Berkeley)

http://www.euronet.nl/users/warnar/leeuwenhoek.html

STUDY QUESTIONS

1. Choose any claim, advertisement, or other propaganda that you find in the media. Present the arguments in the construct of "if and only if" or "Evidence, Logic, Falsification". After your presentation, do you still accept the claim?
2. Choose any news item describing a scientific advance reported in a newspaper, magazine, or on television. Trace the source of the story as far as you can, and analyze the presentation in terms of "if and only if" or "Evidence, Logic, Falsification". Can you identify the controls? What was falsified?
3. Which of the experiments described in this chapter do you consider to be the most convincing? Why?
4. People often say that obesity is a "metabolic problem". From a statistical standpoint, what would you say might be a reasonable indication that the problem was truly medical?

   Which of the arguments presented in the previous chapters as supporting the theory of evolution meet the criteria described in this chapter? Which do not? What would be required to complete the arguments?

CHAPTER 10

THE INDUSTRIAL REVOLUTION, POPULATION POTENTIAL, MALTHUS, SOCIAL PRESSURE, AND COMPETITION

MALTHUS

The 18th and 19th centuries were a period of substantial social pressure in Europe. The growth of the cities and the rise of industrial era influenced the social structure in many ways. In an obvious biological manner, increased communication in the world brought plague, which to a large extent destroyed the feudal system (Chapters 27, page 359, and 28 page 376) and introduced many other communicable diseases. Furthermore, general health of humans, as measured by examination of graves, church records, size and condition of skeletons, evidence of age at beginning of menstruation (menarche), infant deaths, lifespan, and other documents, deteriorated from earlier times. The causes were fairly evident: in a rural environment even peasants have access to a reasonably varied and nutritious diet, but in large cities the sources of food are distant from the people who need the food. Given the problems of transportation, distribution, and storage of food, city dwellers are unlikely to receive the range of foods available to farmers, even if finances are not a problem. When food is acquired not by barter but by exchange of money, the exchange becomes dependent on employment, salary level, and politics. For instance, during the Irish Potato Famine of the 1850's, wheat was being grown in Ireland but for the most part was paid in lieu of cash rent to English landlords, even while there was severe starvation in Ireland. Thus the lower classes in the cities were frequently stunted in growth, generally not healthy, and subject to disease. Since there was no knowledge of dietary requirements or the origin of infectious disease, the problems were not addressed in any effective manner. (For a discussion on the identification of sources of disease, see Chapters 27 and 28.)

Thus there were plenty of problems in the cities, but there was a further, larger one that was the source of considerable difficulty in European societies. Humans in many societies begin to have children between 15 and 20, are capable of having children for 20 to 30 years, and may live 20 years beyond that. Thus even if each couple has two surviving children, there will be six people alive (and eating) for each reproductive group. But humans often have more than two children. In societies in which infant mortality rapidly decreases, as has happened in most large societies in the 20th C, it typically takes approximately one generation before the tradition of having very large families gives way to a tradition of very few children. (In France, the typical ending of a children's book "and they lived happily ever after" translates to "and they had lots and lots of children".) Where humans have plenty of resources, populations expand at very high rates. When Europeans first settled the Falkland or Malvinas Islands, they had an average of eight children per family. If there is absolutely no barrier to survival and reproduction of the children, and assuming that each family has seven children at age 20, all of whom survive, in 100 years a single couple could produce almost 17,000 descendents!

In the growing cities of the 18th and 19th C, health was not very good, but it was tolerable, and with the steady supply of food that agriculture had produced, infant mortality decreased. Thus many families produced several surviving children, considerably more than the slightly over two children per family that it would take to maintain the population constant. This put a substantial financial strain on the families to provide sufficient food for their children, leading to many social means of sending the children elsewhere: to the clergy, military service, or apprenticeships. But while families coped individually with these stresses, the problem was far more structural.

As the economist Reverend Thomas Malthus (1766–1834) realized from the data he accumulated, a modest excess of births over deaths could produce a very rapid increase in population. As is illustrated in extracts from his tables (Tables 10.1 and 10.2) English society at the beginning of the 18th C was reproducing at a rate of approximately 165 births per 100 deaths. At this rate in 500 years a single couple would account for over 8000 descendents. Plague disrupted this expansion in 1709

Table 10.1. Malthus' calculations (Book II, Chapter XI, Table I. When in any country there are 103,000 persons living, and the mortality is 1 in 36.)

If the proportion of deaths to births be as	Then the excess of the births will be	The proportion of the excess of the births, to the whole population, will be	And therefore the period of doubling will be
11	277	1/360	250 years
12	555	1/180	125
13	833	1/120	83 1/2
14	1110	1/90	62 3/4
15	1388	1/72	50 1/4
16	1666	1/60	42
10: 17	1943	1/51	35 3/4
18	2221	1/45	31 2/3
19	2499	1/40	28
20	2777	1/36	25 3/10
22	3332	1/30	21 1/6
25	4165	1/24	17
30	5554	1/18	12 4/5

Table 10.2. Some real numbers that Malthus obtained

Annual Average	Marriages	Births	Deaths	Proportion of marriages to births	Proportion of deaths to births
5 yrs to 1697	5747	19715	14862	10 : 34	100 : 132
5 yrs—1702	6070	24112	14474	10 : 39	100: 165
6 yrs—1708	6082	26896	16430	10 : 44	100: 163
In 1709 &1710	a plague	number destroyed in 2 years.	247733		
In 1711	12028	32522	10131	10 : 27	100: 320
In 1712	267	22970	10445	10 : 36	100 : 220
5 yrs to 1716	4968	21603	11984	10 : 43	100 : 180

and 1710, but then in 1711 the survivors produced far more children, averaging 3.2 children for every death. At this rate 2 would become over 8,000 in 264 years! However, as Malthus insisted in his opening chapter, it is not very easy to increase agricultural production. One can increase the amount of land farmed, but soon the fields will be too far from the city to import perishable foods, or the expense of transport will make the produce non-competitive or unaffordable. Otherwise, the only options are a limited number of dry, storable grains (if they can be protected from rats) or increasing the yield per nearby field, which is not easy to do. By Malthus' calculations, at best production of food can increase linearly, that is, 1, 2, 3, 4, 5. However, when conditions are adequate to good, populations can increase exponentially, that is 1, 2, 4, 8. This leads to an obvious problem. Let's make the very conservative assumptions that couples have their first child at 20, and have four children that survive to adulthood. To make the mathematics straightforward, we will assume that all four children are born simultaneously. Look what happens every twenty years:

Inevitably, the amount of food available for each person will decrease (Table 10.3). Malthus' essay is available online at *http://www.econlib.org/library/ Malthus/malPlong.html* (6th Ed) and *http://www.econlib.org/library/Malthus/ malPop.html* (1st Ed)

Although the example is easy to follow in the case of food, the same principle applies for any resource: water, housing, jobs, heat. Thus, at some point there will be a struggle for these resources. In this struggle, some will win and others will lose. The struggle is likely to be violent, and may be in an organized fashion, such as wars (think how many wars between nations and explorations to conquer new lands have really been about resources) or may be individual, as with the types of thievery, avarice, and murder depicted by authors such as Dickens. The result

Table 10.3. Expansion of populations and food available, assuming linear expansion of food supply and geometric expansion of population

Years	Population	Food	Food/person
0	2	100 lbs	50 lbs/person
20	4	110	27.5
40	8	120	15
60	16	130	8.1
80	32	140	4.4
100	64	150	2.3

is that, by one means or another, violence, famine, or disease will re-establish the balance between available resources and the number of individuals seeking those resources. This is the lesson of Malthus: populations increase exponentially, while resources increase linearly. Each member of the population has access to fewer and fewer resources, which will lead to a struggle for existence. The losers will not survive, returning the population to sustainable levels. To Malthus, the struggle is inevitable.

To be fair, Malthus' calculations, frightening as they are, are not that accurate. By extrapolating his numbers—always a risky proposition, as explained in Chapter 1, pages 11–13—he calculated that the world would run into a catastrophic situation by the middle of the 19th C. Clearly, this did not happen, both because the real expansion of population did not continue along the extrapolation that he predicted and because agriculture improved more rapidly than he expected. However, his essential point was correct, that human populations tend to expand rapidly, and more rapidly than the food supply. In real numbers, between 1950 and 2000, the population of India rose by 5.7% per year. This was the period of the green revolution, which allowed an increase in production of food grains, to 8% per year. In several parts of the world, women average from 6 to as high as 7.5 children per woman. At this rate, assuming that one couple has children starting at age 20 and each lives to age 80, at the end of those sixty years there will be over 140 people to feed.

DARWIN AND MALTHUS

Darwin, with his lively curiosity, had returned from the Beagle with many observations and more questions. He no longer doubted that fossils represented an earlier age of the earth, and that the earth was old enough to have witnessed a gradual transition of earlier organisms to the modern ones. In other words, he was convinced of the fact of evolution. This, however, was not immensely novel. His grandfather had proposed such an interpretation, as had Buffon, Cuvier, and Lyell. The problem was that, without an absolute timeline and without a mechanism, the fact of evolution had no real meaning. If God created all organisms at one time, why should they have similarities suggesting anatomical, geographical, and historical relationships? If species could be formed constantly, what drove the creation? In our terms, the evidence was present, but there was no logic and, without logic providing a hypothesis, there could be no falsification. Darwin was still mulling over these problems two years after he had returned when, for diversion, he read Malthus' book. What he read he recognized immediately, and he realized that he had a mechanism for the creation of species. He had seen the capacity of species to expand when uncontrolled—feral horses in Argentina, goats on the island of St. Helena (Chapter 7), convincing him that what Malthus said about the capacity of the population in cities to grow was valid for animals and plants as well. Thus the struggle for survival would apply in nature as well.

Alfred Russel Wallace had also read Malthus, whose book influenced politicians and philosophers of the time. Thus it was no coincidence that both Darwin and Wallace more or less simultaneously recognized the principles of natural selection. Malthus had described the situation in human terms—actually noting that the rules applied to all living things—and it was not so gigantic a leap to recognize that this struggle could lead to selection. Wallace called Malthus' essay "the most important book I read". Darwin himself, who referred to Malthus as "that great philosopher;" noted that it was "the most interesting coincidence" that the two discoverers of the laws of natural selection had come to their conclusions by reading Malthus. Like the simultaneous rediscovery of Mendel in 1900 (Chapter 13), such apparent coincidences reflect the convergence of data (evidence) with logic. Often great ideas are missed because one or the other is lacking, and ultimately the discovery is consensus of scientific thought as much a function of the consensus of scientific thought as it is an individual breakthrough.

Darwin had also established that all individuals in a given species vary and, though there was no explanation as to mechanism, it was also understood that the variations could be inherited. In fact, the systematists following the Linnaean classification were much discomfited by variation. Each species had its own norm, and variation from the norm was abnormal: ideally, each Dalmatian dog should look like the picture of another Dalmatian, with the same number, layout, and size of spots, but some were less perfect. Part of Darwin's genius was that he recognized that the variations were the heart of the story of the origin of species, rather than an annoyance. New species could arise from the variant individuals. The last step therefore would be to connect the data and the ideas. If, in the struggle for survival, some variants were more likely to survive (or rather, leave their descendents to the next generation) and if the variation that helped them to survive was inherited, then the struggle would become selection for those most adapted to survive, or selection of the fittest. In fact, Darwin connected many more arguments to this hypothesis, including relating the choice of breeders of pigeons to determine which pigeon would contribute to the next generation, and we will return to the major hypothesis in Chapters 11 and 12.

For the moment it is simple enough to understand the general principle. Most fish have swim bladders, air-filled sacs into which most secrete gases. In a few fish, however, the swim bladder is connected to the digestive tract, allowing the fish to swallow or burp air so that they can regulate their buoyancy and thus remain without effort at various depths in the water. Suppose that several variations of a fish live in a pond that, during an infrequent but recurring dry spell, dries up. One of these fish has a variation in its circulation such that it can absorb oxygen from its swim bladder, turning it into a sort of primitive lung so that it can get oxygen from the air rather than through its gills in the water. It will survive a dry spell to leave descendents to the next generation. This variation will also allow it to rest at the edge of the pond, where it may be able to hide from enemies or catch more food. Further selection along this line will eventually enable it to survive on land, where it will encounter no competition, since it is the first vertebrate to get onto land. Thus selection will continue, and the species will evolve.

The industrial revolution, leading to the increasing urbanization and industrialization of western societies and specifically that of England, had resulted in a considerably different pressure on society. As Malthus emphasized and many noted (Dickens' great novels were written mostly around the 1850's) poverty and the growth of slums in the cities were a major problem. Malthus had described it as a struggle for existence. Herbert Spencer, after toying with the issue as early as 1851, in developing the ideas that later were to be called social Darwinism (Chapters 31 and 32), by 1864 had read Darwin and described Darwin's argument as "survival of the fittest". Darwin finally incorporated this term into later editions of his book. As is noted later in the discussion of the social biology of apes (Chapter 29), ideas come from many sources, including social concerns to which scientists as well as others are susceptible. In this case the social conditions of the 19th C led to the recognition of the potential for overpopulation and to the recognition of a struggle for existence. One of the characteristics of genius is the ability to make connections among apparently disparate ideas. The theory of natural selection as the mechanism of evolution derived from the several currents of the exploration of the earth, the economic incentive to better understand the earth, the geological understanding of time, and the social conditions that led thinkers to identify the issues of overpopulation and the struggle for existence.

REFERENCES

http://www.ac.wwu.edu/~stephan/malthus/malthus.0.html (Malthus' complete essay, from Western Washington University)

STUDY QUESTIONS

1. State the major tenets of Malthus' argument and defend the proposition that he is correct.
2. State the major tenets of Malthus' argument and defend the proposition that he is wrong.

3. How many children are there in your family? Choosing the age of your youngest parent at the time that the oldest child was born as the time for one generation (i.e., if the first-born was born when your mother was 30 and your father was 32, choose 30 as the generation time) and assuming that everyone dies at 90, calculate how many of your clan there would be in two generations; in 10 generations.

4. Conduct a similar calculation for the largest human family with whom you are acquainted.

5. For any pet or domestic animal with which you are acquainted, calculate how many young it has in its lifetime. Then calculate how rapidly the population would increase. For instance, assume that a pair of dogs can have 20 pups in a ten-year lifetime. (They could actually have 40 or 50.) Then at the end of 10 years, there would be ten pairs, each of which could have 20 pups, giving 100 pairs.

6. A tree can easily produce 10,000 seeds in a single season and may survive 100 years. Why are there not more trees?

7. Look up in government or political tables the total agricultural production of a specific nation or state over a period of one hundred years, and compare this production to the size of the population over the same period. What do you conclude?

CHAPTER 11

NATURAL SELECTION: THE SECOND HALF
OF DARWIN'S HYPOTHESIS

NATURAL SELECTION

Darwin had read Malthus. He understood Malthus' basic argument and realized that a simple observation of nature demonstrated that Malthus' thesis could be directly applied to the survival of individual plants and animals, and ultimately to species. Malthus, a minister in the Anglican Church, had observed the growing wretchedness of cities and, beginning in 1798, published the following thesis: populations tend to increase exponentially (2, 4, 8, 16, 32) while the food supply increases linearly (1, 2, 3, 4, 5, 6). The observation was simple: if each couple has four children, then for two people in generation 1 there will be four people in generation 2; and if these four (two couples) each have four children, then in generation 3 there will be eight. What this leads to, as Malthus demonstrated by estimates of population size, is that the society will soon run out of food and some will die, by starvation, war, or disease. Thus some will die, and the population will not expand at its full potential. As Darwin saw, this logic obviously applies to the rest of the biological world. If a pair of spawning fish lays 50,000 eggs, but next year the fish population is more-or-less the same, then on average 49,998 of the eggs have died before returning to spawn. This is also true for plants with their thousands of seeds, and even for mammals or birds that have one infant per year but go through several breeding cycles in their lives. Thus nature includes huge levels of mortality for all creatures.

Darwin's extension of this idea to the natural world was just the first part of his great insight. The second part of the insight was that who would live and who would die would not necessarily be completely random. For instance, suppose there was a variation in color of caterpillars, such that some matched the leaves on which they lived and fed better than others. Suppose also the likely scenario that the largest number of the caterpillar deaths was the result of bird predation (birds eating them). The birds would most likely find and eat the caterpillars that looked least like the leaves, ignoring those that most resembled the leaves. This would be selection rather than random loss. Darwin further supposed that the variation itself was not random but was inherited—that is, that the moths deriving from bright green caterpillars produced bright green caterpillars, and those deriving from dull green caterpillars produced dull green caterpillars. If the variation was inherited and some variants survived better than others, then the next generation would consist of a higher proportion of the more favored variant. In each succeeding generation the proportion of the more favored variant would increase, until finally virtually the entire population would consist of the most favored variant. If at an earlier time that variant had been very rare, over time the species would have changed from the less favored to the more favored variant. Wallace proposed essentially the same hypothesis, with the exception that he focused on the existence of populations of varieties, in which one population would survive at the expense of another. He did not question the origin of the varieties. Darwin argued that the variations were individual, leading to the survival of individuals of specific characteristics. Neither had, at the time, a mechanism for inheritance of traits.

This is what natural selection, or "descent with modification" means. All species produce far more young than two per couple, and yet the population sizes remain roughly constant. Individuals vary in many characteristics. If one variant of a characteristic favors survival of an individual, and this characteristic is inherited, then the species will gradually over time evolve to resemble the favored variant— descent with modification.

There is one modest correction that we need to add: We used the term "survival," but all we really need is that the individual achieve reproduction of the next generation. Thus salmon die shortly after laying their eggs, and female black widows and praying mantises eat their mates shortly after they have mated. There are even insects whose young are born by chewing their way through, and killing, the mother. Survival is not the issue; leaving young to the next generation is. Thus the theory of natural selection more specifically says that there is inherited variation in the ability of individuals to leave progeny (young) to the next generation. Those individuals that leave more progeny will in successive generations increase their proportion in the population, until the species would change. Given sufficient time, this natural selection would be able to create new species and even major new types of animals or plants.

SURVIVAL OF THE FITTEST—NOT WHAT YOU THINK IT MEANS!

Two terms cause considerable confusion (and anger) whenever evolution is discussed. The source of the problem is that scientific terminology is by necessity precise, while ordinary speech is not. As we have discussed previously (page 10), words such as "significant" have different meanings to working scientists than they do in casual conversation. In a somewhat similar manner, "natural selection" and "survival of the fittest" are often used in public in a sense broader than, or even in conflict with, the scientific meaning.

"Natural selection" refers to a series of inferences based on some relatively straight-forward observations. Although the observations seem fairly obvious, their implications did not take hold until the social observations typified by Malthus began to be accepted. It had been clear for centuries that humans could change the appearance of species by controlling breeding. Whether one considered the races or varieties of dogs, cats, goldfish, wheat, corn, oranges, apples, or peaches, it was evident that human selection could adapt species in many ways. Dogs could be selected to be hunters, work-dogs such as huskies, burrow-entering badger hounds such as dachshunds, lap dogs such as Shi-tzus or chihuahuas, racing dogs, etc. The new idea proposed by Darwin was that this type of selection could take place by natural forces, and that species could be changed by these forces. (Wallace did not make the connection between human-controlled breeding and natural selection. See page 92.)

The first set of observations was simply Malthusian, noted for the animal and plant world. They were that all species have great potential fertility; population can increase exponentially; but that populations tend to remain stable in size; and that environmental resources are limited. Let's look briefly at these before moving to the inference. Whether we talk about fish, which can lay 10,000 eggs; trees, which can shed hundreds of thousands of seeds; or mosquitoes, which can lay a few hundred eggs and go through a generation in two weeks, it is obvious that most species can easily reproduce enough young to fill any location on earth. Even humans can do this. When humans reach a new, uninhabited but fertile land, as when they first settled the Falkland (Malvinas) Islands, families can average eight children. Allowing for a generation time of twenty years, one couple can produce 64 descendants in forty years, and 512 descendants in sixty years. But it is also obvious that such a population explosion rarely happens. Fish may lay 10,000 eggs, but by-and-large there will be the same number of trout in a stream from one year to the next. Even though a pair of mosquitoes, starting to breed for instance on April 15 could produce 100 quadrillion quadrillion (100 followed by thirty zeros) mosquitoes by October 15, the mosquito population varies only moderately from one year to the next. We understand today that there is not enough food for these astronomical numbers. Other factors may also be limiting. For instance, there might not be very many places to hide in a stream, so that many baby fish would be very visible to predators. These observations: that any species has a potential to reproduce that is greater than the standing population; that population sizes tend to remain stable; and that resources can be limiting, lead to the first important inference:

Inference: There must be a struggle for existence and only a fraction of offspring survive.

This much is relatively obvious, but the next jump requires some observation and thought. The first observation is that individuals vary extensively in characteristics. This is obvious in humans and dogs, but it is also true of all other creatures. Even penguins can identify their mates and their offspring without confusion and, if you cared to make the study, you could find differences among individual ants. The second observation is also somewhat obvious, but must be coupled to the first to build a hypothesis. The second observation is that much of the variation is heritable. We understand this today. Children generally bear substantial resemblance to one or both parents. If we wish to have a Dalmatian puppy that will mature with few spots, we have a better chance if we breed two lightly-spotted dogs than if we breed two dogs with heavy black patches. Although one peach tree might produce better peaches because it is in better soil, overall we will get better peaches by growing trees from peach seeds gathered from the best trees rather than from the worst trees. The inferences, however, are more profound, and consider that nature enforces the same choices that we might make. If we discard the peaches from a rather sickly tree and instead plant the seeds only from the healthiest tree, we have guaranteed the survival of the progeny (young) of the latter tree (plus the tree that pollinated it) and have condemned the former tree to extinction. The proposition is that nature does the same thing.

> Inference: Survival is not random. Those individuals with the traits that fit them best to the environment will leave more offspring.

In the same sense that we chose which peach tree would leave young to the next generation, nature can do by virtue of the fact that a large percentage of the new generation will die. Much will be random: Perfectly healthy seeds will land on rocky or otherwise inhospitable soil, will be eaten by birds or other animals, or will succumb to other uncontrollable events. But for some of the seeds, their survival will not be random. Perhaps one seed can resist a late or early frost a bit better than another. Perhaps its shell is just a bit harder, so that a squirrel cannot bite through it. Perhaps its shape allows it to be carried, by wind, water, or animal, to a more distant location, where there are more sites in which it can grow. Perhaps one of the fish fry (baby fish) is colored just a little darker and is less visible against the sides of the stream, or its markings make it much harder to see against the plants in the stream. It will survive infancy and grow to eventually reproduce, while others will not.

We use the examples of baby fish and seeds because most mortality is in infancy, but the same rules could apply to the adults. One bird's preference for a nesting site might lead it to choose a site that turns out to be far more secure in high winds than the choice of another bird. A male fish might have brighter, more colorful markings that appeal to a female. A bird has a larger beak and can eat larger seeds than other members of its species.

The continuation of this hypothesis states that, for example, in a time of famine only the birds with the larger beaks can eat a different type of seed and therefore survive, the next generation will be the children of the large-beaked birds and will, on average, have larger beaks. This process can continue, with each generation having larger beaks than the previous generation, gradually changing the species.

This latter phenomenon has been seen to occur in time observable to humans. Evolutionists seeking to test the hypothesis have observed that climate conditions during the growing season strongly affect the size and availability of seeds. In several instances, a drought resulting in smaller seeds has led to greater survival of birds with smaller beaks, and the subsequent downward shift in mean size of beaks in the population. Other studies, identifying other quantifiable sources of selection, have established equivalent changes in other characteristics. Such immediate evidences of evolution have been most clearly observed in islands or other isolated populations, where migration to and from other locations or huge variation in resources does not confuse the issue.

> Inference: This unequal ability of individuals to survive and reproduce (SELECTION) will lead to a gradual change in population, accumulating favorable characteristics.

This, then is what natural selection means. In the same fashion that humans can produce German shepherds, Shi-tzus, and greyhounds by selecting characteristics of dogs over several generations, nature could alter species over many generations by selecting for characteristics that give one individual a survival advantage over another. This is what is meant by "survival of the fittest".

There is one other term that we need to define. Most people use the expression "survival of the fittest" to mean the strongest, biggest, or most capable of making money. In the context of evolution, "fittest" does not carry this connotation at all. "Fittest" means ONLY "better capable of leaving offspring to the next generation". This is the only currency in which natural selection works. Any variation that makes it more likely that one individual will leave offspring to the next generation than another individual makes the first more fit. A smaller cockroach, one that can squeeze into a crevice and thus avoid being stomped on; greater tolerance for living in a terrible climate, such as a desert or the arctic; greater timidity, as opposed to a more curious individual, who sticks his nose out while a predator is still in the area; acceptance of a food shunned by other animals—all of these might be examples of greater "fitness". The rule is that, whatever works, works. Any adaptation that improves the possibility of leaving progeny can be selected for. It has nothing to do with beauty, strength, or size. This is why we have many bizarre shapes and lifestyles of creatures—spindly, fragile creatures, creatures that live in very hostile environments, creatures that eat poisonous plants and animals, species in which the male is a puny parasite attached to the female, species such as the black widow and the praying mantis in which the female eats the male after mating, species in which the female is an immobile bag of eggs, and species in which the young hatch by devouring the mother's body and destroying her. LIMITING reproduction may even be a selective advantage, if overbreeding threatens to use up resources. Consider two populations of, for instance, grasshoppers. One population produces 300 eggs per couple, which unfortunately leads to consumption of all available leaves by early August. Two hundred ninety-five nymphs die of starvation before reproducing. The second population lays only 100 eggs per couple, but this population does not use up the entire food supply, and more survive to reproduce. Depending on other circumstances, lower reproduction may be a selective advantage.

Figures 11.1 and 11.2 illustrate some of the truly bizarre creatures—a very small sampling of the many that could be shown—that can be found in this world. The theory of natural selection proposes that for one reason or another each of these

species appeared because its ancestors were better able to survive and reproduce than were ancestors that might have been a bit more "normal".

SPECIES THAT DO NOT "EVOLVE"

A final question that one can raise is the following; if this process of selection operates continuously, how is it that some species do not change? After all, we have many types of animals and plants on earth that we often call "primitive". We consider fish and frogs to be less evolved than mammals and birds, or ferns and mosses to be less evolved than flowering plants. Many individual types of animals and plants have been on this planet for a very long time. We can identify 300 million to 400 million year old fossils that are clearly dragonflies, cockroaches, and ferns, very similar to species seen today (though not actually the same species). Did these creatures simply opt out of natural selection? Is it possible for a species to reach perfection and not continue to evolve?

Most likely the answer to the first question is "no" and to the second question, "sort of" or "in a sense"? We do not really think that creatures achieve "perfection". As we will discuss in Chapter 26, the forces of evolution include interactions with other species, so that, for instance, predators could evolve to get better, forcing the prey to evolve, and in any case, as discussed above, perfection could lead to overbreeding and, ultimately, starvation. The issue seems to be more that, if the environment does not change, then creatures adapted to that environment will not change very much. Sequoia trees would fit this definition. Their ancestors first appeared with the dinosaurs, 180 million years ago, and they were quite common in the temperate, mist-filled climates of the time. Since then, the changing earth (see Chapters 22 and 23, pages 303 and 319) has reduced the area on the earth with that type of climate. Thus most of the sequoia trees have left us. A few remain, very similar to their ancestors, in the few areas on earth that maintain a climate similar to the period in which they thrived.

The same might be said of creatures such as the horseshoe crab (which is more closely related to spiders than to crabs). Although in fact it is different

---

Figure 11.1. A handful of the bizarre creatures found on earth. A. Anteater, South America. B. Spider shrimp, Australia. C. Leaf insect, Malaysia. D. Pelican, USA. E. "Pacman" frog, Argentina. F. The orange creature is a coral, a community of sea anemone-type creatur/es. The red, white, and blue creature is a sea cucumber, related to starfish. The magenta (violet) animal is a type of marine mollusk. Two fish are also visible. Australian Great Barrier Reef. G. An antlion. This insect larva uses its spade-like head to dig a pit in sandy soil. When an ant slips on the edge, the antlion by jerking its head showers its victim with sand so that it loses its grip. When the ant falls to the bottom, the antlion grabs it with its piercing and sucking mouthparts. You can see these common creatures around houses in the United States. H. A duck-billed platypus. It has a duck-like nose (left) and a beaver-like tail. It spends most of its life in water, lays eggs, and nurses its young, and feeds them with milk secreted from sweat-like glands—not true nipples—on its chest. First descriptions of this animal were assumed to be a hoax in Europe. Credits: Ridiculous animals - Antlion: From Swain, Ralph B, 1948, The Insect Guide, Doubleday & Co., Inc., Garden City, NY, illustrated by SuZan N. Swain

Figure 11.2. A. Manta ray, Atlantic Ocean. B. Pacific octopus, Pacific Ocean. C. Baobab trees, Africa. This list can be continually expanded. Credits: Manta - "© Photographer: Harald Bolten | Agency: Dreamstime.com", Octopus - "© Photographer: John Abramo | Agency: Dreamstime.com", Baobab - "© Photographer: Muriel Lasure | Agency: Dreamstime.com"

in many respects from its ancestors, its resemblance to trilobites is striking, particularly considering that the trilobites lived 300 to 400 million years ago (Fig. 3.5).

Horseshoe crabs live near rocky ocean shores, and the physical characteristics of such shores have not changed much since the oceans were formed. The earth has grown warmer and colder, but there was always a range of temperatures such that some seas were warmer than others. As we described above, the criterion for natural selections is "Whatever works, works". A perfectly reasonable corollary would be, "If it ain't broke, don't fix it." A horseshoe crab, primitive though it may be, is a very efficient creature. I once watched a dog try to attack one. Its shell is a dome that very few creatures can get their mouths around, it can flail its tail and do some damage to an attacker, and it can scuttle into the sand quite rapidly. There are few creatures in the sea, sharks included, that can find a way to take a bite out of it. Besides, there is very little meat under all that shell. Whatever works, works. It worked for this group of trilobites, and it works today. If it ain't broke, don't fix it.

You can reasonably challenge this line of argument by asking, "If the trilobites were so good, where are they now?" The first answer is that they were a very large and diverse group of animals, found in many locations on the globe including China, Morocco, Rochester, New York, and Oklahoma. As a group they did very well, surviving for almost 300 million years, three times as long as the dinosaurs. The second part of the answer is that there have been many massive changes in the history of the earth (see Chapter 23, page 319). These changes have led to great shifts in the predominant creatures on the earth, from the amphibian and ferns to the dinosaurs and pine tree-like trees (gymnosperms or conifers) to the mammals and flowering plants. During these periods the trilobites finally disappeared, leaving their descendants, including the horseshoe crab. One species is quite common along the Atlantic Coast of North America, and others are found in Asia, but this group of animals is no longer predominant in the world. The few that have survived, though well adapted to their environment and that environment is not changing rapidly, are different from their ancestors and their descendants will differ from them.

REFERENCES

Browne, J., 2002, Charles Darwin, The power of place, Alfred A. Knopf, New York.

Darwin, Charles, 2004 (1859) The origin of the species, Introduction and notes by George Levine, Barnes and Noble Classics, New York.

Eldredge, Niles, 2005, Darwin, Discovering the Tree of Life, Norton and Company, New York.

Gould, Stephen Jay, 2002, The structure of evolutionary theory. Harvard University Press, Cambridge MA.

[ant lion] Zim H. S. and Cottam, C, (Irving, JG, Illustrator) Insects. A Guide to Familiar American Insects, Simon and Schuster, New York, 1956.

STUDY QUESTIONS

1. Argue for or against the position that Malthus' hypothesis was correct (Malthus' full argument is available online) and that it is also correct for the biological world.

2. Argue for or against the position that Malthus is the true father of the theory of evolution.

3. Are there true evolutionary relics in the modern world? In what way are they true evolutionary relics? In what way are they not?

4. For those creatures alleged to be evolutionary relics, under what conditions do they currently live? Do these conditions differ from conditions in nearby environments?

5. Describe, in your own words, the primary inferences of Darwin's hypothesis. Be prepared to defend each one with appropriate evidence. Is there any evidence to the contrary?

6. Define, in your own words, "survival of the fittest".

7. Identify the strangest plant or animal that you have ever encountered. Can you identify any reason why such a shape, habit, or behavior should enhance the potential of the species to reproduce?

CHAPTER 12

DARWIN'S HYPOTHESIS

DARWIN'S HYPOTHESIS

Darwin's hypothesis consists of two major arguments: That evolution had occurred, and that the mechanism of evolution was natural selection. Neither idea was original, but it was Darwin's linking the hypothesis that animals and plants had evolved to the hypothesis that this evolution was driven by natural selection that provided the logic necessary to interest the scientific community and beyond. Part of the first hypothesis depended severely on acceptance of the evidence that argued that the world was substantially older than 6000 years.

The structure of the world, before the exploration of the New World and the planet, seemed reasonably ordered. It was relatively easy to identify each species and to imagine that one could count all species and completely account for the menagerie on Noah's ark. However, the period of exploration made the definitions less clear. Were the gull of North America, which were lighter in color and had somewhat different markings, the same species as those found in Scandinavia? Questions such as these forced scientists to focus more on the meaning of the variability of species, and even the definition of the term "species" (see Chapter 5, page 68 and chapter 11). Thus the range of variation came under consideration, and with this, acceptance of the idea that, through human choice and control of breeding, one could generate immense variety in the appearance of dogs, chickens, pigeons, horses, cattle, etc. Darwin even devotes an extensive section of the first part of Origin of the Species to a discussion of pigeon breeding. Thus, as early as the 18th C, scientist-philosophers such as Buffon and St. Hilaire had hypothesized that species had evolved or changed with time, and anatomists such as the great Cuvier considered that the common features of the skeletons and musculature of all vertebrates derived from a common ancestry. However, based on the assumption that the world was 6000 years old, and observations of the number of generations it took to effect small changes in the appearance of domestic animals, the hypothesis that all varieties of all animals and plants had evolved from a common ancestor was simply absurd. The derivations must have been part of the act of Creation, or the relationships reflected simply God's (or a Designer's) reuse of the same tools and parts.

HUTTON, LYELL, AND GRADUALISM

However, the search for precious metals, iron, and coal had led to an interest in the structure of the land, as discussed in Chapters 22 and 23. By the end of the 18th C, Hutton had identified various types of soil and rocks, such as old lava and sedimentary accumulations. He had learned to identify different formations— useful for predicting the location of minerals—and realized that they frequently corresponded across all of Europe. Charles Lyell, working on the principle that he promulgated as Gradualism, understood from Steno's principles (Chapter 2, page 27 and Chapter 3, page 35) that upper layers were younger than deeper layers and, based on his observations of rates of erosion and sedimentation, tried to estimate how long it would take to build such layers. The numbers he came up with by far exceeded Biblical time. He even went to the New World to view its spectacular geography. From estimates of the rate at which Niagara Falls is receding or moving upstream, he calculated that it would take 35,000 years for it to have cut back the seven miles from the original face of the bluff. Using the same sort of estimate, he felt that it would have taken the Mississippi River 60,000 years to produce the delta of precipitated mud where it enters the Gulf of Mexico. It matters little that his calculations were substantially off, based on errors in his estimates of the rate of cutting of the falls and the depth of the delta. The point was that, when he published his Principles of Geology in 1830–1833, it had substantial impact. It argued cogently the hypothesis that many others had begun to consider, that the earth was considerably older than the biblical age. Most importantly, Darwin read the book during his voyage on the Beagle. Finding, as expected, fossil shellfish high in the mountains of the Canary Islands and Chile, he tried to calculate, using Lyell's principles, how long it would have taken to lift shoreline to those heights. He came up with figures ranging from the 10's of thousands to millions of years. His estimates, like Lyell's, were not that accurate, but they accomplished something very important: they extended the time over which evolution could have occurred and, freeing the outer boundary from the biblical wall of 6000 years, they raised the possibility of much greater extensions of time.

THE WEIRDNESS OF THE NEW WORLD

So there was time for evolution to have occurred. Was there really evidence that it had?

Exploration raised other questions. For instance, the New World contained strange animals like armadillos, primitive scaly mammals that roll up into a ball when frightened (Fig. 12.1).

Perhaps it was OK that God decided not to favor Europe with strange animals such as these, but what did it mean that there were fossil giant armadillos (the size of a Volkswagen "beetle") in South America? If each species was created uniquely, one might expect that, for instance, armadillos once ranged widely in the world and that they now were confined to northern South America through the

Figure 12.1. Armadillo. This nocturnal creature is common from Texas southward through South America. Inset: Armadillo rolled into a defensive ball

southwest of the U.S.A. Why were the only fossils of these animals found in the same areas that modern armadillos were found? Was it possible that the modern armadillos were related to the fossils? Was it therefore possible that armadillos had evolved only in the new world, and that the modern small animal was a descendent of the larger fossil animal? If this were the case, why should one conclude that the giant armadillo resulted from an act of special creation, rather than that it too had evolved from something else? Georges Cuvier, at the beginning of the 19th C, had clearly stated that, in the fossil record, the further down one went from the surface of the rocks, the less the fossils looked like present-day life. He had suggested that current animals were not exactly like those at creation. Again and again Darwin encountered this problem, whether with all the fauna in the Cabo Verde Islands or with finches in the Galápagos Islands: unique groups of species, found in one location, with apparent affinity to different species from a nearby location. For instance, each of the finches in the Galápagos Islands is a distinct species, but they are clearly finches, and they have some resemblance to finches along the coast of Ecuador. If each is an act of special creation, why was this little group of special creations confined to one specific region? Would it not be simpler to assume that creation was a continuing process, that each variety of finch had not been individually created during the sixth day of creation, but rather that first there had been one type of finch, and that this type of finch gave rise to other types?

THE RELATEDNESS OF ANIMALS

This type of reasoning quickly extends backward to some very provocative ideas. For instance, could finches have arisen from other types of birds? Birds are latecomers in the fossil record, and their bone structure is similar to that of some dinosaurs, while the eggs of birds and dinosaurs are very similar. Could birds be related to reptiles? If this is the case, is it possible that the reptiles themselves came from something else? After all, there was an era in which the skeletons of amphibia (frogs, toads, and salamanders) but not reptiles are found. Etienne Geoffroy St Hilaire had argued that the similarity of vertebrate anatomies meant that they were indeed related, and Jean-Baptiste Lamarck had even suggested that one animal type transformed into another. Lamarck had argued that use or disuse of an organ or limb would determine whether it would grow or atrophy (wither). By the end of the 19th C, this argument of use was disproved and ridiculed, but at the time it made sense—after all, animal structures seem well adapted to the lifestyle of the animal—and Lamarck was, in the last analysis, a topflight and well-respected biologist. Nevertheless, the idea of special creation was so firmly embedded in western (and upper class English) thinking that, as Darwin coped with the implications of this train of thought, he began to realize that to abandon special creation for descent with modification was "like confessing a murder".

SIMPLICITY IN SCIENCE: OCCAM'S RAZOR

There is another element of scientific logic that was known to Darwin and also applies here. This is called Occam's Razor. William of Occam (or Ockham) was a 14th C English logician and Franciscan friar, who argued that, if there were several possible alternative explanations of a phenomenon, the one that required the fewest assumptions was most likely the correct one. This principle would apply to the question of why so much of the structure of one animal was similar to that of another animal. For instance, in limbs as different as those of a frog's foot, a bird wing, a bat wing, the flipper of a seal, a horse hoof, and the hand or foot of a human, the bone structure is very similar (Fig. 12.2) and one can recognize the basis of an original termination in the equivalent of five fingers. By the hypothesis of special creation, it would appear that God reused the basic plan, even though an engineer might have designed more effective support structures for the different uses of the limb. Even worse, one might suggest that the amphibian or reptilian version was a primitive version that was improved for mammals. While it might be flattering to assume that mammalian bone structure is the best possible or perfect one, does it make sense that God would make practice versions? Would the hypothesis of common descent be much simpler? That is, that the tetrapod (four-footed) bone structure appeared once in evolution, and that each of the many types of vertebrates contained a form of that original structure, which had been inherited and modified through the generations (descent with modification)? As Darwin had noted for himself in his notebooks, 'Once grant that species...pass into each other....& whole [Creationist] fabric totters & falls'?

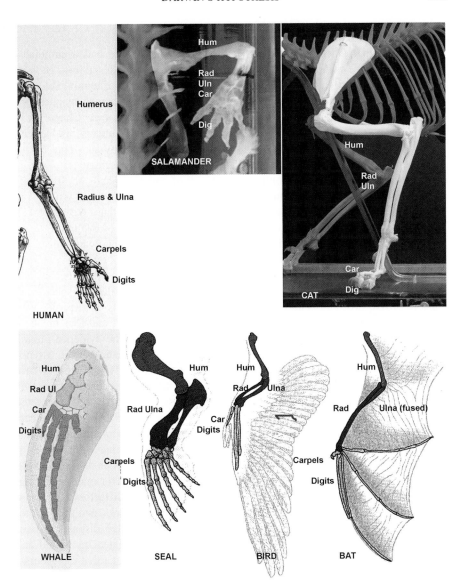

Figure 12.2. Limbs of various vertebrates (human, salamander, cat, whale, seal, bird, and bat). Although they serve different purposes (grasping, walking, swimming, flying) or have evolved independently to serve the same function (seal vs whale; bird vs bat), they all have the same bone structure. Credits: Modified from Gilbert S F, Developmental Biology, 8th Ed, Sinauer Press, Boston

PUNCTUATED EQUILIBRIUM

By the time that Darwin published *Origin of the Species* he had accumulated sufficient evidence—and the logic of Malthus as well as the evidence from geology and physics was sufficiently convincing—that the hypothesis of natural selection was well received by the scientific community. There were of course many questions left unanswered. The mechanism for heredity was unknown (see Chapter 13), leaving a substantial gap in the total logic, and some who were uncomfortable with the hypothesis, including Lord Kelvin, argued against it, but by-and-large by the end of the century there was near-universal acceptance among scientists.

Why therefore is there still controversy? Much of course depends on the beliefs or preferences of individuals, but the argument focuses on two major issues. The first is a misunderstanding of the use of the word "theory" in science (see page 11), but the second is that "scientists still disagree about the theory of evolution".

This latter point deserves some attention. Scientists, and especially biologists and geologists, do not "disagree about the theory of evolution". In essence, everyone agrees that the earth is very old, that species have evolved from other species, and that among the important forces driving evolution is the fact that all species are capable of overbreeding. There is also no doubt that individuals of a species vary in ways that can affect their ability to survive and reproduce, and that much of this variation can be inherited. We will continue to explore the evidence linking species in subsequent chapters, but at this point we can concede that the essence of the theory of natural selection is widely if not universally accepted. Where evolutionists differ is over the relative importance of sexual selection (the means by which one sexual partner chooses its mate), the necessity of species to be separated into two groups (such as on an island and the mainland) for evolution to proceed, the relative importance of predators, as opposed to disease or random mutations, to force selection, whether or not natural selection can operate at the level of genes (a gene can be selected for even if the net effect is bad for the individual), whether or not there is competition such that there is selection for male genes that are disadvantageous to females and vice versa, and other such factors. These are essentially arguments over the mechanics of how evolution works, and they do not challenge the basic premise. Perhaps the most important of the disputes is over the concept of "punctuated equilibrium," as presented by Stephen Jay Gould and Niles Eldridge. These authors, major theorists in evolution, argue that rapid bursts of change interspersed with long periods of stability are the norm rather than the exception. They argue that, for instance in the case of the trilobites, a major shift in environmental conditions—to some extent hypothetical, but backed by substantial general evidence—caused a rapid evolution of the group that produced the trilobites. Once they had appeared, they remained stable, with relatively little change, over the course of their 300,000,000 year history. Gould and Eldridge argue that this is the primary driving force of evolution: a major change in conditions on the earth creates a very difficult time for some or all creatures. Some of the most extreme variants can cope with the new conditions, and selection and evolution occur very rapidly in evolutionary terms (over a few million or tens of millions of years).

Once the tumult is over and the new variants have established themselves, they persist with extremely modest change until the next upheaval. This interpretation has substantial implications as we consider issues such as those of global warming, but it does not fundamentally challenge the theory of natural selection. It merely generates a new hypothesis as to the dominant forces for natural selection. As a scientific hypothesis, it is a good one, since it carries implicitly the appropriate tests to evaluate, such as verification that the evolutionary history of many more species is punctuated, but it does not challenge the evidence that evolution has occurred.

The concept of evolution is distinct from that of natural selection. There is essentially no doubt that evolution has occurred. The theory of natural selection is the best—and an extremely well substantiated—hypothesis describing the mechanism by which varieties that we see today were created. What is under discussion today is the details of the mechanism. As in all good science, one question leads to another, and we continue to burrow deeper into the meaning of the question. This issue is discussed further in Chapter 14, page 191.

REFERENCES

http://www.linnean.org/contents/history/dwl_full.html

http://www.clfs.umd.edu/emeritus/reveal/PBIO/darwin/ (Darwin-Wallace 1858 paper, from Univ. of Maryland)

http://www.darwiniana.org (Essay on Darwin and Evolution, from International Wildlife Museum, Tucson, AZ)

http://nationalacademies.org/evolution/ (Essay on evolution, from National Academies of Science)

STUDY QUESTIONS

1. Restate Darwin's hypothesis in your own words. Which elements are essential?
2. Make a prediction based on Darwin's hypothesis as you have phrased it in question 1. What would you expect to find?
3. What elements from earlier science were essential to Darwin's hypothesis? Explain.
4. Argue an alternative hypothesis as to why, for instance, fossils of armadillos are found only where armadillos survive today.
5. How was it possible for early geologists to argue for an older earth? What might have caused them to come up with numbers that today we consider to be incorrect?
6. What are some of the primary issues today in evolutionary theory? Argue for AND against the statement "Scientists still disagree about the theory of evolution". Do you think that your arguments are valid, even if you are not a trained scientist? Why or why not?

CHAPTER 13

THE CRISIS IN EVOLUTION

THE CRISIS IN EVOLUTION

By the end of the 19th Century, the evidence that species could vary and change was overwhelming, so that one no longer needed the tedium of Darwin's exhaustive documentation. Likewise, although the firmest proof of the age of the earth (radioisotopic dating, measurement of time using the physical characteristics of light, and documentation of continental drift) was yet to come, numerous lines of evidence including astronomy, physics, several arguments from geology, geology, and biology all converged on the conclusion that the earth was at least millions rather than thousands of years old. Thus there was time to produce not only all the breeds of dogs but even to produce mammals, birds, frogs, insects, or grasses, trees, mosses, and ferns. One major impediment to accepting the idea that the earth could change was now resolved, and the idea of evolution began to achieve acceptance. However, this cognizance led to an unforeseen problem. As people began to accept the idea of evolution, they began to explore the mechanisms by which it could occur. And now there appeared a major theoretical problem: by common under-standing of heredity, evolution could not work. We shall see shortly that common understanding was based on denial of the obvious, but nevertheless it was the province of contemporary scientific thinking and therefore an issue that ultimately caused Darwin to doubt his own hypothesis. The argument was as follows:

Throughout history and in many societies, the role of women in heredity has been treated with some disdain, even with completely self-contradictory arguments. "Bring forth men children only" (Macbeth to Lady Macbeth) simultaneously suggests that women have control over the choice and that they betray men by not producing boys, or that they decide the sex of the children, or that they provide fertile or infertile terrain for the development of male children contributed by the father, as was evidenced by many royal marriages being terminated (by one means or another) because the Queen did not produce a son. In the 19th Century, with the observation of sperm, the male-centric interpretation was that the man implanted a microscopic child into the woman (a homunculus, see Fig 13.1), and the woman was a sort of ambulatory flowerpot.

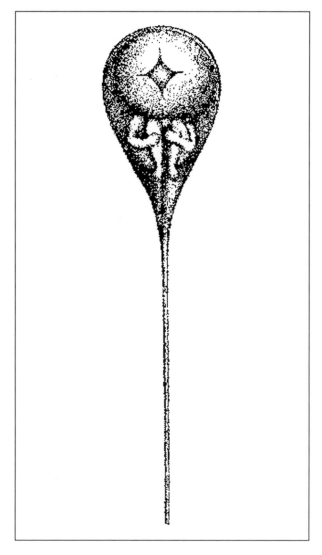

Figure 13.1. Homunculus. Image of a homunculus (Latin: tiny man) as early microscopists believed that they saw in sperm. Sperm had relatively recently been discovered, and even more recently determined to be natural constituents of semen rather than infectious parasites, and they had finally been associated with fertility. Furthermore, microscopes were able to resolve images barely smaller than sperm. In other words, they could distinguish the shape of sperm but not really determine any structures inside the sperm. This did not prevent microscopists from interpreting what they saw in light of the assumptions or prejudices of the time. (Because of the physics of light and the limitations of the human eye, today's light microscopes can more consistently see structures the size of sperm, but the theoretical limit of resolution is close to this size.) To determine structures within sperm or bacteria, some of which are approximately the same size, one needs to use an electron microscope or use any of several elaborate technologies or computer enhancing. Credits: From Nicolaas Hartsoeker's Essai de dioptrique (published in Paris, 1694, public domain (Wikipedia)

THE ORIGIN OF INHERITANCE

Even when thoughtful men conceded some contribution on the part of women to the child (since children could look like their mother), how this contribution got made was a matter of some speculation. The most reasonable idea seemed to be that essences of the parents were distilled into the gonads and were in some manner packaged into what would eventually be recognized as eggs and sperm. After all, children tended to look like parents. If the father had hairy fingers, how would that information get to the child other than by being carried through the blood to the testes and then into the sperm? Likewise, strong and athletic parents tended to have strong and athletic children. The parents were not born strong. Somehow the strength that they had acquired got collected and delivered to the children. Some of this mythology still exists in the rules and classifications in horse breeding, which distinguishes horses that have bred previously from those that have not.

There are two issues here. One is that characteristics are collected and distilled into the children, and the other is that the characteristics (like strength) can be modified throughout life, and the modified form delivered to the child. The modification argument carries a specific testable implication—i.e., it is a testable hypothesis—and it was extensively tested during the latter half of the 19th Century. The hypothesis was the inheritance of acquired characteristics, as represented by the following logic: A giraffe's long neck arose because generation upon generation of proto-giraffes reached ever higher for leaves on trees, their necks grew with constant stretching, and their children inherited the longer necks. Some rodents, such as guinea pigs or hamsters, have short or no tails. The experiment therefore is to cut off the tails of successive generations of mice. Eventually there should be nothing to distill to the babies (or, minimally, the tails have never been used and should atrophy) and babies will be born without tails. Lamarck had specifically proposed this argument, and the experiments were many times repeated, always with the same results: the babies always had full-length tails. Therefore, this element of the argument, that modified (acquired) characteristics could be inherited, went down in crashing defeat.

The other issue is the distillation of characteristics. If one argued the distillation of characteristics, then one would have to deal with the possibility that women could distill characteristics into eggs—and after all, both boys and girls could take after their mother's side of the family—but this created a very dangerous intellectual problem: dilution. In a nutshell, this is the problem: I have a brand new characteristic, one which in Darwin's terms makes me extremely fit. Let's say that I can photosynthesize my own food. My children ought to populate the earth. However, I am, to use 19th C terminology, a "sport," what we would today call a mutant. The characteristic appeared for the first time with me. That means, of course, that the woman I choose to marry does not have the characteristic. Since she contributes to the egg that she builds, let's concede that she contributes half of the characteristics. That means that my child gets only half of my ability to photosynthesize. Since my child will presumably choose a partner from outside of the family, rather than a brother or sister, my grandchildren will have only one fourth of my ability to photosynthesize. In the course of a few generations, my wonderful ability will have been diluted to unmeasurable or ineffective levels. A "sport" or new mutation or new variant cannot be propagated in a population; it will inevitably be diluted into non-existence.

So we have a problem: Evolution is logical, it makes sense, there is evidence that it has occurred, but there is no way in which it can occur. Unless we can resolve this problem, we have to throw out the whole hypothesis. The hypothesis fails the L (Logic) part of the ELF rule.

SOLUTIONS TO THE PROBLEM

Toward the end of the 19th century, an observation gave a hint as to what might happen and, armed with this hint, several scientists set out to see if they could find a solution. The hint came from embryology. August Weissmann, doing a very careful study of how eggs and embryos developed, had come across some very peculiar colored bodies[5] (the literal translation of chromosome) that underwent an elaborate ballet every time cells divided. Not only did they undergo an elaborate ballet, the ballet in cell division that produced an egg or a sperm cell was very different from that when, for instance, a liver cell divided. He called the ordinary division of cells **mitosis**, and the division (actually a pair of divisions) to produce an egg or sperm cell **meiosis**. He was aided in coming to this conclusion by the fact that he had chosen for study some very small worms and insects, so that he could see the chromosomes without having to cut the animals up. In these animals, the different behavior of the chromosomes during meiosis as opposed to mitosis is spectacular (Fig 13.2).

What Weissmann saw was that, for one cell to make two cells, the chromosomes doubled before the cell divided, and then half of the total chromosome population went to each cell. Thus each daughter cell had as many chromosomes as the original, pre-division cell (Fig 13.2a). Very interestingly, in meiosis, the chromosomes doubled once, but the cell divided twice, so that each of the four daughter cells (eggs or sperm) ended up with half the number of chromosomes as the original cell. When the new individual was reconstituted with an egg and a sperm, the original number of chromosomes was restored (Fig 13.2b). So that's how it worked! Each individual consisted of half his mother's chromosomes and half his father's chromosomes. As an adult, this individual would produce eggs or sperm with half the number of chromosomes, and the fertilization would restore the number.

Weissmann had also recognized the cells that give rise to the sex cells (eggs and sperm) and begun the investigation that led to a second very important conclusion. In some animals, the cells that give rise to the sex cells, known as the germ cells, are recognizable in very young embryos, sometimes as soon as they are formed, and they can be followed throughout the development of the embryo. In insects, the

---

[5] The chromosomes are not colored, but they can be readily stained by the dyes that histologists were beginning to use.

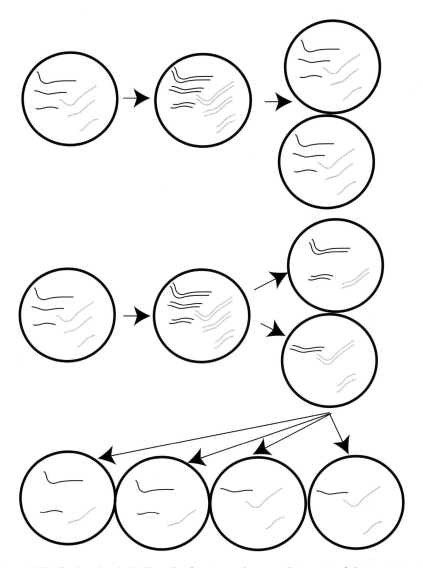

Figure 13.2. Mitosis and meiosis. The cells of most organisms contain two sets of chromosomes, one from the father (black) and one from the mother (gray). The number of chromosomes varies among organisms from one to hundreds. Humans have 23 pairs of chromosomes. Upper figure: The nucleus of this cell contains six chromosomes consisting of three pairs, distinguished by size. When the cell prepares to divide, each chromosome builds an equivalent partner, and the nuclear membrane dissolves. At this time the chromosome is described as consisting of two chromatids (the partners). Each chromosome, with the chromatids still attached, lines up in the center of the cell, and one chromatid of each chromosome is towed into each of the two developing daughter cells. The nuclear membranes reform, and there are now two cells identical to the original mother cell.

In meiosis (lower panel), which occurs in germ cells, The chromosomes double as before, but in the first division one entire chromosome consisting of the two chromatids, from either the mother or the father, moves into each daughter cell. The choice is random, so that each cell will end up with a random mix

germ cells even at one point reside outside of the embryonic body proper (Fig 13.3) Thus in addition to the lack of evidence supporting Lamarck's hypothesis, one could now argue that the germ cells were physically separate from the body, the characteristics of each individual were carried in the chromosomes that resided in the germ cells, and that chromosomes would not migrate from the cells of the body (or soma, from the Greek word meaning, not surprisingly, "body") to the germ cells. Thus there was further evidence against the hypothesis of inheritance of acquired characteristics.

We have progressed to the point today that germ cells can be transplanted from one animal to another, and the results are consistent with the interpretation that they do not change. For instance, if one identifies an easily recognizable characteristic such as body color, and transplants the germ cells from an ebony-bodied fruit fly to an egg from a yellow-bodied fruit fly, the egg will develop into a normal, fertile yellow-bodied fly but will bear young that are ebony-bodied. The inherited characteristics are determined by the characteristics of the germ cells (Fig 13.4).

In terms of the primary problem, that of dilution of mutations, the recognition of chromosomes suggested a possible solution. Chromosomes were not diluted from generation to generation. They duplicated, were divided equally, and recombined to form a new individual. Perhaps it was possible that inherited characteristics could also be preserved intact? But chromosomes could not be characteristics. Humans have only 46 chromosomes (23 pairs) but obviously far more than 23 or 46 different characteristics. Some animals and plants have only 4 or 5 pairs of chromosomes, and there are even some with a single chromosome pair. What was the connection?

THE SEARCH FOR THE RESOLUTION OF THE PROBLEM

Thus in 1900 there was considerable scientific ferment, with the question of how one could resolve the problem of dilution, evidence that germ cells were not changed by their residence within the body, and some suggestion of a means of not diluting characteristics. Thus it was not a lucky or bizarre accident, but a product of the way science works, that three laboratories simultaneously rediscovered Mendel's original paper, which carried the potential solution to the dilution problem. Mendel's paper was not totally obscure. It was published in a reputable journal, but it was mathematical, theoretical, and—to be frank—probably boring to the evolutionary theorists. What Mendel had done was to see if the laws of chance, then being worked out by mathematicians to understand how gambling worked (and for the practical reason of helping casinos, which were then very popular, calculate odds that would ensure a

Figure 13.2. of these chromosomes. The cells then divide again, but in this division the chromatids separate as in mitosis. The result is one duplication plus two divisions, so each of the four resulting cells contains half the DNA of the starting cell, and has only one copy-randomly from mother or father-of each chromosome. These cells are the future eggs and sperm. When a sperm fuses with an egg, the original number of chromosomes is restored

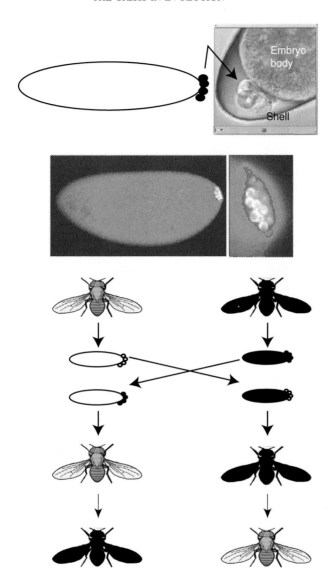

Figure 13.3. Insect pole cells. In most insects, before the embryo has taken shape a small group of cells briefly accumulates at the posterior end of the embryo (left figure, arrow). In the 19th C, André Haget, a French biologist, destroyed these cells with a hot needle and was surprised to find that, although the larvae were normal, the adults were as normal as they could be except that they were completely sterile. Subsequent studies traced these cells ultimately to the gonads (ovaries or testes), and similar, though far more obscure, cells were found in the eggs of vertebrates. Right figure: Pole cells in the embryo of a small wasp. Middle: Pole cells at the posterior end of in a fruit fly (*Drosophila*) embryo. The pole cells have been stained so that they fluoresce green. Lower: When pole cells are transplanted between a normal-colored fruit fly and a black fruit fly, the eggs turn into flies of their host color but produce young of the transplant color http://www.snv.jussieu.fr/bmedia/CoursPCEMDEUG/DocDrosimage/Drosophila%20pole%20cells.jpg. Credits: http://www.snv.jussieu.fr/bmedia/CoursPCEMDEUG/DocDrosimage/Drosophila%20pole%20 cells.jpg

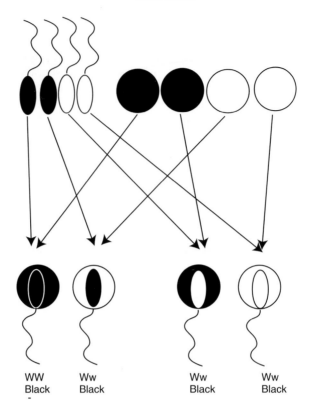

| WW | Ww | Ww | Ww |
| Black | Black | Black | Black |

Figure 13.4. The genetics of color. For any trait such as color illustrated here, each sperm and each egg will be descended from the four cells of a meiotic division, and will have one chromosome for the color trait, either from the mother or the father. The illustration here assumes that the mother and the father both have one black and one white trait. If they are pure-breeding, all the eggs or sperm will be the same. Each of the products of meiosis is illustrated, since the chance of getting either one is random. We also assume that if the resulting child carries any copy of the black trait, its color will be black, since for many genes one version is the absence of another version. Here, the black trait is the ability to make black pigment, and the white trait is the absence of that ability. Depending on which egg combines with which sperm, we will on average get three black animals for every white animal. Two of these three will, like their parents, be able to produce white grandchildren. One of the black animals, and the white animal, will breed true, since it does not contain the other trait. We conventionally describe this situation by designating the black trait as dominant and the white trait as recessive, and geneticists usually symbolize these relationships by giving the characteristic the letter designation of the recessive gene, symbolizing the recessive variant by lower-case type and the dominant variant by upper-case type. Thus the phenotype (how the animal appears) is black or white, as written out. The genotype (what its genetic composition is) is WW (pure-breeding or homozygous black), Ww (black but not pure-breeding or heterozygous), or ww (white and homozygous or pure-breeding).

Note: human racial coloration is far more complex than this and consists of several genes. Also, if for the heterozygous forms the animal containing a single copy of the gene was able to make far less pigment than the homozygous form, the heterozygote might be distinguishable as gray, not black. Flowers may be pink rather than deep red because of such a situation

profit), would apply to inheritance of characteristics as well. In other words, he wanted to see if the chance of getting blue eyes followed the same mathematical laws as the chance of getting two heads in a coin flip or rolling two ones with dice. In 1900, three groups finally realized that this dry mathematical exercise provided the key to the dilution problem. Characteristics could be passed from generation to generation without dilution. The key lay in the way that Mendel did his experiment.

What Mendel did was very simple: Instead of asking, in effect, "Does this girl look more like her mother or her father?" he asked, "Is the color of her eyes that of her mother or her father? Is the color of her hair that of her mother or her father?" In other words, he subdivided general impressions into highly localized or specific characteristics, and only then did he see very clear-cut patterns. Specifically, he saw that some characteristics could hide other characteristics, but that the hidden characteristics could reappear in later generations, unchanged, undiluted, and unaffected by passage in an individual with different characteristics.

All this sounds very abstract, but it can be described in easily comprehensible terms, and terms that scientists of the time might have recognized had they understood that human inheritance was like that of animal and plant inheritance. Throughout the world, but especially in northwestern, northern, and eastern Europe, most scientists had encountered the situation in which a red-headed child was born to a couple with dark hair but in whose families redheads had been seen. It was simply the situation of, "Little Mary has Uncle Ed's red hair!" This was the essence of what Mendel had described. Characteristics (red hair) could be passed hidden from one generation to another, and reappear uncorrupted in a new generation. It should have been obvious to anyone who thought about it, it resolved the conundrum of the dilution problem, and it took 35 years to rediscover the experiment that explained it.

MENDEL SAVES EVOLUTIONARY THEORY

The rediscovery of Mendel resolved that last huge hurdle to the intellectual acceptance of evolution and led to general acceptance in the scientific community of the theory of evolution. This general acceptance and popularization of the theory led to a resurgence of theological and religious challenges that will be addressed later. At this stage it is important to understand what Mendel's experiments and results were, and how they were interpreted. Like all scientific information, what Mendel saw and interpreted has been subjected to some adjustment, as some variations and finer details have come to light, but the essence of his results are as follows. You can read his original paper on the internet at http://www.mendelweb.org/.

Mendel was an Austrian monk who raised peas in his garden. He wanted to see how pea characteristics were inherited and, as noted above, he subdivided the characteristics that he chose to observe. Pea plants can differ in many characteristics: the plants can be tall or short; the peas can be yellow or green; they can be wrinkled or round; the flowers can be purple or white; etc. Rather than treat inheritance as one complex muddle, he asked very simple questions: if he crossed peas with purple flowers with peas with white flowers, would the flowers of the resultant pea plants (the children) be purple, light purple, or white? If he crossed tall peas with short peas, would the children be tall, short, or intermediate? If he crossed plants bearing yellow peas with plants bearing green peas, would the peas be yellow, green, or yellow-green? Would the offspring of plants bearing wrinkled peas and plants bearing round peas be wrinkled, round, or in between? What he found was startlingly simple and unconfusing.

To make this discussion clearer, it will help to use the terminology that geneticists use. The crosses originate between two pure-breeding lines, that is, peas that always produce purple flowers are crossed with peas that always produce white flowers. The peas in this cross are the parental or P generation. The seeds that are produced in this cross become the first filial or F_1 generation. (Students familiar with any Romance language will recognize the fil-root as indicating son or daughter.) These plants are then crossed with each other (there are no laws or customs forbidding brother-sister marriages in plant breeding) and the seeds produced from these crosses are the second filial or F_2 generation. In symbolic form:

$$P_a \times P_b \longrightarrow F_1 \otimes \longrightarrow F_2$$

What Mendel saw was the following: ALL of the F_1 generation looked like one parent, not the other. In other words, in the flower color cross, all of the F_1 plants produced purple flowers. There were no light purple flowers or white flowers. The white characteristic had disappeared. Similarly, in the tall/short cross, all the progeny were tall; in the yellow/green peas cross, they all had yellow peas; and in the wrinkled/round cross, all the peas were round. There were no intermediates, and one characteristic had disappeared.

He then inbred the F_1 generation to get an F_2 generation. In this second generation, the lost characteristics reappeared. There were no intermediates, but there were white flowers, short plants, green peas, and wrinkled peas. Not only did these lost characteristics reappear, but they reappeared in a specific pattern. The lost characteristics reappeared as approximately one fourth of the plants. There were three plants with purple flowers for every plant with white flowers, and so forth. The actual data from Mendel's experiment are shown in Table 13.2.

To Mendel, this was a distribution indicating that the characteristics combined as a matter of chance, since the mathematics was the same as that for flipping coins: He invented a specific description: the traits that appeared in the F_1 generation were "dominating" (today we say "dominant") and those that disappeared were "recessive". Mendel recognized that what he saw was chance recombination. For instance, if one flips two coins, one has an equal chance of getting each of these four combinations: two heads; heads, then tails; tails, then heads; and two tails. If one ignores the tails and counts only the times that one gets at least one head, then one will get at least one heads in three out of four double tosses. Mendel explained that, if the purple trait could hide the white trait, as was seen in the F_1 generation, then the three purple to one white ratio was identical to the "at least one heads" ratio. All he had to hypothesize was that each parental plant contributed at least one "coin" or color characteristic. The purebreeding purple flowers would produce only purple characteristics, and the purebreeding white flowers only white characteristics. Each plant would have two of each characteristic, since the F_1 had to have two. The cross would be as follows:

Purple, purple x white, white $\longrightarrow$ purple, white, which would be purple since purple could hide white.

Here we need a little terminology. The F_1 plant has a **phenotype** (appearance) of purple, since it is purple. However, its **genotype** is hybrid; it has a purple character from one parent and a white character from the other. Thus it differs from the purple parent, which has only purple characteristics, and from the white parent, which has only white characteristics. Since the plant as it grows bears two copies of a color trait, it is **diploid**, and the unfertilized eggs (seeds) and sperm (pollen), which each have only one copy, are **haploid**. The pure-breeding strains, which have the genotypes of purple, purple or white, white, are **homozygous** (from the Greek, "like eggs") and the F_1 plant, which has the genotype purple, white, is **heterozygous** ("different eggs").

If two of these F_1 hybrid purple plants are crossed, each will contribute both purple and white characteristics to the children. Each pollen grain or each seed will contain only one copy of the color characteristic, either purple or white, and the double, diploid, form will reappear when one egg combines with one pollen grain. The cross can produce four possible outcomes, as indicated by the Italics (Table 13.1):

Some of Mendel's actual data were as is illustrated in Table 13.2:

Or, on average, there will be three purple F_2 to one white F_2. The conclusion, therefore, was that characteristics were discrete and not blends or dilutions, that

Table 13.1. The results of genetic crosses

F_1 Genotypes (all Purple, white)	F_1 Phenotypes (each plant produces purple flowers)	F_2 Genotype	F_1 Phenotype
Purple, white × *purple*, white	Purple, purple	purple, purple	Purple
Purple, white × purple, *white*	Purple	purple, white	Purple
Purple, *white* × *purple*, white	Purple	white, purple	Purple
Purple, *white* × purple, *white*	White	white, white	White

Table 13.2. Mendel's data

Plant type	Total Yellow	Total Green	Total Round	Total Wrinkled
315 round, yellow	315		315	
101 wrinkled, yellow	101			101
108 round, green		108	108	
32 wrinkled, green		32		32
TOTALS	416	140	423	133
RATIOS	2.97 to 1		3.18 to 1	

they could be hidden, that they could reappear intact in a future generation, and that they were distributed randomly among children.

INTERPRETING MENDEL

It helps to have a little sense of how this works. In many situations, the recessive form is the absence of the dominant form. For instance, for the purple and white flowers, the plants that make white flowers cannot make the purple pigment. Usually, this happens because the plant carrying the recessive trait has lost the mechanism (an enzyme[6]) to make the pigment. When it is crossed with a plant that can make the pigment, the plant that results now has the enzyme, the pigment is made, and the flowers are purple. It is very much the same as the following: if both you and your spouse have keys to the car, even if you lose yours, you will still be able to drive the car as long as one key remains.

We will call the characteristic a **gene**, and we will use the term in the sense that "he carries a gene for red hair". The gene that can mask another gene is a **dominant gene**, and one that can be masked is a **recessive gene**. For instance, carrot-red hair is typically recessive to truly black hair. If one parent comes from a line of only black-haired people and the other from a line of only redheads, the child (F_1) is likely to be black-haired (heterozygous, carrying both the gene for black hair and the gene for red hair) but could produce red-haired children (F_2) if he or she married someone who similarly carried a gene for red hair.

Although the situation is more complex and we will need some more explanation of the structure of genes, for the moment we will consider that a gene is the information to make something, such as a pigment. The gene itself is DNA (defined on page 193) and carries the information how to make an **enzyme**, which is a protein that can carry out a specific reaction, for instance converting a red pigment to a black one. From this you can see how most dominant and recessive genes work.

Hair pigments are made from chemicals (molecules, which are the individual particles of chemicals) of different colors, in the following sequence (Fig 13.5):
1. A colorless pigment is converted into a yellowish pigment.
2. The yellowish pigment is converted into an orange pigment.
3. The orange pigment is converted into a red pigment.
4. The red pigment is converted into a brown pigment.
5. The brown pigment is converted into a black pigment.

Each step here is accomplished by a specific enzyme (page 184). A red-haired person has the enzymes to complete steps 1, 2, and 3, but lacks the enzyme to complete step 4, and the synthesis stops at that point. Stopping after step 1 would yield a blond.

A black-haired person can complete all five steps. Thus in the F_1 heterozygote, the child of the black-haired parent and the red-haired parent, the gene for the enzyme

---

[6] See further discussion in Chapter 15, page 221.

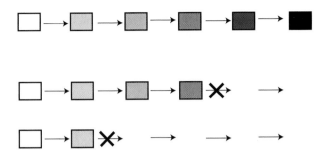

Figure 13.5. Pigment formation. Upper row: The synthesis of a pigment is a several-step process, with each step controlled by a specific enzyme. Thus, in this sequence, an uncolored precursor material is converted successively into yellow, orange, red, brown, and finally black materials. Middle row: The enzyme converting the red pigment to brown is lost, or mutated. Synthesis of pigment stops at this point, and the resulting animal is red rather than black. Lower row: The enzyme converting the yellow pigment to orange is mutated, and the animal consequently has a yellow coat color

for step 4 is missing from the genes of the red-haired parent but is contributed by the genes of the black-haired parent (Fig 13.6).

Thus this child will be able to complete the synthesis of the black pigment and will be black-haired. Should this black-haired child produce an egg or sperm carrying the defective enzyme 4, and this egg or sperm combine with a sperm or egg from a partner likewise carrying the defective enzyme 4, the resulting child would be red-haired.

Most genes operate in more-or-less this manner, and Mendel's laws of inheritance can be demonstrated in all animals and plants. Mendel's interpretation provided the explanation of why new characteristics (mutations, "sports") are not lost in subsequent generations. They are not diluted but are passed intact, even though their effect may not be seen.

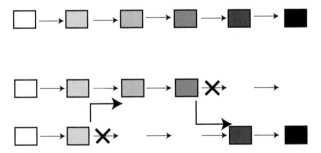

Figure 13.6. Complementation: Lower two rows: The heterozygous child of a red-haired parent and a yellow-haired parent can be black haired, because the defective yellow-to-orange enzyme produced by the one chromosome can be replaced by a good enzyme from the other chromosome, and the defective red-to-brown enzyme of that chromosome can be replaced by the good enzyme from the first chromosome. Thus pigment formation can be completed

Genes are physically very tiny, and we now know that they are lined up on the chromosomes. For instance, there are 20–25,000 human genes, and there are 23 chromosomes, making an average of 1,000 genes per chromosome. Each time a cell divides, whether in mitosis or meiosis, the duplication and movement of the chromosomes carries the genes appropriately into the new cells.

Thus 1900 was somewhat a turning point for the acceptance of the theory of evolution. The evidence for the relationships among animals and plants was abundant; examples of selection could be found almost anywhere one looked; it was now evident that the earth was old enough to have supported the evolution of all the species known; and now it was apparent that characteristics could survive and be passed to future generations. But rather than continue with the history of the social acceptance of evolution, let us first examine, in the following chapters, the evidence for it and the mechanisms by which it works.

REFERENCES

http://www.mendelweb.org/ (Mendel's original paper, from an international resource for the web)

STUDY QUESTIONS

1. Consider the state of knowledge both at the end of the 19th C and today. What are the weaknesses of the theory of evolution, that is, that populations overbreed, there is competition among individuals and selection of the fittest, and that this process gradually changes species? The theory also includes the assumption that such processes could have generated all the life forms that exist or have existed on earth.

2. Some inherited characteristics are not inherited according to the simple rules that Mendel saw. For instance, when one crosses red and white flowers, one might get pink flowers. Can you formulate a hypothesis as to how this might work?

3. Other characteristics also do not form simple Mendelian ratios. For instance, people do not divide into tall, average, and short. Within a range normally of about 5 feet to 6 ½ feet, we find adult humans of all possible sizes. Can you formulate a hypothesis as to how this might work?

4. As noted in this chapter, there are thousands of genes on each chromosome, and the chromosomes move as units into daughter cells. If two genes, say for eye color and for hair color, are on the same chromosome, is it likely that the trait for eye color will separate randomly from that for hair color? Explain.

5. What question did the discovery of chromosomes resolve? What question did the discovery of germ cells resolve?

6. Can you speculate how scientists determined that genes were arranged in a linear manner on the chromosomes? What evidence would they have needed? Do not look at this question as an expectation of your knowledge of detail. Consider how you would go about determining whether or not there was an order to a string of beads, based on how frequently you encountered specific groupings and breaks.

PART 4

THE MOLECULAR BASIS OF EVOLUTIONARY THEORY

CHAPTER 14

THE CHEMICAL BASIS OF EVOLUTION

THE CHEMICAL BASIS OF EVOLUTION: GENES, CHROMOSOMES, AND MENDELIAN GENETICS

Science is an onion. It consists of questions, but each answer opens a new question, much as an onion consists of layer after layer of modified leaves. It is not unusual for a scientist to publish hundreds of papers over a lifetime, yet with each paper insist that he or she is working on the same problem—just delving deeper and deeper into the problem, peeling another layer off the onion. It is much the same issue as realizing the many layers of what seems to be a simple question that a child might ask. For instance, the child might ask, "Why are rabbits brown?" (Since science is concerned with mechanisms, rather than primary causes, "Why" is not an appropriate opening—see page 14—but, since children often use the expression, we will continue along this line.) One answer might be, "Because God made it brown," which is likely to lead to the follow-up question, "Why did God make it brown?" One answer might be, "To allow it to hide from its enemies." At this point the query could go in several directions—talking about natural selection, documenting that color does make a difference, considering the relationship of the color of the rabbit to the color of its environment and the presence of predators, discussing the genetics of pigmentation, or the biochemistry of the synthesis of pigments. If one followed the latter argument, the synthesis of pigments, this could lead to the question of why some molecules are transparent and others have colors, which could lead to an exploration of how atoms are held together into molecules, and how the interaction of atomic structure with light leads in some instances to the light's going through the atom or molecule and in other instances to the light's being absorbed or reflected. This is why the great thinkers of classical Greece, China, India, the Enlightenment, or other cultures did not resolve for all time the questions being asked.

This style is beautifully illustrated by the pursuit of the generalized question, "What is the basis of inheritance?" This pursuit led to the identification of DNA as the genetic material, and subsequently an understanding of how DNA carried information and how this information was transformed into the building blocks of all organisms. This is the story of the rise of molecular biology, surely one of the great episodes in the history of science. It is as abstruse and rarified as any level of knowledge today, but there is no reason why a student cannot understand how it came about. The story illustrates spectacularly well how scientists ask questions and pursue them. For the most part, asking the question and getting an answer was a matter of games and tricks. In the vernacular, "Molecular biology ain't rocket science". It mostly is a matter of cool tricks.

REALLY COOL TRICK #1: HOW TO TURN A NOT-SO-BAD BUG INTO A REALLY BAD ONE. (THE GENETIC MATERIAL IS DNA.)

Once it was conceded that sperm and egg united to form a new individual with the characteristics of its parents, one had to ask, what was in the sperm and the egg that carried the characteristics. The egg contains yolk and nutrients for the embryo. The sperm is much simpler, containing mostly DNA and proteins, but there are many other components as well. At the level of biochemical skill available in the 1930's, even sperm were too complex to use to analyze this question. One needed a simpler model. This model came from microbiologists worried about how diseases were transmitted. In their pursuit of this question, they learned that bacteria could transmit characteristics from one organism to another, and that even dead bacteria could pass on their characteristics to living bacteria. Thus whatever carried the characteristic had to be chemical, and was not a "vital force" or other unique characteristic of living organisms.

The first assumption that everyone made was that the chemical was a protein. This seemed highly logical. Proteins are very complex structures, consisting of a string of a mixture of twenty different building blocks called amino acids, whereas nucleic acids are much simpler, being a string of only four types of their building blocks, called bases. If you consider the amino acids and the bases to be letters in an alphabet, an alphabet with 20 letters can produce a lot more words than an alphabet with only four letters. With an alphabet of twenty letters, we have a language. (This last sentence uses 15 of the letters of the English alphabet.) With four letters, we don't have much: four, fuor, foru, frou, fruo, rouf, ruof, etc. Even allowing words to be different lengths and allowing letters to be used two or three times in the same word (foor, fuur) the language is very restricted. As the logic went, only proteins have the complexity to store all the information needed to build an organism. Unfortunately, however, the first reasonable chemistry began to give a different answer.

The critical experiments, known as the Avery-MacLeod-McCarty experiments (really cool trick #1), were as follows:

Avery and MacLeod, working at what is now Rockefeller University, were studying a type of pneumonia caused by bacteria in mice. They were working with pneumonia-causing bacteria and had isolated a variant (mutant) strain that did not kill the mice. Bacteria can be grown in Petri dishes, in which each single bacterium multiplies and forms a single spot or colony on the dish, much as you may see mold springing up from several isolated spots on a piece of bread. The virulent or deadly form formed smooth colonies that looked like little droplets, while the non-virulent or non-killing form formed colonies that had rough, uneven edges.

We now know that the bacteria that form the smooth colonies secrete a somewhat gelatinous material that both creates the smooth appearance and protects the bacteria from attack by the mouse's defenses, its immune system. The bacteria of the rough variant cannot make this material and are quickly destroyed by the mouse's immune system.

Avery and MacLeod were trying to understand the difference between the rough and smooth bacteria, and what made the difference in the virulence. In one series of experiments, they injected a mouse simultaneously with boiled (dead) smooth bacteria and live rough bacteria. They found, to their surprise, that the mice died. When they took samples from the dead mice and cultured them, they found that what grew in the culture was smooth, virulent bacteria. This was tremendously exciting, because it meant that something from the dead bacteria could convert the rough bacteria into smooth bacteria. Since the experimenters could grow the bacteria and inject them into new mice, which subsequently would die, the rough bacteria had been permanently converted, or transformed, into the dangerous kind. Avery and MacLeod confirmed that they could not grow smooth bacteria from the boiled culture or cause disease if the dead smooth bacteria were injected alone. What this meant was that some chemical in the smooth bacteria survived and transformed the rough bacteria into smooth. It was not simply a question of the chemical coating the rough bacteria and protecting that generation of bacteria, since more smooth bacteria could be grown in the culture and could infect more mice. The rough bacteria truly had been transformed.

Now it was acknowledged that a chemical existed that could carry genetic information and transform one variant of bacteria into another. It now became possible to try to purify this chemical and, using the criterion of transformation, to identify what it was. The scientists pursued this goal and came to a surprising conclusion: the transforming material was DNA, not protein.

Nobody really believed them. DNA was far too dull and uninformative a molecule to carry information (remember our four-letter language). Besides, chemical methods weren't really that good, and even the purest DNA was contaminated with a few percent protein. Obviously, the real genetic material had to be a sort of super-protein that remained during the attempt to separate DNA from protein. The results were not dismissed out of hand, because after all there was nothing wrong with the data or the way the experiment was done, but no-one was really satisfied. The logic was not yet there.

REALLY COOL TRICK #2: BACTERIAL MILKSHAKES

The question still remained alive, until finally a refined means of doing this assessment was developed. This we can describe as the bacterial milkshake, or really cool trick #2. The issue was, can one get really pure DNA and protein, so that it is possible either to identify the "super protein" or confirm that the genetic material really is DNA? This question was addressed in 1952 by Alfred Hershey and Martha Chase. They took advantage of an interesting bit of biology and used a kitchen trick to get an answer. The interesting bit of biology can be summarized in the well known poem, "Big bugs have little bugs/On their backs to bite'm/Little bugs have lesser bugs/And so on *ad infinitum*." What this translates to is that even bacteria have parasites. The bacterial parasites are called bacteriophage, or "bacteria eaters". These are viruses that attack bacteria, eat everything inside the bacterium, and produce new bacteriophage, or phage, that will attack other bacteria. They are quite vicious: if one has a "lawn" of bacteria (a thin layer of bacteria covering an entire Petri dish so that the whole surface is grayish) a single phage and its progeny will kill all the bacteria within range, creating a clear spot or "plaque" in the lawn (Fig. 14.1).

What makes this arrangement so interesting is the manner in which one type of phage attacks one type of common bacteria. This type of phage, which looks like a mini lollipop, attaches to the bacterium, stick end first. It injects something into the bacterium, leaving the shell of the lollipop on the outside. Obviously, what goes into the inside is what is the source of the new phage—in other words, the genetic material. What stays outside plays no further role, being abandoned with the bacterial membrane and wall when the dying bacterium bursts, releasing the new phage into the medium. The whole life cycle takes about 20 minutes. The question then is, what goes inside? To the scientist, the question is how to determine what goes inside.

If that question changes to whether it is DNA or protein that goes inside, there is a way to answer the question. DNA contains a lot of phosphorus but no sulfur, while

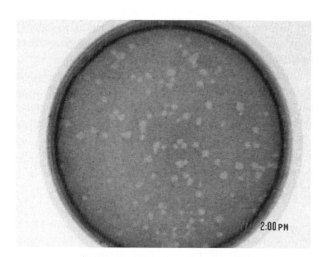

Figure 14.1. Bacterial plaques assay. In this experiment, viruses were scattered on a "lawn" of bacteria, otherwise described as an even coating of bacteria growing on medium in a Petri dish. The bacteria are stained and look dark in the picture. Where a virus has landed, it has infected the bacteria, grown, and reproduced, killing the host and infecting the bacteria next to it. Thus small circles of killing appear, as is marked by the clear areas, or plaques. Each plaque represents the descendents of one virus, or a clone of the virus. Credits: Photograph:–Jeffrey McLean, used with permission

proteins contain a lot of sulfur and not much phosphorus. By the time Hershey and Chase came along, the rise of the atomic era meant that reactors were producing, as byproducts, radioactive sulfur and radioactive phosphorus. Radioactive materials (radioisotopes) can be measured in extremely small amounts or, more importantly, trace contamination can be picked up at levels roughly 10,000 times less than can be detected chemically. So the contamination issue could be addressed, if one could separate what went in from what stayed out. This issue was handled with surprising simplicity. Simply put, the experiment was as follows: Phage were grown in the presence of both radioactive sulfur and radioactive phosphorus. These phage were used to infect bacteria. After a bit of time, but before the phage could kill the bacteria, the infected bacteria were thrown into an ordinary kitchen blender and blended. This knocks the phage off the bacteria. The mixture was then placed in a centrifuge, which spins the mixture at high speed, forcing all particulate matter to the bottom of the centrifuge tube. What comes down is the infected bacteria. What remains in the medium is what did not get inside and was knocked of the bacteria. Hershey and Chase, now having the outside of the bacteria separated from the inside, counted the radioactivity. The answer was unequivocal: The phosphorus (DNA) went in, while the sulfur (protein) stayed outside. The amount of protein that got in could be determined to be less than 0.1%. Thus it became almost impossible to maintain the argument for the "super protein", and the scientific world, reluctantly began to concede that the genetic material was DNA. The question now became, how did it work?

REALLY COOL TRICK #3: MOLECULAR BILLIARDS AND RUSSIAN DOLLS, OR DNA *MUST* BE THE GENETIC MATERIAL

There were many different directions from which one could attack this question, all of which had value, but the next cool trick was a molecular billiards game that made it intellectually necessary for DNA to be the genetic material. This was what is now known as development of the Watson-Crick model for DNA.

The expression "intellectually necessary for DNA to be the genetic material" is a bit of a tough nut to swallow, but it will make sense once we get to the end of the chapter. The first issue we have to address is how the structure of DNA was solved. Although the mechanism involves some of the most difficult aspects of biophysical chemistry, the principle is understandable if you take a little bit of time to think about it.

The chemistry of DNA was known. One can learn enough about chemistry to know that some reactions are possible and others are not. For instance, acids react with bases (vinegar reacts with baking soda) but acids do not generally react with each other. Iron reacts with oxygen to form rust, but gold does not. Using these kinds of arguments, chemists had deduced that DNA consisted of a long chain of sugar-phosphate molecules (deoxyribose phosphate) strung end-to-end:

(1) $\rightarrow$ sugar $\rightarrow$ phosphate $\rightarrow$ sugar $\rightarrow$ phosphate $\rightarrow$ sugar $\rightarrow$ phosphate $\rightarrow$

Simple sugars are made of a few carbon atoms. In the case of deoxyribose, there are five carbons per sugar. To one carbon in each of the sugars was attached a molecule of approximately similar size called a base. Since these did not form the backbone of the chain, they were considered to be side chains:

$$\rightarrow \text{sugar} \rightarrow \text{phosphate} \rightarrow \text{sugar} \rightarrow \text{phosphate} \rightarrow \text{sugar} \rightarrow \text{phosphate} \rightarrow$$

(2)

base	base	base

That was the chemistry. The question was, how did it actually fit together in space? To determine that, one had to get a good crystal of DNA, and then to apply some fairly straightforward tricks and thought to figuring out how it worked. Several laboratories tried to crystallize DNA, with Maurice Wilkins and Rosalind Franklin producing the best crystals.

Now came the question of what the crystal was. By a game of molecular billiards, it became possible to predict that the DNA chain was helical (Fig. 14.2). Linus Pauling had demonstrated a few years before that many proteins, which are chains of amino acids, took on a helical structure (the alpha helix, Fig. 15.1) and he had demonstrated how to recognize a molecular structure. You can understand the principle fairly easily. Throw two stones simultaneously into any suitable body of water, and watch the ripples, particularly where they meet:

As illustrated here, where the ripples meet, they will reinforce each other, producing a stronger ripple. Where the top of one ripple meets the bottom of another ripple, they will cancel each other out. Physicists describe this as the waves being in phase or out of phase. Looking at cross-sections of two waves, they would look like Fig. 14.3. When the waves are in phase, the result is a stronger wave (upper bold line). When the waves are opposite in phase (the trough of wave coincides with the peak of the other), the wave cancels out.

The point of this is that sound and light do the same thing. The "noise canceling" earphones that are sold for airplane travel cancel the sound of the engine by generating waves of sound out of phase with those produced by the airplane, and the shiny iridescence that one sees on puddles after a rain are produced by light reflecting off the top and bottom surface of a very thin layer of oil floating on the water. They can do that because the oil layer is just thin enough to cause the upper

Figure 14.2. A helical structure. DNA is wrapped in a helical structure like this, with the sugar-phosphate chain forming the coils, and the coils held together by the interactions between the sugars and phosphates of one loop and the sugars and phosphates of the adjoining loops

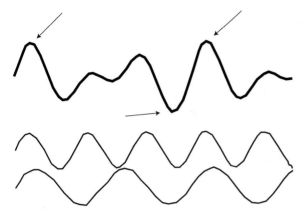

Figure 14.3. When waves meet (lower two lines), their heights sum (upper, heavier, line), so that when they are in phase at the peak or trough the combined wave is at maximum or minimum height (arrows). This phenomenon is noticed, for instance, in tuning a musical instrument. When the tone is close to that of the tuning fork, a beat is heard as the waves occasionally are in phase

and lower reflection to be out of phase. If light waves were the same length as the insides of molecules, we could see the equivalent of iridescence from molecules and be able to measure the molecule.

Unfortunately light waves are much bigger than that. However, the waves of X-rays, which in a sense are a much more intense form of light, are about the size that we expect molecules to be. If we then aim X-rays at a molecule, we might be able to say something about its structure. Specifically, if two successive waves can bounce off of two repeating units of a molecule, they will produce in-phase or out-of-phase reflections, depending on the distance between the repeating units.

This then is the molecular billiards game. The wave length of X-rays is known. When X-rays are shot at the crystal, the rays (billiard balls) bounce off successive repeating structures in the molecule. If X-ray film is placed at the proper point, where the X-rays (balls) hit in phase a spot will be produced, and its position will be a measurement of the distance between repeating units. The type of image that was acquired is illustrated in Fig. 14.4. From it, it was possible to conclude that the pattern was consistent with a helical structure. In other words, the X-rays were bouncing off successive loops of the spiral. The question then became, what did the spiral look like?

Helices can come in many forms, and it was important to understand what this one was. One clue was the density of the crystal. This can be explained as follows (Fig. 14.5): Suppose that you have a bunch of fairly loose springs, like the ones that contact the negative pole of batteries in portable electronic equipment:

You have a box full of them, which will represent your crystal. They can be all scattered loosely, in which case the box will weigh a certain amount, say one pound. It is also possible for the springs to be intertwined with each other. For

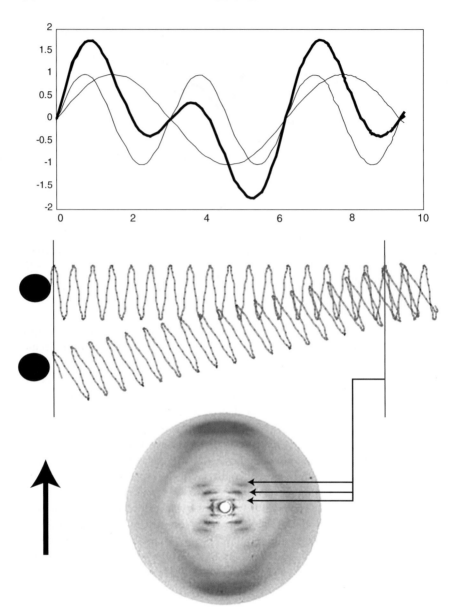

Figure 14.4. The effect of wave interaction. Upper panel: Waves reinforce when they are in phase (aligned with each other) and cancel when they are out of phase with each other (aligned opposite to each other). The heavy line is the resultant, or sum, of the two gray lines indicating waves of differing frequencies.

A crystal has a series of regularly-aligned atoms in it. For an X-ray hitting a crystal (arrow coming from bottom) the reflected waves go in all directions but start from slightly different positions. As they intersect, depending on whether they are in phase or out of phase, they will reinforce each other or cancel each other out. If a piece of X-ray film is set at an angle to catch the reflected rays, in-phase waves will

instance, you could have two springs or three springs in almost the same amount of space that you have one:

In the first case, the box would weigh two pounds, and in the second case, it would weigh three pounds. Since the box size is always the same, the density (weight per volume) doubles and triples. By weighing the DNA crystal and measuring its volume, it was possible to state that the crystal consisted of probably two and possibly three helices intertwined with each other, rather than one or four.

The final question, then, is how they fit together. Knowing the shapes of the sugar phosphates and bases, and the distances between the repeating units, Crick and Watson literally began to assemble models of how the different parts might fit together (Fig. 14.6). Among the various possible structures they found one that matched the numbers quite well. More importantly, it had constraints that led to the conclusion that we started with, that it was intellectually necessary that DNA be the genetic material. The constraints resulted from the measurements of the helix, which indicated that the DNA helix was actually two strands (a double helix). From their knowledge of the way in which the helix was constructed, they could identify both the pitch (distance from loop to loop) and the diameter of the helix. The constraint was imposed by the diameter. Again, based on the X-ray data, it looked like the bases (remember: the side chains) were on the inside of the helix, projecting into the center of the tube. However, the space in the center of the tube was not very generous, given the size of the bases. In fact, there were very few ways in which the bases could fit.

There were two limitations on the way in which the bases could fit. First, the bases are of two general types: a bulky form (purines) in which all the atoms form two rings attached to each other, and a smaller form (pyrimidines) in which all the atoms form only one ring. There was not enough room inside the ring for two purines to sit side-by-side. The only way it would work would be for two pyrimidines or one purine and one pyrimidine to sit side-by-side. Second, molecules can have local charges, vaguely like the north and south pole of a magnet, but here called positive and negative. They work like magnets, in that like charges repel each other and unlike charges attract. Two of the four possible bases are purines and two are pyrimidines. However, because of the way that the charges are distributed on the molecules and the way that they would fit inside the helix, not all combinations are possible. In fact, as Watson and Crick realized, there were only two possible combinations that would work: base A (adenine) across from base T (thymidine)

reinforce each other and produce an exposed spot, and out-of-phase waves will cancel each other and not produce a spot. By knowing the length of the waves and applying suitable mathematics, it is possible to determine from the position of the spot the distance between repeating units, such as the loops of a helix. Lower panel: The X-ray crystalogram produced by Rosalind Franklin that ultimately was interpreted by James D. Watson and Francis Crick as representing a helical structure of DNA Credits: Franklin R, Gosling RG (1953) "Molecular Configuration in Sodium Thymonucleate". Nature 171: 740–741

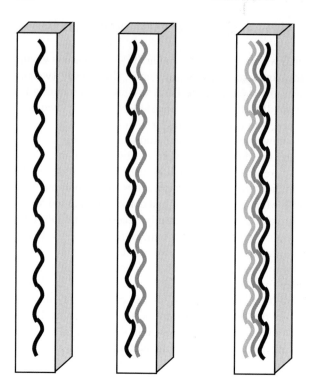

Figure 14.5. Density of crystals. Because helices can nestle in among each other, two or three helices can occupy a volume very similar to the volume occupied by only one helix. However, if one weighs each of these boxes, of course one gets different weights. Thus their densities, or weight divided by volume, are very different. If one can get a crystal of DNA large enough, one can measure its density. Another trick would be to suspend the crystal in liquids in which it will not dissolve, but of different densities. If it is less dense than the liquid, it will float, and if it is more dense, it will sink. This is the same type of analysis that Archimedes used to determine the amount of gold in the king's crown. (He jumped out of his bath and went running to tell the king, shouting "I found it!" [Eureka! In Greek].)

and base G (guanidine) across from base C (cytidine, Fig. 14.7). This would finally explain a curiosity known as Chargaff's rule, which stated that, no matter what the composition of the DNA, the amount of A always equaled the amount of T and G = C. But explaining this riddle was not the important issue. The pairing of A with T and G with C explained how DNA could be the genetic material and made it intellectually necessary for it to be so. In brief, if you pulled the two strands apart and rebuilt, for each strand, a new second strand, then the new second strand would necessarily be a duplicate of the strand that had been pulled off. If strand 1 had an A, then strand 2 had a T, and the new strand 2 (2a) would also have to have a T, while the new strand 1 (1a) built on the old strand 2 would have to have an A. In other words, each strand could create a new strand like the one that it had lost.

This resolved the Russian doll problem. You may know the Russian dolls, or matrioshkas, that come apart, revealing a smaller doll inside; the smaller doll also

Figure 14.6. James Watson (left) and Francis Crick examine the model of DNA that they built to elucidate its structure (1953). Credits: Watson_and_Crick: library.thinkquest.org/C004535/nucleic_acids.html

comes apart, revealing a still smaller doll. In high quality dolls, there may be ten or so different dolls, one inside the other (Fig. 14.8).

In biology, the Russian doll problem consists of the following (supposing that the genetic material is protein, which forms the bulk of our bodies): If protein is the genetic material that carries the information for making (coding for) protein, what codes for the genetic material? In other words, what makes the protein that makes the protein that makes the protein? The Watson and Crick model demonstrated that, according to the structure of the double helix, each strand would serve as a template (mold) for a new strand WITHOUT HAVING TO HAVE FURTHER INFORMATION AVAILABLE. In other words, if the strands could be separated and a new strand could be assembled on each old strand, the molecule could replicate itself. This was the truly important element of the Watson-Crick model of DNA. They had identified a molecule that, by its structure, could be copied without having to have a code for the code for the code for the code for the code... Not only was it now possible for DNA to be the genetic material, because DNA gave an escape from the Russian doll problem, since no other molecule had this property, it was even necessary that DNA be considered to be the genetic material.

The question then turned to how it was possible for DNA to carry the information to produce a human or any other organism. Before we explore that question, however, you may want to note how many sciences ultimately

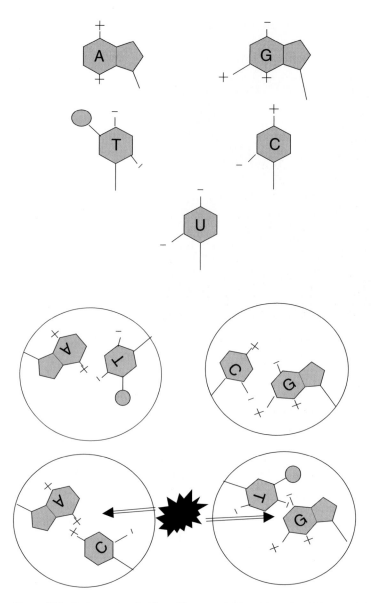

Figure 14.7. Purines and pyrimidines. Upper panel: The purines, A and G, are two-ring structures while the pyrimidines, T and C are smaller one-ring structures. U is used in RNA, while T is used in DNA. Thus the bases of DNA are A, T, G, and C, while in RNA they are A, U, G, and C.

Lower panel: Because of the sizes of the purines, two purines cannot fit across from each other inside the helix. Likewise, since charges on molecules act the same way that magnets do, in that like charges repel each other and unlike charges attract each other, C and A repel each other, as do G and T. T and C, although small enough to fit in the helix, likewise repel each other (not illustrated). This leaves only two possible combinations: A-T and G-C. All of these conclusions derive from the calculations of the size of the helix

Instruction ——⟶ Instruction ——⟶ Instruction⟶Instruction⟶ Instruction

Figure 14.8. Russian matrioshka dolls. Some of them have eight or ten dolls stacked inside each other. The question was, if protein carried the information to make protein, what carried the instruction to make the protein to make the protein that carried the information?

came to bear on this one question. Chemists had established reaction mechanisms and means of calculating and inferring the shapes of atoms and molecules; physicists had understood the wave properties of light and X-rays and how to interpret them; and biophysicists had learned to interpret complex patterns to reveal the structures of molecules. This is typical of any science, that each phase depends enormously on the work of predecessors, even in far-removed fields, and truly convincing arguments are based on the accumulation of data and understanding from many different fields. This is particularly true for the theory of evolution: As is noted on page 94 ff, the consistency of data from many fields is one of the strongest arguments possible for the argument that evolution has occurred.

Another point that you may care to notice is the following: we now can assemble a model of the molecule, and even see it in an electron microscope (another tool that contributes to our understanding). One of the most satisfying experiences in all of science is to see that an intellectual prediction turns out to be true (Fig. 14.9)

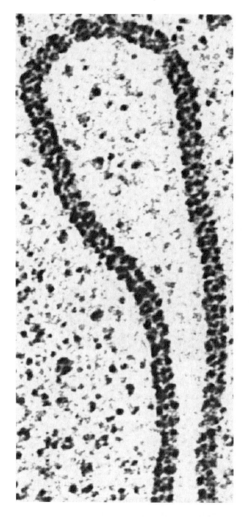

Figure 14.9. This loop of DNA has protein bound to it, as it is found naturally inside of cells, but the helical structure of the DNA strand is nevertheless clearly visible. Credits: Nucleic acids www.biochem.wisc.edu/inman/empics/Protein.jpg

So the question now becomes, "If DNA is the genetic material, how can it possibly carry information?" In other words, how can you possibly get something interesting from a four-color piece of string? Because that's what DNA was, a very long string with four variations. Very boring. Well, it is possible to get something more meaningful out of four variations, if you take the variations in groups. For instance, the Morse code consists of only two variations, dots and dashes, but by assigning values to sequences of one to four characters ("S" = dot, dot, dot; "O"=dash, dash, dash, etc.) one can create an entire alphabet. Another example would be the rhyme heard in the US South, to help people distinguish between the similar-looking deadly coral snake and the harmless milk snake, by the sequence of three colors:

"Red, black, yellow: Dangerous fellow. Red, white, black, that's all right, Jack"

DNA is a string of sugars with bases attached, and proteins are strings of amino acids. So it was logical to assume that the DNA string must somehow represent the protein string. It was already known that the genetic material must be arranged in linear order on the chromosome. This information was determined by very simple logic.

■ Since the number of chromosomes is limited, there must be 1000 or more individual genes per chromosome.

■ If different genes are on separate chromosomes, they will separate randomly, according to Mendelian genetics (pages 134 and 205).

■ If different genes are on the same chromosomes, they should not separate at all, unless the chromosomes can break and rearrange (which they do).

■ If the chromosomes can break and rearrange at random locations, then the closer two genes are to each other, the less frequently they should separate, in the same sense that, in a 1000-link chain, the chance of separating link 671 from link 672 in a random break is 1/1000 or 0.1%, while the chance of separating link 1 from link 1000 is 100%.

■ One can determine the linear order of genes on a chromosome in this manner. In chromosomes that are big enough to analyze, such as those of the fruit fly *Drosophila*, the order is the same as the genetics indicates.

Francis Crick showed mathematically that if this were so, then it would take a sequence of three bases in a row to represent one amino acid. The math is very simple. If one base equals one amino acid, then there can be only four types of amino acids, but in fact there are twenty. If two bases in a row represent one amino acid, then there are sixteen possible pairs of the four bases—close, but no cigar. (Table 14.1) If three bases in a row represent one amino acid, then there are sixty-four possible combinations.

Thus three bases was the minimum number possible for such a coding to work. He even did an experiment to prove it. His hypothesis was that the linear string of DNA coded for the linear string of amino acids, with three bases in the DNA representing one amino acid. He also hypothesized that the code was read only by identifying the first base and stepping by three:

Table 14.1. Possible combinations from different numbers of bases

1 base	2 bases	3 bases			
A	AA	AAA	AAT	AAG	AAC
T	AT	ATA	ATT	ATG	ATC
G	AG	AGA	AGT	AGG	AGC
C	AC	ACA	ACT	ACG	ACC
	TA	TAA	TAT	TAG	TAC
	TT	TTA	TTT	TTG	TTC
	TG	TGA	TGT	TGG	TGC
	TC	TCA	TCT	TCG	TCC
	GA	GAA	GAT	GAG	GAC
	GT	GTA	GTT	GTG	GTC
	GG	GGA	GGT	GGG	GGC
	GC	GCA	GCT	GCG	GCC
	CA	CAA	CAT	CAG	CAC
	CT	CTA	CTT	CTG	CTC
	CG	CGA	CGT	CGG	CGC
	CC	CCA	CCT	CCG	CCC
4	16	64			

THEBADBOYFEDTHEFATCATANDDOGTHEBIGREDBUG
(THE BAD BOY FED THE FAT CAT AND DOG THE BIG RED BUG).

Therefore he proposed that getting out of sequence would be a disaster. It was known that certain chemicals could damage DNA (cause a mutation) by getting tangled in the helix and causing the DNA to add an extra base when it replicates, while radiation and other chemicals could damage a base and cause it to be lost. Therefore he proposed the following experiment: DNA was proposed to code for enzymes, proteins that can carry out reactions such as digesting food. If in bacteria he could cause a mutation by adding a base, the resulting enzyme would be a mess and would not work:

THEBADDBOYFEDTHEFATCATANDDOGTHEBIGREDBUG
THE BAD DBO YFE DTH EFA TCA TAN DDO GTH EBI GRE DBU G..

Likewise, if he could cause a mutation by removing a base, the enzyme would not work:

THEBADBO.FEDTHEFATCATANDDOGTHEBIGREDBUG
THE BAD BOA NDT HEF ATC ATA NDD OGT HEB IGR EDB UG

However, if he could combine the two mutations, he might get an enzyme that would have one area that was a problem, but mostly it would be normal, and it might work:

THEBADDBO.FEDTHEFATCATANDDOGTHEBIGREDBUG
THE BAD **DBO** FED THE FAT CAT AND DOG THE BIG RED BUG

He then did the experiment, and he got an enzyme that wasn't as good as the original, but did work. Thus the evidence supported the argument that the code was a sequence of three bases representing one amino acid, and the question turned to what the code was. To understand how that was done, we need to know a bit more about how it is possible to get the mutations that one wants to use to be able to examine a phenomenon. In other words, how could Crick get bacteria carrying precisely the two mutations that he would need to answer his question? The story of how this was done involves some of the coolest tricks that I know, which is the story of the origin of molecular biology.

REALLY COOL TRICK #4: PLAYING IN THE KITCHEN IS WONDERFUL FOR BABIES AND FUTURE NOBEL LAUREATES

Using mutants to study mechanisms was obviously a good idea—we can find out if this bulb is the blinking bulb in the Christmas string of lights by replacing it with a different bulb—but hoping to find the right mutation was not the way to go. Herman Muller improved the situation by showing that X-rays could cause mutations, and then producing a lot of them in fruit flies, but even fruit flies take two weeks to grow and require a lot of care. Bacteria grow very rapidly and cheaply (anyone who has let a bottle of milk spoil knows that you can get millions of bacteria in a quart of milk). They divide every twenty minutes. One bacterium will become one million in 20 generations, or less than 7 hours. It could theoretically become almost 5 trillion trillion billion ($4.7 * 10^{21}$ or 4.7 with 21 zeros) in a day. It won't of course. It would run out of food. Bacteria also have another advantage: they are haploid, meaning they have only one copy of each gene. Thus any mutation would be immediately obvious, as opposed to the situation for most diploid organisms, in which one characteristic may be hidden by another and be identifiable only by reproduction. For instance, if one of your parents has black hair and the other red hair, you could have black hair but carry the red hair characteristic hidden by the black, and no one would know that you had it unless one of your children or grandchildren was red-haired. However, there were two limitations to using bacteria to study genetics. First, the haploid style may be an advantage but can also be a disadvantage. Since bacteria don't have characteristics such as eye color or wing shape, most mutations that people can identify are the loss of the ability to use something, for instance the milk sugar lactose, as food. If one is looking for new mutations, the clearest evidence that one has found the mutation is that the bacterium has died, in which case of course the mutation has been lost: quite an embarrassment. Worse, in the early 1950's it was thought that bacteria did not recombine sexually. They were considered simply to keep dividing, replicating the same chromosomes over and over again. Thus getting mutations to study how things worked was an exercise in frustration. If you wanted to ask, for instance, whether a bacterium's ability to resist penicillin was related to its ability to resist streptomycin, you could get a mutant that resisted penicillin and you could get a mutant that resisted streptomycin, but you would have no way to get both mutations in the same organism. Edward Tatum posed this problem to a young graduate student at Yale, Joshua Lederberg. Lederberg, literally playing around with kitchen equipment, figured out how not to lose a new mutation. By doing so, he demonstrated that bacteria could recombine successfully, and launched the era of molecular biology.

What Lederberg did was maddeningly simple, in the sense that brilliant experiments usually lead to a "Why didn't I think of that?" response. He made a rubber stamp. What he really did, as the story is told, is that he borrowed a piece of velvet from his wife. When you see a droplet of bacteria growing on a Petri dish or in a jar of jelly, what you are seeing is a colony of clones. One bacterium has landed there, found food, and kept dividing until there are hundreds of thousands or even millions of bacteria, each genetically identical to its parent, siblings, and progeny. If you touch a piece of velvet to the colony, the velvet will pick up some

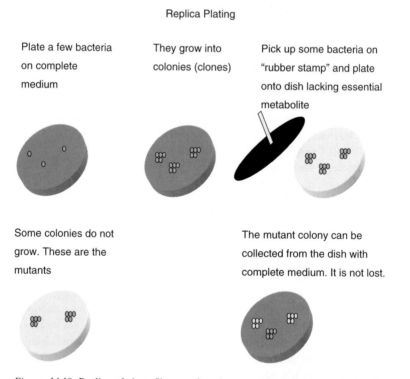

Replica Plating

Plate a few bacteria on complete medium

They grow into colonies (clones)

Pick up some bacteria on "rubber stamp" and plate onto dish lacking essential metabolite

Some colonies do not grow. These are the mutants

The mutant colony can be collected from the dish with complete medium. It is not lost.

Figure 14.10. Replica plating. Since each colony represents the descendents of one bacterium, this technique provides a means of identifying mutants that cannot survive under certain conditions while not losing the mutant because it died. It was the key to the origin of molecular biology

of the bacteria. If you now touch that velvet to another Petri dish, you will leave some of the bacteria on the second dish, much as a rubber stamp leaves ink in the appropriate places on a piece of paper (Fig. 14.10). The genius of this experiment is that you can test for defects, such as the inability to make the amino acid arginine or tryptophan, by raising the bacteria on media lacking these ingredients, but you have not lost the original colony, which is still growing on the original Petri dish containing all possible nutrients. By using this trick, Lederberg was able to identify and collect many types of mutants. Others had suspected that such mutants existed, but had always lost them. The purpose of collecting the mutants was that Lederberg could now ask the basic question, could bacteria recombine sexually? It was the same question as asking if a lion and a leopard could mate and produce young, or if a peach and plum could be crossed to produce a nectarine.

Again, the basic experiment was very simple.

Lederberg had one mutant that could not produce the amino acid arginine. Let's call it arg–. Thus it could not grow in media lacking arginine. He knew that it could back-mutate only very rarely into a form that could produce arginine (arg+). Approximately one in 1,000,000 bacteria could do that. In other words, if he diluted the bacteria in a medium so that there were 10,000,000 bacteria per ml and spread that milliliter of bacterial suspension onto a Petri dish containing medium that lacked arginine, approximately 10 colonies would grow. Likewise, he had another mutant that could not produce another amino acid, tryptophan (trp–), and it could back-mutate at the same rate. He then mixed bacteria that could produce arginine but not tryptophan (arg+,trp–) with bacteria that could produce tryptophan but not arginine (arg–, trp+) and plated them onto a dish that contained neither tryptophan nor arginine. The only bacteria that could survive on this dish would have to be able to produce both arginine and tryptophan (arg+, trp+). This could arise in one of four ways: The arginine-requiring organism could back-mutate; the tryptophan-requiring organism could back-mutate; either one could produce a chemical that the other could use (this was ruled out by other experiments) or they could share genes, such that the arginine-deficient organism could get a good arginine gene from the tryptophan-deficient organism, and vice-versa. How could he tell?

The numbers gave it away. As we noted above, when arg– organisms were plated onto the arginine-lacking plate, only about one in one million could grow. The same result occurred if trp– organisms were plated onto the tryptophan-lacking plate. However, when he mixed the two types and plated them onto a dish lacking both arginine and tryptophan, one thousand colonies grew. In other words, by simply mixing the bacteria, he got a 100-fold increase in conversion. He grew these bacteria to show that this truly was an inherited difference, and otherwise eliminated the hypothesis that this might be chemical replacement of the missing nutrients. By eliminating all other hypotheses or interpretations, he was forced to the conclusion that mixing the two types of bacteria allowed them to exchange genetic material. In other words, bacteria could recombine sexually.

This was not simply a quirk or a silly story to tell at a party. It opened the possibility of moving genes around in bacteria to finally learn what genes were and how they worked, which made it possible to do all the molecular biology that we do today. Furthermore, this recombination is the primary means by which bacteria develop resistance to antibiotics and multiple resistance (to many antibiotics). It also is a major means of viral mutation, and is a significant component in generating cancer cells.

Most of the mutations that Lederberg used were inability to make or digest products that the bacterium needed. In other words, they were failures of enzymes needed to synthesize the product or break it down into usable form. Enzymes are made of proteins, and we return to the question of how DNA carries the information. Once it was possible to produce bacteria with many types of mutations, it also became possible to ask how genes were constructed and how DNA carried information to make proteins. There are many stories about this search—most of which involve really cool tricks to get these molecules to reveal their secrets—but we cannot tell them all, and we do not have to maintain a strict historical sequence. Let us start with the question of learning how to identify the sequence of bases in DNA and learning how to read that sequence.

The first problem that we have to deal with is that there is a LOT of DNA. We have enough DNA to make 1,500,000 genes, though we actually have only 20–25,000 genes (the other 98.4% of the DNA being apparently useless used for instructions on when to be active or other, unknown functions) genes, and frogs have even more (Fig. 14.11). We are still at the level of trying to find out how three bases code for a single amino acid. Where do we start? The tricks that we pick up here are the same ones that will eventually be used for forensic analysis, for tracing the evolution of humans, for determining whether or not Neanderthals are related to us, and for genetic engineering, whether for crop production, repair of disease, or more dubious enterprises. We start by cutting the DNA up into manageable sizes, under controlled circumstances so that we know exactly where we are cutting it. This works because nature does it for us.

REALLY COOL TRICK #5: VIRUSES KNOW HOW TO CUT UP DNA

As we noted above, viruses are sometimes the diseases of bacteria—they infect and kill the bacteria. Sometimes, however, they do not kill bacteria but simply go along for the ride, like a parasite that does not really harm its host. The easiest way that they can do this is to hitch a ride on the bacterial chromosome. The bacterial chromosome, interestingly enough, is a circle; it has no end. What the virus does is open the circle, stick its DNA into it, and then close the circle back up, sort of the way that a magician makes two circles join together and come apart (Fig. 14.12). The bacterium then goes on with its life, dividing on schedule, but also replicating the viral DNA as it replicates its own.

For this to work, the virus has to cut the DNA in such a way that it will not disrupt an important function for the bacterium. It does this by identifying only very specific sites on the bacterial DNA where it will cut. This is accomplished because it has enzymes known as restriction endonucleases. The term "endonuclease" means that the enzyme cuts the DNA in the middle, rather than chewing in from the end, and the qualifier "restriction" means that at a specific sequence of DNA, for instance a specific sequence of four to six bases, for instance GAATTC. Even more interesting, look again at the sequence. It is what we call a palindrome. Palindromes are sentences that read the same backwards and forwards: MADAM I'M ADAM or ABLE WAS I ERE I SAW ELBA. In this case, the palindrome is the opposite strand, which reads backwards exactly like the first strand:

GAATTC →

← CTTAAG

Figure 14.11. There is a lot of DNA in a cell. In this preparation, a mitotic chromosome was spread on the surface of water to allow it to expand. All of the fine strands are DNA. This article was published in Cell, Vol 12, J.R. Paulson and U.K. Laemmli, The structure of histone-depleted metaphase chromosomes, Pages 817–828, Copyright Elsevier (1977). Credits: From J.R. Paulson and U.K. Laemmli, 1977. Cell 12: 817

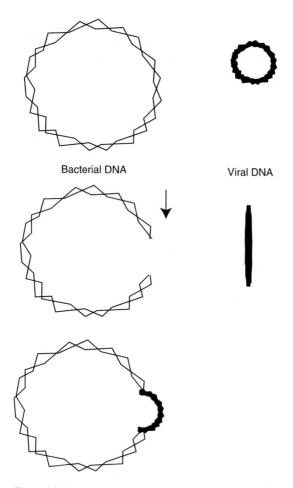

Figure 14.12. Viral insertion into DNA. The DNA of both the bacterium and the virus (phage) are circular. The virus cuts the DNA of the bacterium and simultaneously opens its own DNA into a straight piece of DNA. It then attaches the ends of its DNA to the ends of the bacterial DNA and splices the circle back together. As the bacterium reproduces, it makes a copy of the viral DNA as well as a copy of its own

This is fine, because the strands face in different directions. (In the sugar-phosphate-sugar-phosphate backbone, the phosphate is attached differently to the two sugars. It's rather as if you had two battery holders, each of which took a string of batteries, all facing the same way, but the battery holders were set up so that in one, all the positive poles faced left and in the other, they faced right.) If, for instance, the restriction enzyme cut between the G and the A in this sequence, as the enzyme EcoR1 (for E. coli Restriction Enzyme 1) does, then it will cut both strands, leaving a little bit dangling over. The dangle will become very important in a bit.

Consider what this means: The probability of finding an A next to a G is 1 in 4; the probability of finding AA next to G is ¼ × ¼, or 1/16. The probability of finding the entire sequence is ¼ × ¼ × ¼ × ¼ × ¼, or just about 1/1000. Fruit flies have about 122 million bases and 14,000 genes, while humans have about 3 billion bases and 20 to 25,000 genes. This means that this one enzyme might cut up human DNA into 3 million pieces. If we can use it, it would be like taking a very unwieldy book with no punctuation and cutting it into pieces by cutting every time we found the ending "-ation". (It would be even more meaningful if we cut it every time we encountered the word "chapter".) If we can separate the DNAs from different chromosomes or by other characteristics—this can be done—we can get a manageable number of fragments to analyze. The restriction endonuclease is the first of several tricks in this bag. It is now used commonly in forensic medicine. This is how it is used:

There are many regions of human DNA that are very variable, so much so that they are nearly unique for every person. If we can analyze that region, which we can identify by another trick, we can distinguish one human from another. Restriction endonucleases come into play because the piece of this variable region will be the same size only if the two pieces we are comparing are identical. Look at the following sentences, from which we will cut a piece by cutting only after the string of characters "and the":

The buffalo and the *prairie dog are characteristic of the plains. The cockroach and the* pigeon are characteristic of the city. (69 characters)

The buffalo and the *prairie dogs are characteristic of the plains. The cockroach and the* pigeon are characteristic of the city. (70 characters)

The buffalo and the *prairie dog are characteristic of the plains. The cockroach and pigeon* are *characteristic of the city. XXX* (≫103 characters)

The buffalo and the *prairie dog are of the plains. The cockroach and the* pigeon are characteristic of the city. (deletion: 57 characters)

The buffalo and prairie dog are characteristic of the plains. The cockroach and the *igeon are characteristic of the city.* Xxxxxxxx (?? characters)

Thus it is clear that, if we can separate pieces of DNA by size, we can identify them, analyze them, or at least distinguish which are identical and which are not. How to separate them by size is the next cool trick. It depends on the same principle that can be seen in mid-afternoon in major cities: middle school children get through the subway turnstiles much faster than big or even obese people. That's really cool trick #6.

REALLY COOL TRICK #6: GETTING DNA TO RACE

Everyone is familiar with gels such as gelatin ("Jello" ®) and you may have seen agar on a Petri dish. These gels are made of long strands of molecules (proteins or carbohydrates) that are tangled among each other, rather like a bowl of spaghetti. However, they also hold onto water very well, so that the structure of the gel is water suspended among the strands as if it is in a sponge. What is interesting about this is that the meshwork of strands leaves holes (water passages) about the size of

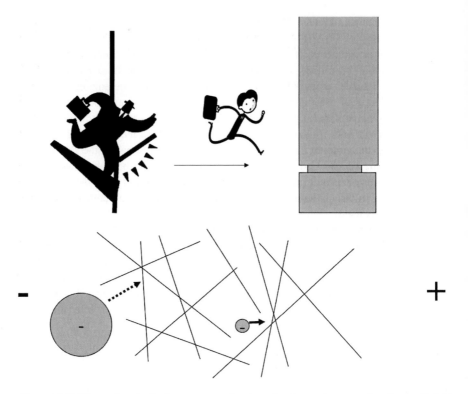

Figure 14.13. Electrophoresis. In the same way that, in racing for a subway car (gray) a small child can squiggle through the gate faster than a much larger individual (upper panel), small molecules can squiggle through gates faster than large molecules (lower panel). A gel is made such that it has holes equivalent to gates, and the holes are approximately the size of molecules. Since most proteins are negatively charged or can be made to be negatively charged, they can be attracted to the positive pole (anode) of an electrical field. However, since they have to cross the gel to get there, the smaller molecules move faster. Thus, proteins can be separated, and eventually identified, by size

molecules. By increasing or decreasing the amount of the material to make the gel, we can get gels with holes of different sizes. We can use this to make gates for the pieces of DNA.

Happily, DNA is an acid (deoxyribonucleic acid) and a characteristic of acids is that under the right conditions they are negatively charged, or negative ions. If you place ions between the positive and negative poles of a battery, negative ions will move toward the positive pole and positive ions toward the negative pole. So, if we put DNA between two electrical poles, but interpose our gel between them, the DNA molecules will move toward the positive pole, but the smaller ones will move through the gel faster, while the bigger ones get tangled in the mesh. This is electrophoresis (Fig. 14.13).

This would be fine if we could get all the DNA we wanted, but often it is hard to come by, from crime scene evidence, or a dinosaur bone, or from a newborn mouse that we would like to identify but that we do not want to kill to get its DNA. Because DNA of interest was often in very short supply, Kary Mullis searched for a way to get more of it, and developed really cool trick #7, the polymerase chain reaction (PCR).

REALLY COOL TRICK #7: LOOKING FOR CRAZY BACTERIA

If you have ever tried to untangle two springs, a Slinky® toy, a hose, electrical cord, yarn, or a braid, you know that rotating one strand causes the other to rotate as well. It actually is quite a complex trick to unwind DNA, but in order for DNA to replicate, the strands have to separate so that a new strand can be built on each old strand. All organisms do this by using a complex set of enzymes collectively called DNA polymerase. The polymerase recognizes single-stranded DNA and builds a new strand on it. DNA will also unwind at high temperatures, so theoretically you could use that unwound DNA to make new DNA, but unfortunately the polymerase is cooked at that temperature. Proteins including enzymes are not stable at high temperatures. They get permanently deformed and they precipitate, as the white of an egg (mostly the protein albumen) does when you cook it. Thus we have a problem: We can get the DNA unwound, but we cannot use it at the temperature at which it is unwound.

There is, however, a solution. There are organisms (bacteria) that live in hot springs, such as those at Yellowstone Park. Some bacteria live in water hotter than 90° C (194° F)! If they live at that temperature, then it follows that they reproduce at that temperature, meaning that their DNA polymerase can survive and work at high temperature, the temperature at which most DNAs naturally unwind. Using their polymerase (Taq polymerase, from the bacterium *Thermophilus aquaticus*: Translated from the Greek and Latin roots, the name means "heat-loving creature in the water"), we might be able to synthesize new DNA from the strands of the unwound DNA. That is the "P" (polymerase) of "PCR". But it only doubles the amount of DNA. Doubling the amount of a vanishingly small amount of DNA does not help much. That's where the "CR" (chain reaction) comes in. If we run the cycle once, the 2 original strands become 4. If we run the cycle a second time, the 4 strands become 8. If we run it a third time, the 8 strands become 16. In other words, something that doubles each time increases at a geometric rate. If we run the cycle 10 times, we have increased our original DNA 1000-fold. If we run it 30 times, we have increased it one billion-fold. This is the normal procedure for PCR: by an automated procedure, a trace amount of DNA is run through approximately 30 cycles of polymerase reaction, thereby creating enough DNA to work with and study. Of course, everything depends on having a really clean bit of original DNA at the start and not getting fingerprints, dust, or bacteria into the preparation. You really don't want a one-billion fold amplification of that hamburger grease that was on your fingers.

Finally, to study the evolution of animals, one very important tool today is to compare sequences of DNA from different animals to see how closely they are related. For instance, we know that the gene for our hemoglobin is extremely similar to that of chimpanzees, less similar to that of other mammals, but more similar to that of mammals than to that of birds, etc., and we can trace resemblances all the way to fish and beyond. Using the DNA, we can determine where whales came from (hippopotamus-like animals) and where vertebrates arose (from starfish-like animals). We can trace human migrations, and, making some assumptions about how fast mutations arise, use the number of mutations as a molecular clock (see Chapter 9 page 118). To do so, we need to be able to sequence the DNA. It turns out that this is really quite easy. We can do it by electrophoresis as we described above. Of course, we need a couple of tricks.

REALLY COOL TRICK #8: IF THE WATER MAIN IS WORKING AT 1ST ST. BUT NOT AT 3RD ST., THEN THE BLOCKAGE MUST BE NEAR 2ND ST.

DNA polymerase works by adding one base at a time to a growing chain bound to the intact chain. As we have discussed, the backbone of the chain is sugar-phosphate-sugar-phosphate-sugar-phosphate...The phosphate links to the sugars through oxygen on the sugar. (A simple sugar, like grape sugar, consists of 6 C, 12 H, and 6 O. Table sugar consists of 12 C, 24 H, and 12 O. To make a chain of sugars, the phosphate links to an O on one end of one sugar to an O on the other end of the next sugar.) If one of those oxygens is missing, the phosphate cannot link like a series of hook-and-eye links, with one hook missing, and the extension of the chain will stop. There is a synthetic form of base like this, called a dideoxy base (deoxyribose, the sugar of DNA, already lacks one oxygen; dideoxyribose also lacks the oxygen to which the phosphate would bind). A chain terminated by the addition of a dideoxyribose will be shorter than normal, and will run faster in electrophoresis. We could therefore recognize its existence, but how do we know what it is, and how do we read sequences?

There were several efforts to resolve this problem, but the one that has worked very well is this: It is possible to make a dideoxybase fluorescent and, better yet, make each of the four (A,T,G,C) fluorescent in a different color. What we now do is to prepare our DNA-synthesizing mixture with DNA polymerase and the DNA we wish to sequence. To this mixture we add a mixture of the fluorescent dideoxy bases, but not enough to stop the reaction totally. Let's see what happens. We will assume that the strand synthesis begins with a G. Some of the strands (there are actually millions of separate strands) will incorporate the fluorescent dideoxy G (ddG) and therefore end. Thus the shortest strand will have the fluorescent ddG plus the base that it was attached to. Other strands will incorporate a normal G and go to the next base—let's say it is an A. Some strands will incorporate the ddA and stop. Therefore the second shortest strand will be the unknown base-G-ddA and be three bases long. Other strands will incorporate a normal A and continue. After a while, we will accumulate a series of newly synthesized DNA, each type of strand being one base longer than the previous, and each ending with a fluorescent dd base. Let's electrophorese this, and put a sensor (light meter) along the path. The light meter is capable of distinguishing the colors of the bases. It now records the strands as they pass by: G, A, ... In other words, the fluorescent bases are establishing the sequence for us! In reality, it is not possible to separate more than about 1000 bases at a time. To sequence entire genomes (all the genes) of animals, the DNA is broken into small pieces by restriction endonucleases, the sequences of the pieces are read, and the continuation of one piece to the next is identified by overlapping pieces, as you might fit together a torn-up newspaper by pairing partial letters from one piece with partial letters on the next piece.

The final really cool trick is described more because it is headline news than because of its relevance, but rearranging genes can be used to study evolution. It is the principle of genetic engineering, and we are aware that virus invasion of our chromosomes has changed our inheritance. The viruses use this trick, and it is the basis of all the stories of headlines. It relies on the palindromic sequences described above that some restriction enzymes use.

REALLY COOL TRICK #9: WHEN ALL THE PUZZLE PIECES ARE THE SAME

DNA can be damaged in many ways—by sunburn, X-rays, heat, and many chemicals. In order to survive, all organisms must have means of recognizing and repairing the damage where possible. One of the means of repairing DNA broken at the sugar-phosphate bond is an enzyme called DNA ligase. It can be isolated, purified, and used in the laboratory.

Now look at the palindromic sequence produced by EcoR1:

GAATTC →

← CTTAAG

The cut comes between the G and the A, leaving the two strands as

xxxxxG   AATTCxxxxxxxx

yyyyyCTTAA   Gyyyyyyyy

Note the AATT loose ends. Remember that any cut that the endonuclease makes will produce these loose ends. These loose ends, or overhangs, can still stick together, with the A's and T's still associating or binding, and the DNA ligase can repair that kind of break. But suppose that the strand on the right side came from a different piece of DNA?

xxxxxG.AATTCwwwwwww

yyyyyCTTAA.Gzzzzzzz

The DNA ligase would not be able to distinguish between the "good" DNA and the "fake" DNA; all it would do would be to repair that break, and the DNA would be a hybrid of the original piece on the left and the original piece on the right—an engineered piece of DNA. This is the heart of "genetic engineering". Different pieces of DNA can be attached to each other so that, for instance, a gene conferring resistance to frost can be inserted into a crop plant, allowing the plant to be grown in more northerly areas. Some of the better-known agricultural uses today include adding growth hormone genes to farm-raised fish to increase their growth rate and causing some crop plants to automatically produce insecticides normally produced by other plants or bacteria. Medical uses include production of usable quantities of hormones by cells grown in culture, production of highly specific and highly sensitive diagnostic reagents and production of specific proteins, sometimes deliberately altered, to fight specific diseases or cancers. The bulk of the most exciting advances in biomedical research today are based on the use of animals and plants with manipulated genes. There are as yet no cures based on "correcting" genes in individuals, because it is one thing to get a cell in culture to produce a specific protein, but it is much more complex to assure that one can place a specific cell in the body, at a specific location, so that it will produce the desired protein product only when it is needed and will distribute it only where it should go. There are also dangers inherent in altering organisms, mostly related to their potential to escape and compete with the native forms. Laboratory animals typically not only carry the desired altered DNA but are bred so that they cannot survive in the wild, but it is not guaranteed that all agricultural restrictions are so stringent.

However, the dangers are often greatly exaggerated. Essentially no food that you eat today resembles its wild form. All have been manipulated by selective breeding, deliberate induction of mutations, and cross-breeding, to improve the size, palatability, or appearance of the food. The original tomato was much smaller and berry-like, similar to its relative the nightshade. Potatoes, also members of the nightshade family, could be very toxic if they were allowed to become green or to sprout, but the toxicity has been bred out of them. Corn was similar to a tassel of grass. No apple or peach that you eat today resembles the original crabapple-like fruit but instead is a sterile hybrid, propagated by grafting onto other roots. All the variations of oranges have a similar history, as is indicated by names indicating human oranges: tangerine (from Tangiers); Clementine (after St. Clement); mandarin (from China). The large and almost always successful production of wheat in the western world depends on the judicious choice of insect-resistant strains when insects are a major problem, rust-resistant strains when this fungus spreads, and wheats specifically selected to emerge early in the spring or to grow late into the fall. Genetic engineering is a more efficient means of doing what we have always been doing, but does not represent a theoretically or morally new direction in human behavior.

REFERENCES

Campbell, Neil A. and Reece, Jane B.(2004) Biology (7th Edition), Benjamin Cummings, Boston, MA.

STUDY QUESTIONS

1. For any of the examples given above, describe the ELF logic on which it is based. Are there any flaws or limitations to this logic?
2. Make your own diagram of the several steps necessary to isolate and sequence a specific piece of DNA. Explain these to a classmate.
3. Make your own diagram of the several steps necessary to introduce a new piece of DNA into another piece of DNA. Explain these steps to a classmate.
4. Which of the "really cool tricks" do you find to be the most intriguing? Why?
5. Do you think that "really cool tricks" were used to build other sciences? Why or why not?
6. Describe a situation in which you or a friend or relative worked out a clever or ingenious means of solving a particular problem. Was this solution effectively different from the "cool tricks" that eventually led to Nobel Prizes? Why or why not?
7. Some scientists claim that, "For every difficult experiment, there is one organism that will be perfect to conduct the experiment." Does the story of molecular biology support or contradict this claim?

CHAPTER 15

THE STUFF OF INHERITANCE: DNA, RNA, AND MUTATIONS

THE CHEMISTRY OF INHERITANCE

To understand what evolution is, what selection is, and how it works, we need to look at the physical mechanisms by which it occurs. This is basically the same style as distinguishing between the statement "the oven doesn't work" and addressing the problem. A stove—let's say, a gas oven—is a pretty simple device. Gas flows in through a pipe from a pipeline or tank; turning the regulator opens a valve that lets gas into the oven, and an ignition mechanism (heated electrical element, continuously burning flame—pilot light—from gas admitted through a small, continuously open valve) ignites the gas. The regulator valve also usually has a thermostat that will decrease or stop the gas flow when the desired temperature is reached. If the oven "doesn't work" we have to determine if the gas source is providing gas, if the ignition mechanism is working, and if the thermostat is functioning and allowing gas to enter. In a similar fashion, to understand what evolution is, in a sense somewhat more complete than "rabbits are brown to hide from their enemies," we need to have a sense of the components of evolution. In other words, we need to know what makes rabbits brown, and how evolution can create white and brown rabbits. Therefore this chapter addresses the following points:

- The pigment of a rabbit is made by enzymes.
- Enzymes are proteins
- Proteins are linear arrays of amino acids
- The body carries information on how to make these linear arrays of amino acids in genes, which are located on chromosomes.
- The genes are also linear arrays of molecules, but instead of being proteins, they are DNA.
- Enzymes, proteins, and DNA are macromolecules.

Most biological activity is carried out by the use of giant organic, carbon-based molecules. Because they are giant, they are called macromolecules. For instance, the blood pigment hemoglobin contains approximately 3,300 carbon atoms, over 6,000 nitrogen atoms, over 1,000 oxygen atoms, approximately 550 nitrogen atoms, and four iron atoms. Macromolecules are typically built by organisms by linking, usually end-to-end, a sequence of smaller molecules. The smaller molecules that are linked have many variations but some features in common.

221

PROTEINS

Hemoglobin belongs to the class of macromolecules called proteins, which are defined as macromolecules consisting of chains of amino acids. An amino acid is a small carbon-based (organic) molecule that has both a group called an amine group and a group producing an acid. An amino acid might look a bit like this, with the amine and the acid both attached to the carbon:

$$\text{Amine} - \text{Carbon} - \text{Acid}$$

Carbon, however, can have up to four different atoms (things) attached to it, and in amino acids there is at least one major other component attached to the carbon. The other component is very variable. There are twenty of these other components. Some dissolve easily in water, while others do not. Thus each different amino acid has unique properties such as solubility. In spite of these unique properties, they all have the same basic structure:

Variable group

$\updownarrow$

Amine $\leftrightarrow$ Carbon $\leftrightarrow$ Acid

Proteins are long chains of these amino acids linked together through their amine and acid groups, with the variable groups sticking out of the chain. (In the diagram below, the carbon is symbolized simply by a "C" and the variable groups by V1, V2, etc.

V_1 V_2 V_3 V_4

$\updownarrow$ $\updownarrow$ $\updownarrow$ $\updownarrow$

Amine$\leftrightarrow$ C$\leftrightarrow$ acid$\leftrightarrow$ amine$\leftrightarrow$ C$\leftrightarrow$ acid$\leftrightarrow$ amine$\leftrightarrow$ C$\leftrightarrow$ acid$\leftrightarrow$ amine$\leftrightarrow$ C$\leftrightarrow$ acid$\leftrightarrow$ amine

All the differences among proteins—between hemoglobin, steak, egg white, finger-nails, hair, digestive juices, saliva, wool, and the gristle of meat—derive from the differences in the variable groups.

CARBOHYDRATES AND FATS

Proteins are one major class of macromolecules. Other major classes include the **carbohydrates**, which are linkages of sometimes many thousands of sugars (small molecules consisting most commonly of six carbons, twelve hydrogens, and six oxygens). In carbohydrates, the manner in which the sugars are linked is important. Cellulose (wood, paper, or cotton) and starch have the same sugars, but they are linked differently. **Fats** have far fewer oxygens than other macromolecules and, by definition, dissolve in oils, gasoline, and similar substances. Finally, **nucleic acids** such as DNA are linear sequences of small but complex molecules called nucleotides. Thus, in general and in the simplest version, macromolecules can be symbolized as follows (M = Monomeric unit such as a sugar or amino acid):

$M \leftrightarrow M \leftrightarrow M \leftrightarrow M \leftrightarrow M \leftrightarrow M \leftrightarrow M \leftrightarrow M \leftrightarrow M \leftrightarrow M \leftrightarrow M \leftrightarrow M \leftrightarrow M \leftrightarrow M \leftrightarrow M \leftrightarrow M \leftrightarrow M$

As you might guess, these chains do not simply lie there in straight rows. They can bend and wrap around each other. One of the most common structures of proteins that we know is a spiral; another is a pleated version (Fig. 15.1).

Straightening hair or curling it consists of using heat and water, or chemicals, to disrupt the natural spiral form of the hair protein, stretching the hair, and letting it settle into the pleated form, much as if you overstretched a spring. Eventually, if there is enough moisture around, as water or humidity, it will recurl into the alpha helix. This is why straightened hair becomes curly in humid weather, or curled hair restraightens. (Hair straighteners contain reducing agents that reduce (break) further links called disulfide bonds that strengthen the helices by binding amino acids across the loops. These bonds can then reform in other ways, maintaining a more permanent disruption of the helix.)

Some of these proteins have the ability to vastly speed up reactions. When they have this ability, they are called enzymes. For instance, even sterile beef, with no bacteria present, will eventually break down into amino acids, though this will take many, many years, even thousands of years. If a digestive enzyme such as pepsin or trypsin is present, the complete digestion of a mouthful of beef will take an hour or so. The enzymes can do this because they can fit very tightly against the protein molecules in the beef and bring the reacting parts together, much as when a key matches the tumblers in a lock, allowing the cylinder to rotate (Fig. 15.2).

Now we can deal with the question of the resources of evolution. An enzyme is needed to create the brown pigment. (Actually, many are needed, but for the

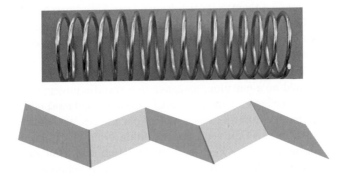

Figure 15.1. Many proteins can form a helix, as in the top figure (it is a particular type of helix, called alpha helix since Linus Pauling, who worked out which one it was, postulated several possible helices, which he described as alpha, beta, gamma, etc (A, B, and C in Greek). They also can form a pleated structure, called a beta pleated sheet, as shown in the lower figure. Hairs are long sequences of many of these proteins, aligned in the hair. Styling hair consists of stretching the normal form of the hair protein, the alpha helix, into the beta form. Heat and moisture can allow it to return to the alpha helix form, and so resume its original shape

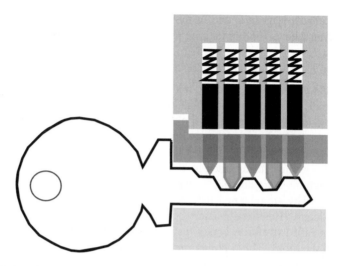

Figure 15.2. As a key must be specifically shaped to push the tumblers so that they are aligned (the cylinder turns at the junction between the gray pins and the spring-mounted black pins above them), a substrate must fit perfectly into an enzyme for the enzyme to be able to take it apart or be able to attach it to something else. If the enzyme is altered, which one could picture as having a big piece of dirt under one of the pins, the fit will not be perfect and the enzyme will not work

illustration one will suffice.) Because of the lock-and-key arrangement by which they work, a very small change will prevent an enzyme from working, as a newly-cut key will not work if there is a small burr or metal fragment left from the cutting. Such a very small change could be the substitution of one amino acid for another. Hemoglobin consists of four chains linked together, two pairs of identical chains. (Picture a four-stranded braid, with two brown strands and two blond strands.) Of the total of 584 amino acids in hemoglobin, changing two of them (one in each of the two identical strands) will create the disease called sickle-cell anemia. We know of several other instances in which the change of a single amino acid can change the character of the protein.

This, then, would be a mutation: a change in a specific protein that produces an identifiable difference between the individual carrying it and most other individuals. We would describe the difference as a mutant phenotype. We could even lose the protein altogether, for any of several reasons. In the simplest case, a visible change or mutation is caused by the switch of a single amino acid. How does this occur? The body builds, or synthesizes, these proteins from their building blocks or amino acids. Somewhere in the body an instruction manual must exist that tells the body in which order the amino acids must be added. This instruction manual is called the genome, the collection of genes that are instructions for the manufacture of individual proteins. As is described in more detail in Chapter 16, the genes, located on the chromosomes, are composed of DNA. DNA is a macromolecule consisting, like proteins, of a linear array of subunits. In this case the subunits are nucleotides.

A nucleotide is itself made of subunits, a base and a sugar called deoxyribose (the "D" in DNA"). The deoxyriboses are linked together by phosphates (combinations of phosphorus and oxygen) and form the backbone of the chain, with the bases sticking off the side like the variable "side chains" of the proteins.

base   base   base   base   base   base
↕   ↕   ↕   ↕   ↕   ↕
dRibose ↔ P ↔ dRibose ↔ P ↔ dRibose ↔ P ↔ dRibose ↔ P ↔ dRibose ↔ P ↔ dRibose

The sequence of bases in the DNA is the information (code) for making the sequence of amino acids that will be the final protein. Suffice it to say that the change in a single base can result in a change in a single amino acid, and a change in a single amino acid can produce a noticeable difference in characteristics of an animal or plant. In fact, most mutations are alterations of a single protein, such as sickle-cell disease, or the ability to make brown pigment. So now we have an understanding of what will happen in evolution. If for any reason the DNA is altered so that the enzyme needed to make a brown pigment is abnormal, the rabbit will have a different color or perhaps no color at all. The alteration of the DNA can be as small as the change of a single base. In a very profound sense, evolution depends on the chemistry of DNA.

REFERENCES
STUDY QUESTIONS

1. If the question, "Why do monarch butterflies fly south in the fall?" is not a good scientific question, how would you rephrase the question to make it better? Explain.
2. How does an amino acid differ from a protein?
3. What is a macromolecule?
4. Proteins have the peculiar property that they are soluble only at certain levels of acidity, and precipitate in more acid conditions. What might you suspect that bacteria do to milk to make it curdle?
5. Because of the strength of the bonds that hold proteins and lipids together, biological lipids tend to melt at approximately 104° F (40° C), while protein structure falls apart at approximately 122° F (50° C). Do you think that this has anything to do with the fact that people tend to hallucinate when they have high fevers? How hot does water have to be for you to be burned by it?
6. If a protein can contain 1000 amino acids, why should the change of a single amino acid make a difference in how it functions? (Hint: wrap something like a flat electrical cord carefully around a suitably wide structure such as a broomstick. Wrap the wire so that each loop lies side-by-side to the loops next to it. Now do the same thing but attach something the size of the plug to the middle of the wire. What happens?

CHAPTER 16

THE GENETIC CODE

THE BILINGUAL DICTIONARY IS tRNA

The final trick consists of being able to translate the code. After all, it does not help much if, in a war, you have intercepted an enemy's message and you recognize that it is encrypted (in a coded language) if you cannot read it. The bulk of our functioning body is protein. How do we get from the DNA code to the protein? By the 1960's, this was a critical question. Beyond Crick's hypothesis of a triplet code and experiment (see page 205) various scientists attempted to find evidence that the hypothesis was correct. For instance, it became possible to get the amino acid sequence of readily available and easily purified proteins such as hemoglobin. Normal and sickle-cell hemoglobin were analyzed, and it became apparent that the two differed by only one amino acid. The changed amino acid in sickle-cell hemoglobin is much less soluble in water than the normal amino acid, making the hemoglobin less soluble as well and causing the sickle-cell hemoglobin to precipitate in the red blood cell under certain circumstances. The red blood cell is then deformed and catches in the smallest blood vessels, causing clogs and clots that can cause considerable pain and damage.

This is not an analysis of sickle-cell disease (but see Chapter 32, page 425) but for geneticists there were two very important lessons to be learned: first, that a mutation could be as small as one amino acid, which might theoretically result from a single base change. (The single base change was subsequently confirmed, see page 247 and page 231). Second, a severe change in characteristics (phenotypic change) could be produced by the change of a single amino acid.

To understand how the DNA was decoded, we need to know a little bit about how proteins are made, and we can explain this by the use of a few analogies. The problem is that the DNA is in the nucleus of a cell, separated from where the proteins are made, in the cytoplasm. So the first question is how we connect the two. The analogy is as follows: Everyone has seen the stockboys or stockgirls in supermarkets. They are the ones who bring materials from the storerooms to the shelves for the consumers. This is not quite the image that we need. A better image is from factories, or at least from factories in which manufacturing is not fully automated. For instance, let's describe how a fender of a car might be made. It starts as a flat sheet of metal that is placed into a large machine called a press. The press does exactly that: A large and very heavy upper part moves downward and presses the sheet against a mold or template, bending the sheet into the form of a fender. The *press* is non-specific; it is simply a machine that exerts enormous pressure on a sheet of metal. It could bend the metal into any shape desired, depending on the shape of the *template* (Fig. 16.1). Such a machine of course is very heavy and is not movable. To make the fenders, *stockboys* bring to the machine and its operator a continuous supply of fresh metal sheets and take the finished fenders to the next station, where, for instance, holes might be cut for lights. This image now includes all the components that we need: The *press*, which is a complex collection of molecules in the cytoplasm called a *ribosome*; the *stockboys*, which are small molecules called *transfer RNAs*, which serve to bring fresh amino acids (unbent steel) to the ribosome (press) so that they may be linked together to form proteins; and the *template* for the press, which is a molecule called *messenger RNA*. The messenger RNA is what carries the information from the DNA in the nucleus to the ribosome. To picture what is happening, we need two other terms: *transcription* and *translation*. To transcribe something is to copy it as you hear it, without necessarily understanding. For instance, suppose that you are in France and want directions to the train station. You ask a native, making gestures and sounds to imitate a train, and the native tells you, "Vous allez au coin, tournez à droite, et la gare est à deux cent mètres sur votre gauche." You dutifully *transcribe* what you hear: "Vou zalley zo kwan, tourney za drwat, eh la gar eh tah duh sont

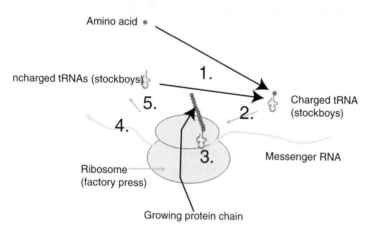

Figure 16.1. Protein synthesis. 1. An amino acid is attached to a specific tRNA , which has a specific anticodon and accepts only one type of amino acid. It acts as the stockboy. 2. The charged or loaded tRNA moves to the ribosome (equivalent to the press) which has bound mRNA (equivalent to the template). 3. The tRNA binds to the mRNA codon that matches its anticodon. Thus it is in a position to move its amino acid onto the end of a growing protein chain. Its act of transfering its amino acid ratchets or nudges the mRNA along the ribosome so that the next codon will be in position to donate an amino acid. 4. Once the tRNA has released its amino acid, it detaches from the ribosome and returns to pick up a new amino acid. Ultimately, the mRNA, chugging along the ribosome, will present a codon for which there is no code and no tRNA is attached (a "stop codon"). The additions will cease and the finished protein will be released

metruhs seur vohtra gosh." This is not very helpful to you, but you take it to an English-speaking friend who knows French well, and she looks at it and *translates*, "You go to the corner, turn right, and the station is six hundred feet on your left." Your step was transcription: not changing the language, but putting the French into written form. Her step was translating, converting the meaning from one language to another. Similarly, messenger RNA is made from DNA in a base-pairing manner very similar to that in which a second strand of DNA is made. This is transcription. We are still in the language of nucleic acids. RNA differs slightly from DNA; its sugar is ribose, not deoxyribose (ribonucleic acid, not deoxyribonucleic acid) and instead of the T (thymidine) in DNA, it has U (uridine) (Fig. 16.2). Messenger RNA, or mRNA, is copied from the DNA strand, carrying the code, and is transported from the nucleus to the cytoplasm, where it serves as the template or mold on the ribosome press. The translation is handled by the marvelous stockboys, or transfer RNAs (tRNAs). There are approximately twenty tRNAs, one for each amino acid. They are marvelous because one end of each tRNA has a triplet codon that will match a three-base sequence on the mRNA (and therefore resembles the original DNA), while the other end is specific for a single amino acid. The transfer RNA is therefore the bilingual dictionary. On one end of the molecule it has the French word (vous, mRNA) and on the other the English word (you, protein). The tRNA translates from nucleic-ese to protein-ese (Fig. 16.2).

We are telling this story in a linear sequence, but of course the role of the tRNA could not be understood until coding was better known. Many scientists were trying to identify the code, using many different tricks. One of the most original was a mathematical biologist name Martinas Ycas, who reasoned as follows: To get the code, one would have to have a protein with a highly unusual composition, so that one could determine from the unusual composition of the RNA what the code was. Certain moths produce a silk for their cocoon that is made almost exclusively of only two amino acids, glycine and alanine. Ycas presumed that the base ratio of the RNA would be highly distorted. He flew to Africa, collected the caterpillars, and extracted the RNA from their silk glands. However, the base ratio was not distorted, leading Ycas to suggest that the coding RNA must be a very small portion of the total. He was correct: the bulk of the RNA is the press, rRNA, and mRNA makes up only about 1% of the total. Meanwhile, laboratories in France, England, and the US were producing biochemical evidence for the existence of mRNA. However, the code was still not known.

The first breakthrough was almost accidental, in the sense that, as Pasteur said, "In the field of experimentation, fortune favors the prepared mind". Marshall Nirenberg and H. Matthaei was studying how an enzyme called ribonuclease digested RNA. To have a clean and easily measurable material to work with, he made a synthetic RNA consisting of only one base, U (the equivalent of T in DNA). Poly-U, a chain of riboses containing nothing but U, could not be confused with DNA, which would have to have T. To determine what happened to his synthetic RNA, he decided to test if it had any biological function, since it was known that adding RNA to a mixture of several other components would allow the mixture to synthesize protein.

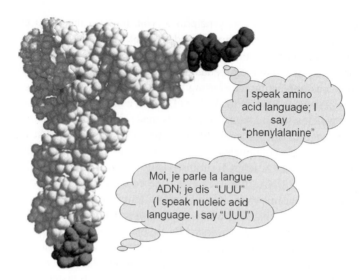

Figure 16.2. Transfer RNA is the translator. At one end it speaks amino-acid-ese, and identifies a specific amino acid. At the other end, too far away to strongly influence the other end, it speaks nucleic-ese, and carries a specific anticodon. Thus the lower end binds to mRNA at a specific location, and the upper end offers a specific amino acid to be incorporated into the protein, in the same sense that a good bilingual dictionary will give the English equivalent of a foreign word. Here the anticodon AAA on the tRNA binds to the UUU on the message, while the other end of the tRNA binds phenylalanine. Credits: Wikipedia: This image has been (or is hereby) released into the public domain by its author, Vossman. This applies worldwide

Therefore they added their undigested poly-U and their digested poly-U to this protein-synthesizing mixture to see how well they would work. They found, to their disappointment, that something precipitated or settled out from the mixture. Precipitation in an experiment like this generally means that something has gone wrong (see comments on stability of macromolecules in Chapter 17). However, Nirenberg and H. Matthaei decided to find out what had happened by analyzing the precipitate. What they found was a considerable surprise. The precipitate was a single compound, an artificial protein consisting of nothing but the amino acid phenylalanine. (Phenylalanine is a very poorly soluble amino acid, and therefore the chain was insoluble.) Nirenberg and H. Matthaei realized that they had conversed with molecules: They had spoken to the protein synthesizing machinery, saying "UUU, UUU, UUU, UUU...", and the machinery had responded, saying, "Oh, I see, phenylalanine-phenylalanine, phenylalanine, phenylalanine..." This was the first codon identified.

To say that there was a race to get other codons is putting it mildly. Nirenberg and H. Matthaei had made the announcement at a meeting of biochemists in Moscow. He stopped in Europe to repeat the talk two or three times, and a month or so later gave the talk at Harvard, in Cambridge, Massachusetts. At the talk, a gentleman not known to the audience stood up, literally with a laboratory notebook in hand,

and read from the notebook, saying that they had identified several other codons by synthesizing other simple nucleic acids. He had flown from New York to Boston to confront Nirenberg and H. Matthaei with the announcement. We learned later that Severo Ochoa, a biochemist from New York, had been at the Moscow meeting and had telephoned his laboratory—not an easy or inexpensive task in those days!—to report how Nirenberg and H. Matthaei had done it, and they had immediately begun the experiments to determine other codons. We now know that, of the 64 possible codons, some are punctuation. A triplet code for which there is no tRNA is called a "stop codon" meaning that the amino acid chain ends where this codon appears. Some of the other codons are "degenerate," meaning that more than one codon can be used for the same amino acid. Thus it took a little time to identify them all correctly. Once the codons were known, other scientists returned to the sickle-cell mutation and a few other similar mutations. Sure enough, there was a one-base change in the sickle-cell DNA, and it was exactly the change to produce the abnormal amino acid of the mutation. This served as an independent verification of the hypothesis of the genetic code. We can now produce synthetic DNAs to change specific amino acids at will, in a process called "site-directed mutagenesis".

The mechanism of coding should now be clear. Because of how we handle it, we tend to call the mRNA sequence the code. Therefore, the DNA anticode sequence AAA produces the code sequence in mRNA, UUU. The tRNA anticodon AAA binds to the template UUU, and the amino acid carried by what we will now call phe-tRNA, phenylalanine, is contributed to the growing protein chain (Fig. 16.1).

There is a final very important point to be learned from the coding hypothesis. Protein-synthesizing kits are now sold to research scientists. They consist of ribosomes, tRNAs, and other necessary ingredients; one needs only to contribute the mRNA to produce the protein encoded by the mRNA. mRNAs from many animals, plants, bacteria, and viruses have been tested. With very few and relatively minor exceptions, the code is universal. In other words, if UUU = phenylalanine in humans, then it also = phenylalanine in frogs, sequoia trees, bees, mosses, bacteria, and viruses. Physically, this does not have to be so. The tRNA molecule is big enough that what it has on one end, where it binds the amino acid, does not impose any requirement on the other end, where it carries the three-base sequence called the anticodon. You can perhaps picture this more effectively if you imagine a set of keys, each of which has been tagged to indicate to which lock it corresponds. There is no reason why the tags on the keys cannot be switched around. Thus the tag AAA on the tRNA corresponds to the amino acid key phenylalanine, but an impish biologist could switch the tag so that you try to insert the key in the wrong lock. In fact, artificial tRNAs have been constructed, in which the anticodon has been altered (Fig. 16.3). For instance, if the AAA of the phe-tRNA is altered to GAA, the anticodon will recognize the codon CUU, which represents the amino acid leucine. Sure enough, the synthetic tRNA with the anticodon GAA dutifully inserts a phenylalanine into a protein where a leucine belongs. It would obviously be very dangerous to have a mutation in a tRNA, and it is not surprising that the coding has been conserved throughout evolution. But the take-home lesson is more

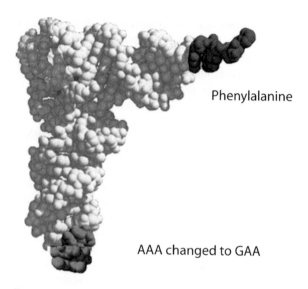

Phenylalanine

AAA changed to GAA

Figure 16.3. Proof of the hypothesis. The first proof was that the DNA sequence that produced the change in a single amino acid that was responsible for sickle-cell anemia was analyzed, and it proved to be a change in a single base in DNA. The change was exactly what was predicted to produce the change of the amino acid. A more elaborate proof consisted in altering a tRNA such that, while it still bound the amino acid phenylalanine, its anticodon was altered so that it bound to inappropriate locations on mRNA. The protein made by this construct incorporated phenylalanine in these incorrect locations. The experiment was done in a test tube. As you might expect (see study questions) such a situation would be catastrophic for a living animal or plant. Credits: Wikipedia: This image has been (or is hereby) released into the public domain by its author, Vossman. This applies worldwide

profound. Basically, the fact that the code is universal would be the equivalent of Europeans arriving in the New World, stepping off the boat, and realizing that the Arawak and Carib Indians addressed them in perfect Spanish! In other words, the genetic language is universal, even though there is no physical reason why it should be. This is the strongest argument we have that all life currently on earth came from one original living type. This is not to say that others did not start and fail—we have no evidence that this did not occur—but that today's living creatures have a common source. So what does this story have to do with evolution? It is of profound interest for the one very simple and straightforward reason mentioned above: The genetic code is universal, but it does not have to be. We could imagine mechanisms to synthesize proteins other than the rather complex DNA → mRNA → export from nucleus → ribosome + tRNA → protein system described, but we do not find other mechanisms. Even where there are differences, for instance in the structure of ribosomes of bacteria compared to the structure of ribosomes of eukaryotes, the similarities are far more striking than the differences. Furthermore, the code is universal: if the code TAC means the amino acid methionine in a bacterium, it means methionine for a sequoia tree, a moss, a fish, a bee, and a mouse. We know that this code is carried by the bilingual dictionary, tRNA, and that the structure of tRNA does not force the association between the base triplet AUG (on the mRNA) and the specific binding of methionine. Therefore, as far as we yet know, there is no physical reason to assume that TAC could not be leucine, or phenylalanine. In fact, we can make artificial hybrid tRNAs, and they can work perfectly well in a test tube, incorporating the (wrong) amino acid they carry at the location specified by the anticodon on the other end. Therefore, the argument is as follows: If there many possible codes, but we find only one, then all life on this earth derives from one source. This does not claim that there was only one origin to life—only that, if there were other attempts at starting life, only one survived to present. One may attribute that origin to any source: God, natural causes, arrival from another planet, etc., but mechanically the descent is the same. We can also conclude that all this life is related, in the same sense that one might reasonably conclude that a freckle-faced, red-haired Caucasian child in a central African village is the child of the freckle-faced, red-haired European couple working in the village rather than the child of the other residents. All life is related because it all looks alike. It is very difficult to understand why this should be so if each form of life was a special, individual creation. Thus we can conclude that the universality of the genetic code offers extremely strong support for the theory of evolution.

CONSERVED GENES AND HOMEOTIC GENES

Equally startling, though perhaps of a different order of urgency, is the frequency of conserved genes. As is described immediately below, these are found in quite unexpected circumstances, and they provide remarkable documentation of evolutionary connectedness. The basic argument is as follows: a gene (a DNA sequence that codes for a particular protein) evolves in a lower animal or plant. The function of this protein is very important to the survival and reproduction of the organism. In the course of time, random events cause mutations, or changes in base sequence that translate to changes in the amino acid sequence of the protein. However, because the protein is very important and because the function of the protein depends heavily on the presence of particular amino acids in defined sequence, almost all mutations prove to be very deleterious or even lethal, and the bearers of these mutations do not survive. Thus the only individuals that produce young for the next generation are those that maintain the gene intact or nearly intact. The gene survives the evolution of new species over a very wide and long period.

The survival of the gene indicates not only the importance of the gene. Where conserved genes are found, one can trace the lineage among organisms, and detect relationship where it is otherwise not obvious. In this manner one can identify an evolutionary history that connects insects and humans and some of the lowliest threadworms or roundworms and humans. These relationships were quite unexpected and are best explained by illustration. We will use three examples, all of which illustrate important principles of evolution. The examples include a group of enzymes known as caspases; genes controlling the development of eyes; and a group of genes known as homeotic genes.

Caspases: There are many ways to make an enzyme that can digest other proteins. We can digest proteins in our stomachs and intestines; inside of cells one can degrade damaged or improperly formed proteins; and in other cases some proteins are intentionally made to be quickly used, rapidly degraded, and built anew. The decomposition of a dead animal ultimately is a process of digestion of the proteins by bacteria and molds. There are well over 100 different types of protein-digesting enzymes (proteases), and there is no particular reason to assume that others could not be designed. Thus it came as a considerable surprise when Junying Yuan and H. Robert Horvitz, looking for a mechanism by which cells in a roundworm commit suicide, identified a gene that produced a protein-digesting enzyme. It was gratifying but not particularly surprising to find the protease—this would be an effective way of destroying the cell—but what was really amazing was that they identified its function because it was very similar to an enzyme found in humans! Further research quickly revealed that not only were the enzymes similar, they had similar, previously unknown functions in controlling the death of cells in humans and other mammals. (This group of enzymes is now called caspases, a technical name that describes to the initiated their function and structure. Cell death is a very important aspect of normal development and physiology, and many diseases including congenital (present at birth) abnormalities, cancer, AIDS, Alzheimer's Disease, Lupus, rheumatoid arthritis, and others at least in part result from derangements in patterns of cell death. There are now over 200,000 papers in the field, and Horvitz was awarded a Nobel Prize in large part because of these discoveries.) Think of it: at least 300 million years separate the lowly threadworm (the miniscule wriggly strings that you sometimes see in stagnant standing water) from us, and yet we use the same enzymes, and the same mechanisms of controlling cell death, to assure the appropriate placement and survival of our cells. Since one could imagine an infinite number of other means to assure proper development, the only possible conclusion is that the first evolutionary appearance of the caspase-based means of regulating cell survival proved wildly successful, and all organisms that derived from that first creature that used it have depended on maintaining it intact.

Perhaps a less abstruse example is that of *eye development*. Insects and vertebrates have gone their separate evolutionary ways almost as long as threadworms and vertebrates have and, although many insects can see very well, their eyes are extremely different from ours (Fig. 16.4). Our eyes are designed like a camera (or rather, a camera is designed like our eyes): a lens focuses light on a retina (film or light sensor); the lens adjusts to change focus; and the orientation of the eye changes to take in different views. It is a very complex instrument, and its complexity has been occasionally used as an argument against evolution. In fact, there is ample documentation for the evolution of the eye from a simple light-sensitive tissue into its present form (Fig. 16.5). An insect's eye is very different. It is more like a bundle of fiber-optic cables, each fiber capturing and carrying a fragment of the entire image. The LED (light-emitting diode) traffic lights and advertising signs that are now appearing in many cities give a sense of the image captured by insects:

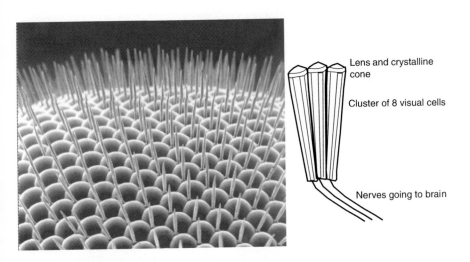

Lens and crystalline cone

Cluster of 8 visual cells

Nerves going to brain

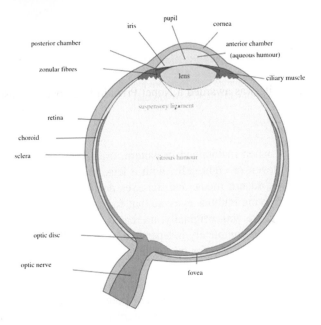

Figure 16.4. Insect eyes and vertebrate eyes. Upper: an insect compound eye. Each unmovable lens provides a fragment of the total image to a cluster of visual cells. No focus is possible, but near images are resolved fairly well. In contrast, in the vertebrate (here mammalian) eye, the lens takes in the entire image, and focus is adjusted by muscles that change the shape of the lens. Also, the shape of vertebrate eyes is maintained by fluid- and gel-filled chambers, which is not the case for invertebrate eyes. Credits: http://remf.dartmouth.edu/images/insectPart3SEM/source/26.html; (Wikipedia)

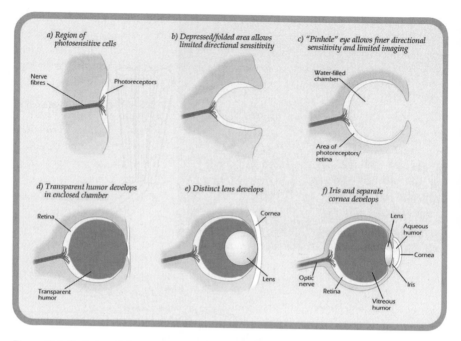

a) Region of photosensitive cells

Nerve fibres

Photoreceptors

b) Depressed/folded area allows limited directional sensitivity

c) "Pinhole" eye allows finer directional sensitivity and limited imaging

Water-filled chamber

Area of photoreceptors/ retina

d) Transparent humor develops in enclosed chamber

Retina

Transparent humor

e) Distinct lens develops

Cornea

Lens

f) Iris and separate cornea develops

Lens

Aqueous humor

Cornea

Optic nerve

Retina

Iris

Vitreous humor

Figure 16.5. Evolution of the vertebrate eye. There are living creatures that have eyes similar to each of these stages. Credits: Created by Matticus78 (Wikipedia)

a series of dots that, when maintained in pattern, create an overall picture. Each "dot" is a group of seven or eight cells, with a lens, that captures a fragment of the image. The lenses do not focus, and the eyes do not rotate. Insects get their breadth of vision by having bulbous eyes, so that each mini-lens covers a different territory (Fig. 16.4). There was, originally, no reason to assume that the evolution of an insect's eye was not completely independent of the evolution of a mammal's eye. That is, there was no reason for this assumption until genetics got better and it was possible to sequence genes.

There exists, in laboratory fruit flies, a mutation called "eyeless". As one might presume, the bearers of this mutation have very poorly developed or absent eyes. In the laboratory setting, where food is readily available, the flies can get along using their other senses, but they are blind. Conversely, it is possible to make the good form of the gene turn on in other parts of the fly's body; in this case extra eyes pop up in the weirdest places. They are non-functional, since they do not connect properly to the brain, but they are otherwise structurally normal eyes. This mutation was known for quite a while. In the sometimes annoying structure of the nomenclature of genetics, the name of the gene is the effect that appears when it is not functional. Thus the eye is missing when the gene is mutated, and the normal form of the gene "eyeless" is responsible for making the eye.

During the 1980's and 1990's, it became possible to sequence genes, to read their base sequences. As the databases became larger, governmental agencies in the US, England, Switzerland, and Japan pooled the information so that researchers could compare sequences and look for common themes, in the quest to understand how genes worked. One, again startling, discovery was the realization that the fruit fly eyeless gene was structurally very similar to a human gene called aniridia. Again following the naming convention, there is a rare mutation in humans in which the eyes are exceptionally small and often non-functional. In one version of the mutation, the most noticeable feature is absence or near-absence of the iris, leading to the name aniridia (absence of iris). The normal form of the gene is necessary for the proper development of the eye. It was certainly provocative to realize that genes controlling the development of the eye in a fruit fly and in humans were similar in sequence.

Since the sequences of both genes were known, it was possible to isolate the normal form of the human aniridia gene and insert it (see Chapter 14, page 191) into fruit flies that had mutated eyeless genes and were therefore blind. This experiment was done, with the result that the human gene was able, at least in part, to restore the development of the fruit fly eye! Thus eye development in insects and humans, even though the eyes are very different, was so genetically similar that the genes were nearly functionally interchangeable! Again, there appears to be no rational explanation except for the argument that, before insects and vertebrates existed, an ancestor common to both evolutionary lines had established a genetic mechanism for building a light-sensing organ. This capability was so important—it is obvious to imagine the value of being able to detect light and darkness—that all ancestors preserved this genetic mechanism, even as they evolved into insects, lobsters, octopuses, fish, birds, and mammals. It is a further and extremely strong argument for our common evolutionary heritage.

A final and likewise important example of common lineage is the remarkable story of the *homeotic genes*. This group of genes is important at many other levels as well, since their appearance may have been one of the bases for the Cambrian explosion (Chapter 20, page 281). The homeotic genes are responsible for establishing the basic layout of the body: why our arms and legs are located at the appropriate positions; why the vertebrae of our chest connect to ribs, whereas those of the neck, lower spine, and sacrum and coccyx do not; why the heart and lungs are in the chest and the stomach, spleen, liver, pancreas, and intestines are in the abdomen. Their name comes from the description of some peculiar mutants seen in fruitflies, in which the parts are mixed up. In one, the poor fly has, where the palps or little feelers around the mouth should be, legs instead of palps. In another, the wing-bearing segment of the thorax is repeated (Fig. 16.6).

In the 1970's and 1980's, Walter Gehring in Switzerland analyzed the genes that were responsible for some of these mutations, and he realized that they all contained a very similar short sequence. This nucleic acid sequence coded for a sequence of amino acids that would take a shape such that it would readily bind to DNA. In other words, these proteins were of a type that would be able to regulate the activity of DNA, exactly what would be needed if one were to determine that

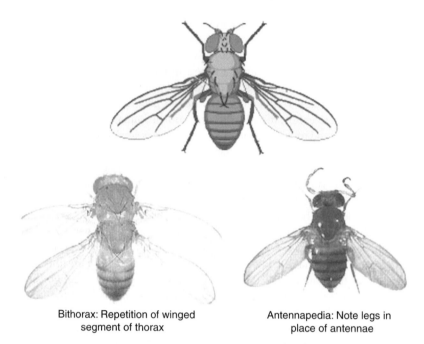

Bithorax: Repetition of winged
segment of thorax

Antennapedia: Note legs in
place of antennae

Figure 16.6. The effect of mutations of homeotic genes. In the upper picture is a normal fruit fly. The fly on the lower left is a bithorax mutant. In this fly the wing-bearing segment (the second thoracic segment) is repeated, creating an extra set of wings. On the lower right is a fly with the mutation antennapedia. In this fly the antennae, which are the anteriormost appendages, are converted to the more posterior legs, which are properly found on the thorax

one region, for instance, should be the head and another the thorax. Furthermore, Gehring and his coworkers found something else quite startling and still not well explained: these similar genes were lined up on the chromosome in the order that they functioned in the body. The ones that determined what would be head came first, followed by those that determined what would be thorax, followed by those that determined the abdomen (Fig. 16.7).

At this time the structures of the six-legged insect or the ten-legged lobster, with their nerve cords along the stomach and their hearts along the backside, were so obviously different from the structures of the four-legged vertebrates, with their hearts on the stomach (ventral) side and their nerve cords along the back (dorsal) side, that there was no assumption of evolutionary connectedness. However, again referring to databases and pursuing the issue, Gehring and many others quickly realized that, not only were close relatives of the fruit fly homeotic genes found in vertebrates, they were arranged on the chromosome in the same sequence as those in the fruit fly! Not only were they similar and their arrangement similar, mutations of them in mice demonstrated that, as in fruit flies, these homeotic genes in vertebrates also established the anterior-to-posterior axis of the mouse. Thus, once again, the existence of common base sequences with common functions—conserved

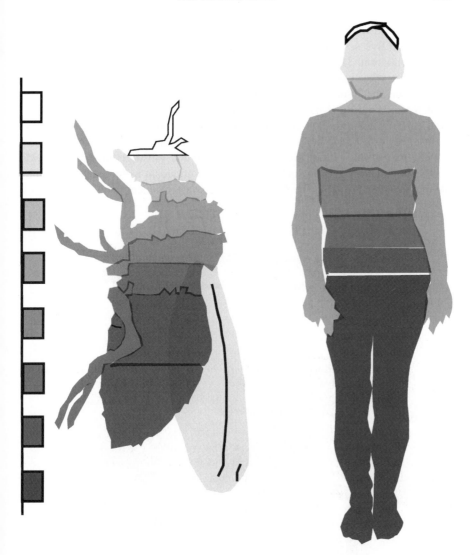

Figure 16.7. Conservation of homeotic complexes from invertebrates to humans. On the left is illustrated the sequence of eight homeotic genes on a fruit fly or human chromosome. On the models of the fruit fly and human the regions that these genes delineate are indicated in corresponding shades. In both humans and fruit flies, the alignment on the chromosome is the same as the anterior-to-posterior realm of action in the body

genes—in wildly different creatures not only establishes the importance of the genes and their functions, it provides a deeply compelling argument for common ancestry. An early precursor of both the vertebrates and the higher invertebrates found a means of differentiating its body parts, and this mechanism was so valuable that it was altered only at the bearer's peril.

Vertebrates differ from insects in that they have more than one set of these homeotic genes, up to five sets, with each presumably adding a greater level of subtlety to the differentiation. This is an entirely different story, but it does provide one more argument for why animal life rapidly expanded approximately 400 million years ago. An earthworm has a sort of a brain at its front end, and it goes in one direction, but if you cut off the front end, the remainder of the worm really has no obvious direction or layout. Similarly, the most primitive vertebrates (we call them "chordates" or "hemichordates" because they don't have bony vertebrae) are not very impressive: picture a longish fish like an eel but without fins, cut off its head, and substitute instead a filtering mouth through which it can suck in microscopic organism as food. This would be one of these creatures (Fig. 16.8). Now picture a fast, efficient fish with keen eyes,

Figure 16.8. Upper: a lancelet. Its head is to the right. This inconspicuous creature (it is only about an inch long) has very little anterior-to-posterior (head to tail) differentiation. It has no elaborate brain or eyes, no skull or teeth. It draws in water and small organisms through its filtering mouth. This is the general appearance of a chordate that does not have a full set of homeotic genes. Lampreys and hagfishes are much larger but only slightly more elaborate. The lampreys clinging to the trout may look like eels, but they have no true fins or jaws. Credits: Amphioxis - Wikimedia Commons. Branchiostoma lanceolatum. Photo by Hans Hillewaert. Lamprey - Sea_Lamprey_fish.jpg (81KB, MIME type: image/jpeg) Licensing This image is a work of a United States Geological Survey employee, taken or made during the course of the person's official duties. As a work of the United States Government, the image is in the public domain. (Wikipedia)

well developed and specialized fins, a true skull with strong and specialized teeth, and clearly differentiated parts of the body. A trout, tuna, or barracuda would be a good example. Perhaps the most important difference between these types of creatures is the presence of one or more full sets of homeotic genes. Tracing back the lineages of these genes (Chapter 9, page 118) we can surmise that they appeared and duplicated approximately 350 million years ago. The obviously much higher efficiency of having highly specialized regions of the body, controlled by the activity of the homeotic genes, leads to the argument that the appearance of the homeotic genes created the basis for the high variety among creatures that led to their rapid expansion in the Cambrian era (page 281). Certainly the appearance of these genes was very important to the incipient animal lineages, as the genes are highly conserved, but homeotic genes do not exist in plants, which organize their structures in very different ways.

REFERENCES

Halder, G., Callaerts, P., and Gehring, W.J. (1995), New perspectives on eye evolution. Curr. Opin. Genet. Dev. 5: 602–609.

Halder, G., Callaerts, P., and Gehring, W.J. (1995) Induction of ectopic eyes by targeted expression of the *eyeless* gene in *Drosophila,* Science 267: 1788–1792.

Land, M. F. and Fernald, Russel D. (1992) The evolution of eyes. Ann. Rev. Neurosci. 15: 1–29.

STUDY QUESTIONS

1. If you were to design a machine for making proteins based on information stored as a linear sequence of DNA, how would you do so? Why?
2. What do we mean, "The code is universal"?
3. Imagine an animal in which a tRNA has been mutated so that it gives a false translation. What would happen to the animal? Why?
4. Occasionally we encounter an organism that makes a unique amino acid, somewhat similar to another amino acid, but with a distinct difference. How might this come about? (Hint: consider the possibility that it is possible to modify an amino acid. At what steps might this occur?)
5. Can you think of any means by which the base sequence at the anticodon of the tRNA could determine which amino acid is attached?
6. If the code is universal, could you for instance use ribosomes and tRNA from bacteria and mix them with mRNA from humans and expect to get a human protein synthesized? Why or why not?
7. If you were to find a creature from another planet, would you expect to find a similar means of making protein and a similar code? Why or why not?
8. If you found an organism that did NOT use the universal code, but still used the tRNA, mRNA, ribosome synthesizing system, would you consider it related to everything else on earth? Why or why not?

PART 5

THE HISTORY OF THE EARTH AND THE ORIGIN OF LIFE

CHAPTER 17

THE STORY OF OUR PLANET

THE STUFF OF INHERITANCE: DNA, RNA, AND MUTATIONS

Knowing how DNA works and how proteins are synthesized can tell us that all life is related, but it does not tell us how life arose. Strictly speaking, this is not an issue for the story of evolution, since the story of evolution begins with the postulate that, once life appeared, natural selection generated the variety that it became. Whether life appeared on earth as an act of God, as dust delivered from another planet—which begs the question of how THAT life arose—or appeared as an entirely chemical process does not seriously affect the hypothesis of natural selection. Nevertheless, it is possible to speculate on the mechanics of creating life. After all, science is about the mechanics, or how things work. The only difference between a scientific analysis of how life might have begun and a religious viewpoint is that the scientific argument is, "If God created life, this is how He might have done it." As the great late Medieval Spanish Jewish philosopher Maimonides put it, "A miracle is not something that could never have happened. It is something that always could have happened, but did not without the intervention of God." Maimonides was one of the scholars who returned Aristotelian philosophy to Europe, and he was widely respected by Christian, Jewish, and Islamic philosophers. Aristotle of course emphasized mechanical explanations for phenomena.

There is much evidence to tell us about many of the steps that presumably transpired, but at present we cannot assemble the entire story. Based on evidence and logic, we can hypothesize several of the steps that probably occurred. The story is constructed as follows:

LIFE REQUIRES CARBON AND WATER

The reason that projects such as planetary exploration search other planets for water and "organic molecules" is that virtually all scientists agree that any form of life must be based on these materials. There are many reasons for this agreement, based on the properties of "organic molecules" and the properties of water. Organic molecules are defined as complexes based on the element carbon. Carbon is unique among the over one hundred known elements in that individual carbon atoms can bond to other carbon atoms (that is, carbon can form chains of molecules) and that the carbon-to-carbon bond or linkage is flexible. Atoms are the smallest intact particles of elements, such as oxygen, carbon, sulfur, or chlorine. Molecules are combinations of different atoms, such as water (two hydrogen atoms combined to one oxygen atom), carbon dioxide (two oxygen atoms combined to one carbon atom) or propane (three carbon atoms and eight hydrogen atoms linked together). Other atoms, such as silicon, can also bond to each other and form chains, but the chains are not flexible; they tend to form very rigid structures such as sand (crystals of linked silicon and oxygen). The flexibility means that a carbon-based molecule can be built in almost any shape imaginable—including giant accumulations of atoms called macromolecules (including thousands of carbon atoms, as well as more thousands of hydrogen, oxygen, sulfur, nitrogen, and phosphorus atoms), providing the enormous variety that is essential for life. For instance, cotton, wood, sugar, and starch are all forms of carbon-based compounds called carbohydrates; fingernails, hair, antlers, steaks, egg white, milk, the blood protein hemoglobin, silk, and skin are all members of another carbon-based family, the proteins; and butane lighter fluid, gasoline, candle wax, animal fat, and cooking oil are members of the carbon-based family called lipids. No other single type of atom will do this. For this reason virtually all scientists are in agreement that life must be based on carbon, and complex carbon-based molecules are granted their own special branch of chemistry. Complex carbon-based molecules are called organic molecules, and the study of them is called organic chemistry.

The properties of carbon-based molecules also determine some conditions of life. For instance, most carbon-based molecules break down if they get too hot. This is what happens to the white of an egg when you cook it, or to your skin if you burn it. Most carbon-based molecules are unstable above 50° C (122° F), setting an upper limit to the temperature at which life could exist. (A very few creatures, such as bacteria living in hot springs and thermal vents, have made special adaptations to their proteins and DNA, to allow them to survive at temperatures near boiling. These, however, are quite rare.) The instability of carbon-based molecules is even the reason why people become delirious if the body temperature rises above 104–105° F (approximately 40° C). The properties of carbon are illustrated in Fig. 17.1.

The other crucial element is water. Water is a wonderful solvent, dissolving at least a little of a vast range of molecules ranging from salts to proteins and lipids. (The fact that you can taste the gasoline if water has come in contact with it indicates that a little of the gasoline has dissolved in the water.) It absorbs and holds onto a lot of heat, which is why coastal areas have warmer autumns and cooler springs than inland areas; it interacts quite well with many molecules, allowing them to react with each other. When it freezes, unlike most molecules, the ice is lighter than the water and floats, preventing deeper water from freezing. But most importantly, it is liquid at temperatures at which carbon-based molecules are stable and can react. Atoms do not move much in crystals such as ice, and so reactions cannot take place. (Imagine making a cocktail or any other drink that must be shaken if the whole mixture is frozen into one big chunk of ice.) Thus water is extremely useful, even essential, for life, and the freezing point of water determines a lower limit

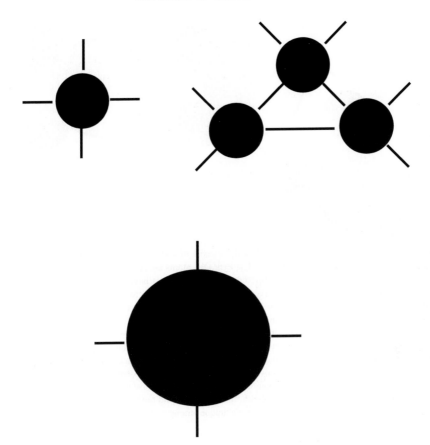

Figure 17.1. Upper left: The atomic structure of carbon consists of an atomic nucleus (black) and a shell of electrons surrounding it, of which four electrons can interact and form bonds with other atoms (lines). Because of the size of the nucleus relative to the shell, the electrons can interact with each other, in effect bending to form bonds in various positions (upper right). The molecule illustrated would be called cyclopropane. Because of this property, carbon atoms bonding to other carbon atoms can form molecules (complexes of atoms) with an infinite variety of shapes and forms, a requisite for being the basis of the enormous complexity that allows us to live. Also, because the bending forces the distortion of a relaxed state, different molecules store energy rather like cocked springs. The molecule cyclopropane is flammable, capable of releasing its energy as heat. (Lower). The atom silicon is similar in configuration to carbon but is much larger. Its electrons are too far apart to interact, like a cartoon character who has blown up like a balloon so that his arms and legs no longer can reach each other. The consequence is that silicon can form only rigid, linear or cuboidal structures like sand

at which life can comfortably exist. We therefore can identify three characteristics that we feel should be present if life is to exist: complex carbon-based molecules, water, and a temperature somewhere above freezing and below about 100° F. This is why probes of other planets, looking for signs of life, explore for traces of liquid water and for complex organic compounds. The structure of water is indicated in

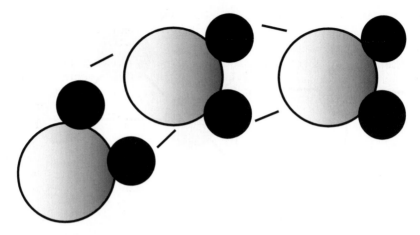

Figure 17.2. Water molecules. Water molecules consist of one oxygen atom (larger circle) with two hydrogen atoms (smaller circles) attached. The hydrogen atoms are separated at an angle slightly larger than a right angle. Water is not ionic. In other words, it does not have a charge and it will not move in an electric field. However, it does have a polarity like a magnet (gradient across the molecule). The oxygen end is slightly negative, and the hydrogen end is slightly positive. Thus different molecules can interact weakly with each other, again like magnets separated by slight distances, holding themselves together and making water a liquid at room temperature. By contrast, methane, a molecule of similar size but not polar, is a gas at room temperature, with its molecules not interacting at all. Approximately one in ten million water molecules do break apart, allowing water to react with many other substances. Because it can do this, water can dissolve many different types of substances. Because it is a liquid at temperatures in which carbon-based compounds are stable and because it can dissolve and interact with many substances, it is a very special molecule for the mixing and building of carbon-based compounds. It has several other properties that make it nearly unique among molecules, leading scientists to conclude that it is essential for life

Fig. 17.2. In terms of the origin of life on earth, we can assume that water and carbon dioxide were present, thus supplying the two primary ingredients. We need to ask if we can get more complex organic molecules by processes other than the metabolism of living creatures. The answer is yes.

FORMATION OF ORGANIC MOLECULES CAN TAKE PLACE IN THE CONDITIONS PREVAILING DURING THE EARLY DAYS OF THE EARTH

There are many means of analyzing atmospheric and climatic conditions in the early earth. One can interpret the composition of the atmosphere by the chemical composition of the rocks. You are familiar with the fact that iron rusts when left in air. In fact, elemental (native) iron reacts relatively easily with oxygen to form ferric oxide (rust). If one breaks open a rock that has lain undisturbed since it was formed, and finds rust inside this rock, one may safely conclude that oxygen was present in the atmosphere the last time the iron was exposed, perhaps in lava coming out of a volcano. If the iron is metallic or in some other form, then no oxygen was present when the iron was last at the surface. It is possible to make this assessment more quantitative, since different reactions occur at different concentrations of oxygen in the atmosphere. Similarly, the shapes of crystals are different if they are formed at different temperatures or under different pressures. Diamonds, graphite, and coal are all primarily carbon, but each is formed under different conditions. Thus one can interpret the temperature at which other rocks formed.

From considerations such as these, most geologists have concluded that the early earth was hot, an argument fully consistent with the hypothesis that the earth originated from the sun. Eventually the surface cooled to below the boiling point of water, allowing incessant rain that accumulated on the earth as hot water. Because of the physics of raindrops, there must have been considerable lightning during this period. The moon was much closer to the earth, generating violent tides hundreds of feet high that swept many miles inland. The rain would have extracted soluble salts from the surface of the earth, carrying the salts to the forming sea. (The concentration of the salts in the sea at this time can also be estimated by the types of crystals that were formed: it was approximately 0.2%, about ¼ of what it is today.) All the oxygen available had already reacted with other molecules, producing sulfur dioxide, water (dihydrogen oxide), carbon dioxide, and other oxides, and there was no free oxygen in the atmosphere. There was, however, sulfur dioxide, hydrogen sulfide (the odor of rotten eggs), and ammonia.

These conclusions had been reached by 1953, when Stanley Miller and Harold Urey*explored the possibility that such conditions could create the first steps toward life. They put together the molecules hypothesized to be present in the early atmosphere, together with a salt water solution. They removed all free oxygen from the mixture and maintained it at high temperature, together with electrical discharges through the flask, simulating lightning, and allowed the mixture to "cook" for several weeks. At the end of this period, they opened the flask and analyzed the solution. What they found were the simplest molecules presumed to be necessary for life: the building blocks of proteins (amino acids), nucleic acids (nucleic acid bases), sugars, and small lipids. In other words simple organic molecules could be formed from inorganic molecules in the conditions presumed to be present in the

- Interact with hydrogen, oxygen, and water

- Mostly unstable at temperatures much above 40°C (104°F)

- Easily deformed by extremes of acidity or alkalinity

- Easily deformed by changes in salt concentration

Figure 17.3. Properties of organic molecules. The structure of carbon-based molecules determines that they can exist only under specific conditions of temperature, salinity, and acidity. Thus, since life depends on carbon and water, all living organisms must function within these limits. Living organisms cannot exist at temperatures at which carbon-based molecules burn or water is converted completely to steam, and organisms cannot function (and can only survive as spores or the equivalent) if the water in their bodies is solid (ice)

early earth. This would be the first step in the creation of life. The properties of organic molecules are illustrated in Fig. 17.3.

LARGER ORGANIC MOLECULES CAN BE FORMED FROM SMALL ORGANIC MOLECULES

Macromolecules are polymers—end-to-end chains, occasionally branched, of these building blocks. If these building blocks are brought together, they can interact to link to each other. We have encountered macromolecules earlier (page 244). They include long chains of amino acids, which are proteins; long chains of nucleotides, which are the nucleic acids DNA and RNA; chains of sugars, which we know as starches, glycogen, complex sugars, and cellulose; and chains of small two-carbon molecules that form fats. More recent research has established that, under appropriate conditions of heat, salt concentration, and sometimes electricity, small molecules can interact to form macromolecules, Thus, although we are still a very long way away, we have the beginnings of something that could eventually acquire the complexity required for life. Based on evidence such as that described in Chapter 17 and elsewhere, we believe that the early seas were warm enough, salty enough, and subjected to enough lightning strikes to recreate these conditions. The reactions would have taken place very slowly. Living systems have developed enzymes, or proteins capable of speeding up reactions, to make the process much more efficient, but in the early earth, enzymes would not have existed, and since enzymes also degrade macromolecules—bacterial enzymes are what causes formerly living materials to putrefy—it would have been possible to accumulate large quantities of these organic molecules. One other element that could cause these molecules to decay was also missing. You are aware that many types of carbon-containing and some inorganic materials, of which wine and iron are primary examples, will "spoil" if exposed to air. This is because many types of molecules react easily with oxygen, turning into different types of molecules. As we will see in Chapter 18, there is also solid evidence that there was no free oxygen in the atmosphere of the early earth. Thus this limitation to the accumulation of organic molecules was also absent. Many scientists therefore feel that it would have been possible in the early earth for large amounts of organic molecules to accumulate. This of course is primarily speculative, but the point is that the mechanism is possible, and thus there is no intellectual impediment to this part of the hypothesis.

THE LIPIDS FORMED CAN NATURALLY FORM PARTICLES THE SIZE OF BACTERIA; THESE PARTICLES HAVE A MOST STABLE SIZE AND CAN GROW AND DIVIDE

In the most precise interpretation of the sentence, the sea is not "living". It contains an immeasurable number of living organisms, but the complex salt-water solution, containing also some soluble macromolecules from plants and animals that have died and burst (sea foam, often seen when a coastline containing a lot of algae is agitated, is created because the carbohydrates—agar, as is used in biological research—are dissolved in the water). Nevertheless, it is not alive; if all living organisms are removed from the water, it will sit there more or less indefinitely. One can argue that the essence of life is control. As the father of physiology, Claude Bernard, said, "The constancy of the internal environment is the basic condition for a free life[7]," meaning that, in order for an organism to function independently of the environment, it must be able to make the conditions inside its body constant—constant body temperature, salt content, concentration of nutrients, oxygen level, etc. The reason for this is that macromolecules are delicate things. They work only when conditions are just right; otherwise they may not work, or even be damaged. Life cannot exist at temperatures at which macromolecules are not stable. Thus we become delirious at the temperature at which the membranes of our nerve cells begin to melt, and we receive burns at the temperatures at which egg whites become solid (the proteins are "denaturing" or precipitating from solution). In order for a living thing to exist, it has to isolate its macromolecules from the rest of the world, and maintain them in a medium with a constant, ideal, amount of salts, acidity, and other supplies. Again, proteins precipitate when conditions become too acid (milk curdles because of acid production by bacteria) or too alkaline. To maintain the constancy, a living creature basically has to keep the proteins inside of plastic bags into which it can allow entry and exit of only limited amounts of the salts, acids, and supplies. These plastic bags are what we know as the membranes surrounding the cell, the nucleus of the cell, and the organelles of the cell. The question is, where did cells come from?

INTERESTINGLY ENOUGH, ORGANIC SOLUTIONS CAN SPONTANEOUSLY FORM PARTICLES WITH THE APPROPRIATE PROPERTIES

Everyone is familiar with the "shake well before using" admonition on prepackaged salad dressings, spray paints, and other items. The object of the shaking is to temporarily suspend particles or one liquid in another liquid in which it is immiscible (it does not dissolve). If the particles are small enough, they may remain nearly permanently, as in homogenized milk. In the case of milk, it is called a colloidal suspension, in which the droplets of fat are suspended in the watery milk. They basically are very tiny droplets of fat, so small that they do not settle out. The fats can form another type of suspension, in which the droplet of fat is not a full droplet, but rather a little bubble, a membrane of fat separating two watery solutions. In this configuration, a suspension of little bubbles is called a coacervate.

Coacervates have very interesting properties. First, as you can readily imagine, the conditions inside the bubble can be different from conditions outside, as a balloon may contain helium while existing in air, which is primarily a mixture of nitrogen

---

[7] La fixité du milieu intérieur est la condition même de la vie libre.

Figure 17.4. The molecules in a coacervate become crowded if the size of the bubble is too small. If the bubble gets too big, they are stretched apart, weakening the structure. Therefore coacervate particles are most stable at intermediate sizes. Smaller particles will fuse to grow, and larger particles will split to shrink

and oxygen. Second, they can have an optimum stable size. The fat molecules cling to each other more firmly than they cling to anything else, and they have defined shapes. If they pack too tightly, they crowd each other, while if the bubble is too big, it is not very strong (Fig. 17.4). Because of this property, in a coacervate suspension consisting of different-size bubbles, tiny bubbles will fuse to become larger and approach the more stable size and, surprisingly, larger bubbles may split to become smaller. This splitting has the appearance of a crude and simplistic type of cell division. Furthermore, the optimum stable size turns out to be approximately the size of bacteria and small cells. A coacervate even has a further interesting property: If anything can move across the membrane, it can attract some molecules, which interact well with the fats, into the inside of the bubble, causing the bubble to grow until it is larger than optimum size, whereupon it splits into two. Thus, although it is speculative, we can imagine the creation of something that would look like a dividing cell. It can even create a stable situation in which the interior of the bubbles, or coacervates, is consistently different from the outside world.

THESE PARTICLES CAN DIVIDE, BUT CAN THEY REPRODUCE?

We would consider this process to be interesting, complex, but non-living. It is purely chemical, rather than living, because it does not control its own reproduction.

One of the defining characteristics of life is its ability to reproduce itself. Thus it is not sufficient for chemicals to tend to accumulate in preferential distribution: A living creature must define that distribution, and expend energy to maintain and reproduce that distribution even when conditions are not favorable. Thus, though it may not have been the first mechanism invented, ultimately life had to invent enzymes, or proteins that can efficiently and rapidly assemble and disassemble other molecules. In the world of physics and chemistry, everything in the long run tends to deteriorate. Living creatures capture energy to rebuild molecules and ultimately reproduce themselves. The question is, how do we move from the interesting complex chemistry of the coacervate to the directed reproduction of a living creature?

Proteins are very special molecules, the spontaneous assembly of which is highly improbable and which are therefore characteristic of living organisms. As we have seen in Chapters 14 and 15, proteins are made by a very elaborate mechanism involving DNA and RNA. One intellectual breakthrough came when in the 1980's Thomas Cech and Sidney Altman demonstrated that RNA, a macromolecule of relatively simple structure, could catalyze reactions—as Leslie Orgel, Francis Crick, and Carl Woese had suggested almost 20 years before (evidence finally coming to the aid of logic). A catalyst brings two molecules into close proximity so that they can interact, and thus speeds up reactions. The catalytic converter in automobile exhausts binds both sulfur compounds and oxygen so that the oxygen can inactivate the sulfur compounds and render them less harmful. An enzyme is a catalyst. The discovery of enzymatic RNA (ribozyme) brought a new idea to the concept of the origin of living creatures. Rather than having to establish a complex process of DNA having to code for RNA so that the RNA could make protein before life could become self-reproducing, Orgel and others suggested that if RNA could act as an enzyme, it might be possible for RNA to reproduce itself without other components. Following the logic of Occam's Razor (pages 50 and 170) this hypothesis required far fewer assumptions and was therefore inherently more appealing. The primary problem was the same as the Russian doll problem (page 200–203): if proteins are required to synthesize RNA and DNA, how do you get RNA, DNA, and proteins together spontaneously? On the other hand, if RNA were spontaneously formed, it might be able to replicate itself, as it does in certain viruses, with the aid of host energy supply and raw materials, and also and also to alter some proteins in enzyme-like mechanisms. Evidence has since accumulated that RNA can indeed act as an enzyme. It also may have been possible to build RNA under primitive earth conditions, perhaps using clay as a catalyst, much as powdered metals are used as catalysts for the catalytic converters that remove impurities from exhaust fumes.

Much of this of course is highly speculative, but is supported at least by the logic of the argument, with some evidence adduced in support of the logic. The subject is somewhat tangential here, but is discussed clearly by scientists such as Orgel (http://www.geocities.com/CapeCanaveral/Lab/2948/orgel.html).

HOW WAS THE DNA→mRNA→PROTEIN SEQUENCE CREATED?

We know little about how we got to the DNA→mRNA→protein sequence, other than the fact that, although chemically it is easier for RNA to be made than for DNA to be made, DNA is a more stable molecule in the chemical conditions that may have existed in the early history of the earth. Thus the hypotheses are as follows:

1. RNA and protein were formed by non-living chemical means.
2. RNA could interact and alter other RNA and proteins. In other words, it was an enzyme.
3. In some of these reactions, RNA, which could form paired double helices like DNA, made copies of itself, or reproduced.
4. Occasionally the reproduction of the RNA would not be perfect and would produce mistakes. These mistakes are mutations, passed along to the next generation. Some of these mutations would improve the ability of the RNA to control its environment and reproduce more efficiently. Thus this mutation would be selected for.
5. Some of the properties of the RNA, or the proteins altered by the RNA, required specific concentrations of salts and acids. Thus there was selection also to structure coacervates such that the RNA and protein were inside, and other proteins became gates controlling what could pass through the lipid (fatty) membranes. These ultimately became the highly specific membranes that we know as cell membranes.
6. Since proteins are structurally more variable and malleable than RNA, and DNA is more chemically stable than RNA, gradually selection worked to make proteins primarily responsible for the many reactions carried out by cells, and DNA responsible for preserving the genetic information.
7. This type of selection would lead to a primitive living organism, perhaps similar to bacteria, which have a cell membrane through which they can import and export specific materials and exclude others. The DNA is naked inside the cell. In many types of bacteria, the DNA is replicated with no sexual exchange between individuals. Some bacteria have little "hairs" by which they can actively move, and some can even sense light, but otherwise there is little other structure or organization. Nevertheless, by being able to control its internal environment, to make new proteins when it wishes, and to reproduce its DNA at appropriate intervals, this type of creature has vastly greater ability to survive and propagate than any inanimate aggregate. It is alive.

Incidentally, although viruses are considerably simpler than bacteria, they are not considered to be the earliest forms of life, for the simple reason that they cannot survive without living cells. They use the mechanisms of living complete cells to manufacture new virus, as is described in Chapter 28.[8]

---

[8] This is similar to a story told in religious circles about scientists creating life:
God is sitting in Heaven when a scientist says to Him, "Lord, we don't need you anymore. Science has finally figured out a way to create life out of nothing. In other words, we can now do what you did in the 'beginning'.

REFERENCES

http://www.geocities.com/CapeCanaveral/Lab/2948/orgel.html (Orgel's essay on Origin of Life)
http://www.ncseweb.org/icons/icon1millerurey.html (The Miller-Urey experiment, from National Center for Science Education)

STUDY QUESTIONS

1. Argue for or against the proposition that all properties of life are properties of the element carbon.
2. Argue for or against the proposition that life can exist in liquids other than water.
3. What would you consider to be the most accurate means of determining the composition of the early earth? Why?
4. What steps would you consider to be the most crucial for life to begin? Why?
5. What characteristics make a living creature, as opposed to an exceptionally adaptable molecule or collection of molecules? Why?
6. Could you imagine a form of life in which DNA directly constructed proteins, without the requirement for RNA? What would it be like? Why do you assume that it would work?
7. Could you imagine a form of life in which RNA copied itself without the aid of DNA? What would it be like? Why do you assume that it would work?
8. Can you imagine a form of life in which protein reproduced itself, without the aid of nucleic acids? What would it be like? Why do you assume that it would work?

"Oh, is that so? Tell me..." replies God.
"Well", says the scientist, "we can take dirt and form it into the likeness of you and breathe life into it, thus creating man."
"Well, that's interesting. Show Me."
So the scientist bends down to the earth and starts to mold the soil.
"Oh no, no, no..." interrupts God, "Get your own dirt."

CHAPTER 18

THE APPEARANCE OF OXYGEN

THE ORIGIN OF BIOLOGICAL (ORGANIC) MOLECULES

There are various snippets of evidence by which we think that we can identify the first signs of life on earth. For instance, some types of marine bacteria grow in peculiar large clumps. These clumps are remarkably similar to unusual rock formations, called stromatoliths, found in various fossil rocks (Fig. 18.1). Also, most carbon is tied up in chemical forms such as carbon dioxide (a gas) or calcium carbonate (chalk), with pure carbon being rather difficult to come by in the test tube. Living organisms can decay to combinations of carbon and hydrogen (oil) or to nearly pure carbon (coal), and many investigators believe that deposits of oil or coal indicate earlier life. Furthermore, living organisms carry out some chemical reactions in a decidedly non-chemical way. Since enzymes must fit perfectly around other molecules to assemble them or take them apart (see page 224), a reaction catalyzed by an enzyme can distinguish subtleties that a powdered-metal or clay catalyst cannot. For instance, because of the way that atoms bind to each other, many organic molecules are constrained to specific shapes. Some are constrained to shapes like a left-handed and right-handed glove, in which one of the three axes (up-down, left-right, back-front) is reversed relative to the other two. Nonenzymatic chemical reactions do not distinguish between the two, but enzymatic reactions do. Many molecules, such as amino acids (the building blocks of protein) and sugars, display this property, and some of them can be preserved almost indefinitely in old soils and crude petroleum. Living organisms vastly prefer, use, and manufacture the left-handed forms of amino acids, whereas chemical reactions make no distinction (Fig. 18.2). The presence of a preponderance of left-handed amino acids suggests the former presence of life. Finally, although it is very subtle, one isotope (see page 104) of carbon, carbon-13, is ever so slightly larger than the more common form, carbon-12. Enzymatic reactions can very slightly distinguish the two, preferring carbon dioxide containing carbon-12 as a substrate to make sugars over a form containing carbon-13. A compound containing a higher ratio of carbon-12 to carbon-13 than is found in surrounding rocks is an indicator that life was present. When these data are all taken together, there is some consensus that life originated on Earth approximately 3.5 billion years ago—astonishingly close to the time that the Earth was sufficiently cool enough to support life, but quite distant from the time that life really began to explode on the earth, approximately 500 million years ago (see page 277). What could have happened to cause these two transitions?

257

Figure 18.1. Stromatoliths. (Upper) Fossil stromatoliths from Precambrian period, found in Germany. (Lower) Contemporary stromatoliths, formed by bacteria on the west coast of Australia. Credits: Fossil stromatolith - de.wikipedia.org/wiki/Stromatolith, Australian stromatolith - http://upload.wikimedia.org/wikipedia/commons/1/1b/Stromatolites_in_Sharkbay.jpg

There are several hypotheses. These hypotheses assume very highselection pressures favoring those who could versus those who couldn't. The first transition is defined by the need for energy. One absolute requirement for life is the ability to capture energy. According to the Law of Entropy coming from Physics, everything

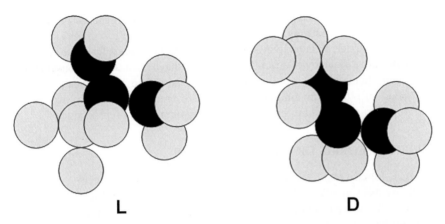

L   D

Figure 18.2. L- and D-amino acids. (The "L" and "D" are from the Latin names for "left" and "right". These molecules contain the same atoms, but because of the way that the atoms are attached they resemble each other as left and right gloves resemble each other. Because biological reactions take place with one molecule fitting into another like a hand in a glove, biological systems usually readily distinguish between the two forms, whereas chemical reactions usually do not. The L form is commonly used in biological systems. The D form is not except in cell walls of bacteria

tends to deteriorate. Entropy is essentially disorder, and it always tends to increase. That is, order decreases. Helium contained in a balloon—ordered by being confined to a specific location—will dissipate, thus destroying that order. A watch can fall apart, but it will not spontaneously reassemble. A sand castle (a highly ordered arrangement of sand) will gradually fall apart, whether by waves, wind, or gravity, but it will not spontaneously appear. But life creates order: an egg becomes an embryo, and the embryo an animal. A seed turns into a tree. The reason that this can happen is that living creatures can use energy, which at one point is contained in a highly ordered arrangement of atoms, to build a different but new arrangement. Wood burns to create heat that causes a machine to bend a sheet of metal into an object. Turning the metal into an object creates order, but in the larger picture the destruction of the wood was a greater loss of order, the conversion of the cellulose into carbon dioxide and water, and the difference was lost as heat. Overall, the Law of Entropy holds, but for the machinist and the object, order has been created. A major adaptation of early life would have been to trap a source of energy to reproduce itself. If the early Earth contained, as we assume, some coacervate/cell structures formed amid a soup of relatively similar organic molecules, then breaking down these organic molecules would have yielded energy that the coacervate/cell (organism) could use. Many organisms continue to do this at present. Yeast can famously take a molecule such as sugar and, breaking it down to ethanol (alcohol), carbon dioxide, and water, can extract enough energy to build new yeast. Bacteria routinely perform similar conversions, often producing foul-smelling products such as butyric acid (the odor of rancid butter). Thus these reactions are possible and may have sustained the earliest forms of life. Virtually all organisms preserve this capability, which generically is known as fermentation.

However, sooner or later living creatures would have used up these resources, since with any efficiency of reaction they would have been able to consume them faster than chemical processes could generate them, and living creatures would have faced an ultimatum: find another source of energy or die. They would have been forced to manufacture their own means of support and reproduction.

THE FIRST FORMS OF PHOTOSYNTHESIS

However, luckily there is another readily available source of energy: the sun. If you focus the sun's rays through a magnifying glass onto a container of water, you raise the water to boiling, and you use the steam to drive a motor, you have captured the energy represented by the light and heat of the sun and used it to do work, to drive the motor. If the motor manufactures an object, you have created order from the energy of the sun. It is possible to capture even more of the energy from the sunlight by not letting any light escape. You can do this by adding carbon to the water so that it is an opaque black. All pigments do essentially the same thing. They absorb some of the sunlight, reflecting back the colors that we see. Thus a green leaf absorbs red and blue, while it reflects back green. It does so by containing special pigment molecules called chlorophyll. (The name "chlorophyll" translates to "the green from the leaf".) The chlorophyll is special because normally we cannot bottle sunlight, but chlorophyll not only absorbs the light but hangs onto it long enough to move the energy elsewhere (Fig. 18.3). This is the first step of photosynthesis, capturing the energy of sunlight and using that energy to manufacture complex organic molecules from simpler molecules. In the best known instance, plants capture carbon dioxide from the air and combine it with water to form sugar:

$$6CO_2 + 6H_2O \rightarrow C_6H_{12}O_6 + 6O_2 \,^9$$

It takes energy to combine simple molecules into larger, more complex ones, and the genius of plants is that they can capture the energy of sunlight to do so. Animals live entirely as parasites on the plants, using the sugar of the plants as a source of energy to build their own molecules. There are a few organisms that dwell alongside volcanoes at the bottom of the ocean, capturing energy from high-energy molecules produced by the volcano, but other than these latter, all energy used by life on this planet originates in the sun.

Capturing energy from the sun does not mean that the chemical reaction is necessarily the one listed above. What is required is that, first, the sunlight be

---

[9] In chemist's symbolism, $6H_2O$ means six molecules of water, each consisting of two atoms of hydrogen and one atom of oxygen. $C_6H_{12}O_6$ means one molecule of sugar (here glucose), consisting of six atoms of carbon, twelve of hydrogen, and six of oxygen. Note that in the equation the numbers of atoms of hydrogen, carbon, and oxygen are equal on both sides of the equation.

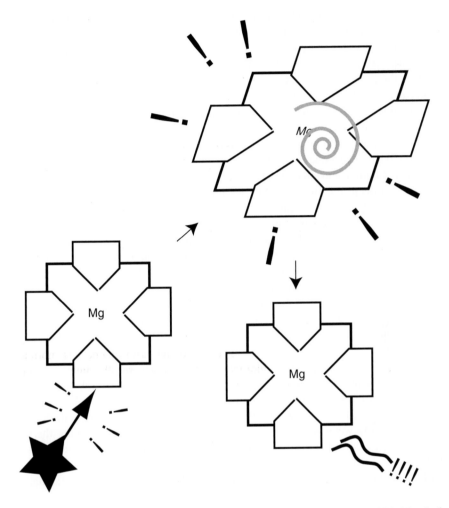

Figure 18.3. The role of chlorophyll in converting instantaneous (light) energy into storable (chemical) energy. A molecule like chlorophyll can be pictured as a large rubber sheet. When a particle of light slams into it, it stretches in a recoiling action. Moments later, it bounces back, releasing most of the energy in a form that can be captured and used to convert carbon dioxide and water into sugar

captured and second, that it be converted into a useful chemical product. The first step requires this special chlorophyll-type molecule which can, fortunately, be assembled spontaneously if conditions are right. Chlorophyll is a compound very closely related to the molecules that animals use to handle the oxygen (hemoglobin and cytochromes) used to burn the sugar, or carry it in the reverse reaction to that illustrated above back to carbon dioxide and water. This type of molecule is very good at moving energy. Its trick, as mentioned above, is to hang onto the energy long enough to use it for something else.

We all know that sunlight can do damage by, for instance, bleaching colored garments. What happens is that the pigment molecules absorb the energy, which breaks the molecules apart and causes them to lose their color. What is special about molecules like chlorophyll is that they can absorb the energy but then bounce it around, rather like a hot potato, until it can be released again without shattering the molecule. Light hitting the molecule is rather like a very fast bullet or ball slamming into a somewhat elastic net or cage. It bounces around, ricocheting around, until it loses enough energy to dribble out of the cage. You can see this property fairly easily. If you extract the chlorophyll from something like spinach into an organic solution like ethyl acetate (fingernail polish remover) and, in a dark room, illuminate the green solution with blacklight (long wavelength ultraviolet light), the solution will glow a deep red. Besides being an amusing trick for Halloween, it illustrates chlorophyll releasing the higher energy of the blacklight as lower energy visible red light. The absorbed photons are released in multiple steps, so that the total of energy gained equals the total of energy given back. So, the first success of organisms was to learn to use the properties of chlorophyll and similar molecules to capture light energy and hold it long enough to use it for constructive purposes. This is the 'photo' part of photosynthesis.

The second step is to turn the energy into a synthesized chemical product such as sugar, which can be stored and used for other purposes. This is the 'synthesis' part of photosynthesis. This would be rather like a bullet slamming into a paddle wheel and turning it—if you can capture the energy, you can turn it into useful work. For a plant, this means being able to use the energy to attach two molecules together.

We can interpret from among reactions that bacteria use today as well as from the geological record what the synthesis step must have been when photosynthesis was first achieved. We know that there was considerable hydrogen sulfide, H_2S, dissolved in the water at the time. Hydrogen sulfide is a gas producing the odor of rotten eggs, but otherwise is chemically similar to water, which differs only by having oxygen, O, in the role played by sulfur, S. Some bacteria, known as the sulfur bacteria, carry out photosynthesis using the following reaction (actually a series of reactions):

$$6CO_2 + 6H_2S \rightarrow C_6H_{12}O_6 + 12S$$

They use the energy sunlight to attach the hydrogen to carbon dioxide, synthesizing sugar. The leftover product is sulfur. Some of the bacteria alive today that can form stromatoliths, as seen in very ancient fossils, also can carry out this reaction. The resulting products are the highly ordered sugar—which can be used as fuel for other syntheses—and the element sulfur, which is a solid powder at normal temperatures. We also know that the sulfur deposits that are mined today are in very ancient rocks. These observations lead us to conclude that the earliest photosynthesis was of the sulfur-producing variety, and that it generated both the energy needed to sustain the rest of life and created the great sulfur deposits of the Andes and elsewhere in the world.

Sulfur photosynthesis, however, suffers from several embarrassments. First, the sulfur accumulates. In shallow lake beds, for instance, the accumulation of sulfur may eventually fill the lake, and the sulfur can react with water or other materials to produce truly noxious (poisonous) materials such as sulfuric acid. Second, hydrogen sulfide may be convenient for a while, but ultimately there is not much of it. If you ever have a chance to climb to the crater of an active volcano, you will encounter the foul smell of hydrogen sulfide, and you will retain a disagreeable but memorable recollection of the experience. Other than volcanoes and hot springs, thankfully H_2S is rare in our atmosphere.

PHOTOSYNTHESIS USING WATER INSTEAD OF HYDROGEN SULFIDE

This is the point. Hydrogen sulfide is generated in a few places and, once the original widely-distributed supply is gone, the marvelous invention of photosynthesis will be useless. Unless one can find a substitute for H_2S. Finding a substitute was, as both modern genetics and geology indicate, the next big achievement of plant life. They substituted water for hydrogen sulfide and produced the reaction that we know today:

$$6CO_2 + 6H_2O \rightarrow C_6H_{12}O_6 + 6O_2$$

(Despite the differences between the yellow powder sulfur and the gas oxygen, the two elements react chemically in many similar ways.) Water as a reactant has several advantages over hydrogen sulfide. Water is far more abundant than hydrogen sulfide, and the waste product oxygen does not precipitate at the bottom of lakes and convert to disagreeable products. It is a gas that bubbles away into the atmosphere. At the very low levels of oxygen accumulated by the first oxygen photosynthesizers, this must have been an ideal solution to the problem of diminishing amounts of H_2S, and the organisms that had achieved this success must have proliferated rapidly. It was the first major worldwide pollution, and it changed the planet. Microscopic examination of the most ancient rocks in Australia, approximately 2.8 billion years old, reveals structures very similar to modern cyanobacteria. The cyanobacteria are so named because they are bluish in color. The blue color is a form of chlorophyll, the primary molecule that can trap light to initiate the reaction. At approximately this time, chemical products in the rocks suggest that oxygen appears in the atmosphere.

One of the most startling indicators of this activity is the so-called banded iron formations or redbeds (Fig. 18.4). These ancient rocks show striping alternating between red and the base color. The red is iron oxides (rust). In the uncolored regions, iron is present but is not iron oxide. What appears to have happened is that alternately there was free oxygen, which would react with free iron, just as iron today will rust, and alternately the oxygen disappeared. It has been speculated, though it is highly unlikely, that this alternation was seasonal. Today there is so much oxygen in the atmosphere that any iron left exposed for any length of time turns to rust, but as the waste product oxygen was first being produced, much

Figure 18.4. Banded iron formation. The lighter colors are red, representing back-and-forth shifts of the availability of atmospheric oxygen. Such patterns are characteristic of only a limited period of the earth's history. Today evenly distributed iron would be completely oxidized. Credits: http://upload.wikimedia.org/wikipedia/commons/5/5f/Black-band_ironstone_%28aka%29.jpg This is a file from the Wikimedia Commons

of it may have immediately reacted with iron and other materials in the soil and water, and never reached the atmosphere. During this time the amount available may have waxed and waned with the seasons and the ability of microorganisms to maintain photosynthesis, as similar-sized bands in bacterial formations similar to stromatoliths are formed by seasonal fluctuations, or there may have been other mechanisms generating the cyclicity. We are not convinced that oxygen began to build in the atmosphere until approximately one billion years ago.

THE IMPACT OF AN OXYGEN ATMOSPHERE

The accumulation of oxygen in the atmosphere changed the earth in many crucial ways. First, as you know, oxygen is highly reactive: hence all the warnings about not smoking or creating sparks when high levels of oxygen are present, since it reacts with many types of molecules. Second, it releases a lot of energy when it reacts: things burn or explode. We can snuff out fires by depriving them of oxygen, or make them burn more vigorously by adding more oxygen. The energy that is released can be captured and used, whether to drive a motor or to drive chemical reactions that living things can use. Third, oxygen can provide an important shield for living creatures. Let's consider these in reverse order.

Every fair-skinned person is aware of the danger of sunburn, and even people with greater pigmentation can burn. Ultraviolet light can be absorbed by molecules, including macromolecules, and it can shatter them. If you imagine a molecule to be a balloon, hitting the molecule with a photon (particle) of ultraviolet light would be the equivalent of stomping on the balloon. Clothes fade in sunlight because the ultraviolet light shatters the pigment molecules. We use ultraviolet lamps to sterilize small areas because the UV shatters the DNA, RNA, and proteins of the bacteria, thus killing them. Some individuals, afflicted with a disease that prevents them from repairing damage from the sun, dare not expose their skin even briefly to sunlight, but must remain indoors and fully clothed at all times. This condition demonstrates that most people routinely suffer damage to the skin (shattered molecules) from sunlight, but repair virtually all of it. Imagine what it would be like to live on an earth that received somewhere between 50 and 500 times more UV. (The number is imprecise because, owing to cloud cover, elevation, and latitude, the amount of UV that reaches the earth at any one time varies.) In the early atmosphere, water could absorb UV and protect life a few centimeters below the surface of the water, but the land surface was under continuous bombardment by an extremely powerful germicidal lamp. Life on land was not possible in the early atmosphere.

Fortunately, when oxygen reaches the upper atmosphere, it can form an unusual form of oxygen, known as ozone, and the ozone absorbs the UV very effectively. It acts essentially as a colored filter, allowing only light less energetic than UV (of longer wavelength) through, much as a red filter will not let blue light through. The black-and-white photographs with dramatic dark skies were produced by using red filters, which prevented the blue light of the sky from reaching the film, while the white light from the clouds remained. Ozone is far from a perfect filter, but by blocking 98% or more of the dangerous UV, it made the land habitable. Land offers many opportunities for life, and this was a very important transition.

It was also a very important transition because products can be burned using oxygen. Cellulose, the main constituent of wood or paper, is a macromolecule consisting of a chain of sugar molecules. If it reacts with oxygen in a fire, it is converted into carbon dioxide and water, releasing the energy that we identify as fire and heat. Photosynthesis has been invented, accumulating sugars, and some organisms can become parasites on the photosynthetic organisms, taking their sugar for themselves. We know these organisms as non-photosynthetic bacteria, fungi, and animals.

These organisms have a few means of using sugar. One of these is simply rearranging the sugar molecule. The simplest version of this is the conversion of sugar, $C_6H_{12}O_6$, into two molecules of lactic acid, $C_3H_6O_3$, or to two molecules of ethanol (grain alcohol, C_2H_6O) and two molecules of carbon dioxide (CO_2). Note that in either case you still have 6 carbons, twelve hydrogens, and six oxygens. This is rather like releasing the spring on a small toy to make it jump. The sugar represents the toy in its cocked position, and the ethanol or lactic acid the toy with the spring released. You have released some energy, which you could use to perform some work. But let's make the toy plastic. If you burned it (reacted it with oxygen), it would release a lot more heat, which you could use to do more work. As a matter of fact, you know that the byproduct of rearranging molecules, ethanol, readily burns, releasing substantial heat. This means that there is still a lot of energy in the molecule. Burning ethanol is simply reacting it with oxygen: $C_2H_6O + 3O_2 \rightarrow 2CO_2 + 3H_2O$, and it releases a lot of heat, converting all of it to carbon dioxide and water. If we could do that, we could get much more energy from the sugar, actually about 24 times as much. Think of it: organisms that can ferment sugars get only 4% of the energy available in the sugar. Once oxygen becomes readily available, the organism that can learn to use it has a huge advantage. Thus some organisms managed to adapt a modified version of the chlorophyll molecule, which can absorb energy in the same fashion as chlorophyll but in this case not using light—think of it as containing an explosion—to carry out this reaction in slow motion. The energy that is released in burning sugar to carbon dioxide and water is the same whether one burns cellulose (paper, cotton, or wood, a polymer of sugar) or whether we "digest" sugar to carbon dioxide and water. We measure it in the same fashion, in Calories or Joules. A calorie is the amount of heat needed to raise one liter of water (about one quart) one degree centigrade (about two degrees Fahrenheit). Sugar releases approximately 4 Calories per gram whether it is burned with a match or digested. The heat of our bodies is the same heat as that produced by a fire. The difference is that living creatures do the burning in slow motion and controlling it so that the body never gets too hot, and as much energy as possible is captured to be used for such purposes as making new molecules (synthesis). Again, for obvious reasons, these organisms were enormously successful. They were so successful that, as the bacteriologist Lynn Margulis suggested, other creatures decided to hang out with them.

THE INVENTION OF THE EUKARYOTIC CELL

Eukaryotic cells are those that contain true, membrane-bounded organelles such as nuclei, mitochondria, and chloroplasts ("eukaryotic" means "with a good nut [in the center]", based on the appearance of a cell and its nucleus when viewed in the microscope). The organelles serve specific functions, since different enzymes work under different conditions and the membranes create compartments in which ideal conditions can be maintained for each process. Mitochondria, which are responsible for respiration, and chloroplasts, which conduct photosynthesis, are peculiar. First, these organelles are enclosed by not one but two membranes, with the inner membrane deeply folded so that it has far more surface area (Fig. 18.5). Second, although proteins can be synthesized inside these organelles, the ribosomes (page 228) on which the proteins are synthesized are unusual. Most of the ribosomes of eukaryotes are a bit different from those of bacteria, and some antibiotics are more toxic to bacteria than to us because of this difference, but the ribosomes of mitochondria and chloroplasts are more similar to those of bacteria than to the other ribosomes inside the cell. Third, as was finally understood in the 1960's, mitochondria and chloroplasts have their own DNA and divide when the cells divide, so that each

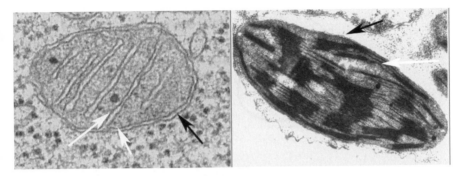

Figure 18.5. Mitochondria, chloroplasts, bacteria. Upper: Mitochondrion (left) and chloroplast (right). The outer membranes are indicated by black arrows and the inner membranes by white arrows. Credits: Mitochondria - http://web.mit.edu/esgbio/www/cb/org/mito-em.gif origin unknown, Chloroplast - http://www.nrc-cnrc.gc.ca/education/images/bio/gallery/pl_chloroplast.jpg

daughter cell has similar numbers of these organelles. Also, most mitochondria are similar in size to bacteria. Finally, some primitive eukaryotes lack mitochondria, but have symbiotic bacteria living with them. Putting all this **evidence** together, in 1966 Margulis produced a **logical** if very startling hypothesis, that mitochondria and chloroplasts were descended from respiring and photosynthetic bacteria that larger cells had taken inside of themselves to benefit from their activities! This arrangement is known on larger scale and called a mutualistic arrangement, as lichens are combinations of algae and fungi, and most animals cannot live without the products of helpful bacteria that live in their digestive tracts. However, in this case Margulis proposed that ancient cells took up these bacteria into vacuoles, or membrane-bound organelles, where the cells delivered precursor products and the bacteria, which lived happily inside the vacuoles, returned their products for the benefit of the larger cell. See Fig. 18.6. Since Margulis first proposed this hypothesis, it has become possible to sequence DNA, and the similarity of mitochondrial and chloroplast DNA to that of certain bacteria is so close that the relationship seems extremely convincing. This later evidence **falsifies** almost all alternative hypotheses. Mitochondria and chloroplasts are nevertheless very different from their original forms, and cannot survive outside of cells, but most scientists are convinced of their origin from bacteria. The mutualism was so successful that the world is now dominated by eukaryotic creatures. The origin of the nucleus is far more speculative and far less certain, but remains an interesting scientific question.

THE ORIGIN OF MULTICELLULARITY

There are many obvious advantages for the cooperative interaction of many cells. One very important advantage is efficiency. In some of the simplest organisms such as a colonial alga, some cells specialize in the photosynthesis that provides food for the organism while others take over the primary responsibility for producing the

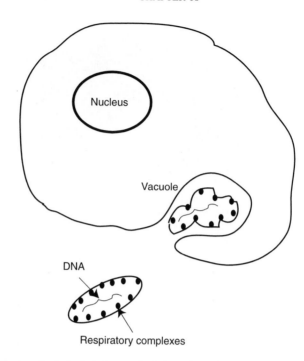

Figure 18.6. The hypothesis is that photosynthetic bacteria, and bacteria that could handle oxygen, lived commensally with larger cells, ultimately evolving to be come completely dependent on the host organism as, respectively, chloroplasts and mitochondria. Because oxygen is so reactive, molecules handling it are embedded into the cell membrane of bacteria, where they can be held in place and their reactions controlled. The bacteria are engulfed by cells that cannot handle oxygen, but they are not eaten. Their cell membranes ultimately become multiply folded to increase their capacity to process oxygen, but they retain their DNA and ability to divide within the cell. The outer membrane is the old membrane of the vacuole, and the inner membrane is the old bacterial cell membrane. Ultimately, some of what they need is produced by the host cell, and they no longer can live independently

new generation (Fig. 18.7). In an example of inefficiency, the simple multicellular organism Hydra, a freshwater relative of jellyfish, has partially specialized cells. What makes its muscle is not a true muscle cell but a cell that contains muscle components but also carries out other functions more typical of cells on the outside of the animal, called epithelial cells. The cell is called a musculo-epithelial cell. As is seen in Fig. 18.8, the cell can contract, but it has to drag along the non-contractile bag of the epithelial component, which incidentally is attached to other cells. It is a rather awkward arrangement. Thus animals and plants that have truly specialized cells, which can perform complex functions (digest, contract, send signals, excrete) very efficiently, but which depend on other cells for their support, have many advantages over multi-purpose, jack-of-all-trades cells. Also, larger size creates the advantage of having more resources to deal with and survive intermittent problems such as lack of food or water. Thus larger, multicellular creatures can do better in variable environments without having to go into deep hibernation or

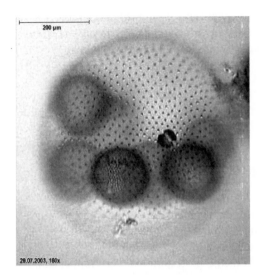

Figure 18.7. Volvox, a colonial alga. Each spot on the larger sphere is a single cell of the organism. The inner balls are collections of cells that have separated from the outer sphere to become the reproductive cells for the next generation. The outer cells will eventually die. This is one of the simplest structures for a multicellular organism. Credits: Micrograph of Volvox aureus. Copyright held by Dr. Ralf Wagner, uploaded to German Wikipedia under GFDL

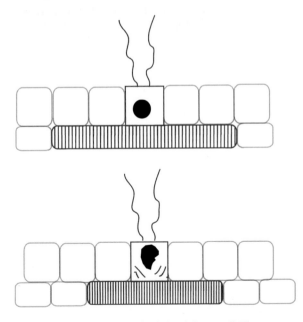

Figure 18.8. Musculoepithelial cell in a hydra. The dark unit is one cell. The upper part faces the outside of the animal and has cilia to stir up water and draw in prey. The lower part works like a normal muscle. When the muscle contracts (lower picture) the epithelial part gets pushed around and is under strain. Higher organisms have separated functions for greater efficiency

make other drastic arrangements. Presumably because of these advantages, although single-celled organisms, particularly bacteria, are enormously numerous, multi-celled creatures are now very prominent in our world.

REFERENCE

Orgel, Leslie (1998) The origin of life. A review of facts and speculations. Trends in Biochemical Sciences 23: 491–495.

STUDY QUESTIONS

1. How do living organisms not violate the laws of entropy?
2. Is there more than one type of photosynthesis? Explain.
3. Does photosynthesis differ from fermentation? Explain.
4. How similar are photosynthesis and respiration?
5. What is the advantage of respiration over fermentation? What is the disadvantage?
6. What is the evidence that mitochondria and chloroplasts descended from bacteria? What is the evidence against the hypothesis?
7. Would you think that any other cellular organelles could have arisen by symbiosis? Why or why not?
8. Do you think that there might be evidence today for a commensalism that might be evolving toward complete symbiosis? What might that be?
9. Is there any advantage to being multicellular? What might that be?

CHAPTER 19

THE CONQUEST OF LAND—EVERY CRITERION FOR THE CLASSIFICATION OF THE MAJOR GROUPS OF ANIMALS AND PLANTS REFERS TO ADAPTATIONS FOR LIFE ON LAND

If the land held such potential for the evolution of all forms of life, why did it take so long (approximately 9/10 of the earth's history) for organisms to conquer the land? There are many hypotheses, but most of them in one way or another relate to the presence of oxygen in the atmosphere. We have many means of estimating how much oxygen was in the atmosphere at various ages of the earth, and we can interpret the appearance and physical characteristics of the land in the presence or absence of oxygen. The principle for estimating oxygen in the atmosphere is very straightforward. If you leave a piece of iron in the air, particularly in humid conditions, it will rust. Rust is the combination of iron with oxygen. If you keep the iron in an atmosphere of pure nitrogen, or prevent oxygen from reaching it by means of paint or oil, it will not rust. Iron rust, frequently seen as red earth, means that iron was exposed to oxygen. Other metals can also rust, turning different colors. For instance, one For instance, one equivalent of rust equivalent of rust for aluminum is a black powder. Also, crystals of stones form differently in the presence of different amounts of oxygen. Each of these reactions occurs at a different threshold, meaning that below a certain level of oxygen the reaction will not occur. By reading these various chemical reactions, we can establish the approximate levels of oxygen in the atmosphere at any given time. For instance, today the amount of oxygen in the atmosphere is a bit above 20%. It began to accumulate quite slowly and then increased rapidly.

Oxygen began to build in the atmosphere approximately 2.5 billion years ago and reached approximately 50% of current levels about 500 million years ago (Fig. 19.1).

What is particularly interesting is what is called banded iron formations. In various ancient sedimentary rocks, ranging from 3 billion years to a bit less than 2 billion years old, there is a peculiar fine striping, black or red rust between lines of sedimentary rock containing unrusted iron (Fig. 18.4). What must have happened is that the amount of oxygen varied during the time that the sediment

271

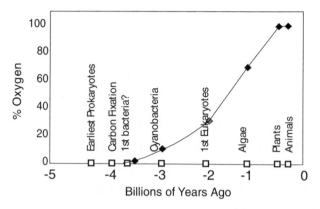

Figure 19.1. O$_2$ in the atmosphere. Chemical indicators suggest that oxygen first appeared in the atmosphere over 3.5 billion years ago. Large organisms did not proliferate on earth until atmospheric oxygen levels had increased to levels nearly equivalent to those seen today (21% of the atmosphere)

was deposited. One can speculate on the mechanisms producing this effect, but the most logical and consistent one is the following: Oxygen would get into the atmosphere only via photosynthesis, meaning that photosynthesis was functioning at that time. The banding pattern represents a seasonal (or most likely other) cyclical variation in photosynthesis. During alternate seasons or periods, photosynthesis was lower, and the available oxygen was used up or diluted further into the atmosphere. Finally, approximately 1.8 billion years ago, oxygen was definitively in the atmosphere but the raw iron available in the sea water and sediment was essentially fully oxidized and the banding ceased. Whether or not this interpretation is correct, the banding is remarkable and indicates that there was an active photosynthesis three billion years ago. In other words, there were living photosynthetic bacteria.

Photosynthesis offers a further advantage, in that both alcohol and lactic acid are soluble in water and will eventually make the water toxic to life—beer and wine stop fermenting when the alcohol level poisons the yeast that produced it— whereas carbon dioxide can escape into the atmosphere. Therefore, if organisms can use oxygen, they have the potential to do more, producing far more energy with lower requirements for food and less risk of poisoning themselves, than organisms that cannot use oxygen (Chapter 18). There is a cost, since oxygen is still a very reactive molecule and can react with (or burn) other molecules in the body. Antioxidants are used in paints, to protect the metal underneath, and in many manufactured foods, to prevent the food from developing a "stale" taste. The argument that oxidation creates damage is the primary argument on which the manufacture and sale of nutritional antioxidants is based. (As a practical matter, ingested antioxidants rarely reach the sites in the body where they might do some good.) If organisms can manage to avoid the damage produced by oxygen, they have the potential of living far more efficiently than organisms that cannot use oxygen.

In any event, oxygen is accumulating in the atmosphere, and all organisms must evolve means of dealing with it. A few organisms do not. The bacteria that cause food poisoning, tetanus, and gangrene depend on living in the complete absence of oxygen and are rapidly killed by exposure to it. Thus oxygen in the air provides the possibility of life capable of producing far more energy and thus living at a more energetic pace.

PROTECTION FROM UV

The other effect of oxygen is readily seen by extrapolating from other situations that most of us have seen. We are generally familiar with the fact that human skin can be burned by excess exposure to the sun, and most people are aware that it is the UV part of the spectrum that does the most damage. (Fig. 19.2). We also know that a risk of sunburn is a type of cancer known as melanoma. Melanomas are produced by mutations of DNA, from which we learn that UV can damage DNA. We deliberately exploit this possibility when we use UV lights to sterilize surgical suites or surfaces that we need to keep sterile. The UV kills bacteria by destroying their DNA.

The extrapolation is that the amount of UV reaching the earth at present is 1% or less of the UV that could reach the earth, and our UV lamps produce miniscule amounts of UV compared to the sun. The UV coming from the sun is substantially blocked by oxygen, which is excited (activated) in the upper atmosphere to form a version of oxygen called ozone. The ozone absorbs UV light and prevents it from reaching earth. This subject is commonly in the news, since human-produced chemicals can also react with and destroy the ozone. You can imagine what the surface of the earth would have been in an oxygen-free atmosphere. It would have

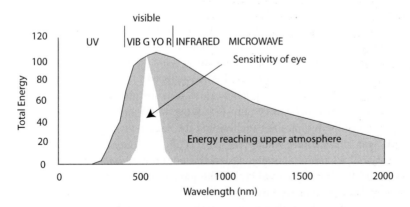

Figure 19.2. Light reaching the earth. The sun emits a wide spectrum of energy. Primarily because of the ozone in the upper atmosphere, most of the destructive high energy emission (short wavelengths, to the left) that reaches the upper atmosphere is blocked. Approximately 99% of the gray area to the left of the white area, which indicates what we can see, does not reach the surface of the earth

been one vast sterilization chamber, bombarded every day by intense ultraviolet radiation. Anything that ventured onto land would have been, quite literally, fried. Life on land, based on DNA, was impossible until sufficient oxygen built up in the atmosphere to block the UV. Although the amount of UV, particularly the most damaging form of UV, UV-B, that is absorbed varies with time of day, elevation, latitude, and cloud cover, today approximately 99% of it is blocked by the ozone, and an even more damaging form, UV-C, does not reach the surface of the earth. Life on land was not possible before photosynthesis.

One of the most interesting arguments that we can make is that we can almost always group the animals and plants that were first defined by the Cambrian explosion (page 277) into categories that we know today, such as arthropods (jointed-footed animals such as crabs, lobsters, and insects), vertebrates and, among vertebrates, fish, amphibians, reptiles, birds, and mammals. In plants, we recognize such groupings as algae, mosses, ferns, evergreens, and flowering or broad-leafed plants. Many of the finer distinctions arose much later, but they represented continuation of trends established by the earliest groupings of animals and plants. What you may not appreciate is that essentially all of the criteria that we use to distinguish among these categories describe one or another adaptation to life on land. Compare a large alga such as a kelp—the large-leafed seaweed—to a spruce tree. The kelp's life is relatively easy. Though one end may attach to a rock, it floats in the water. It needs no serious support structure. Since the leaf is thin, everything that it needs can diffuse into each cell and all of its wastes can diffuse out. To reproduce, it releases its gametes (a type of egg or sperm) directly into the water, where the different gametes will hopefully find each other. Since it exists in a solution of unchanging salinity (the ocean) it does not have to worry about having too much salt or too much or too little water. Contrast this with the situation for a spruce tree. To reach the sunlight, it must have a rigid structure to support itself against gravity. This will require a fairly bulky structure, which means that nutrients and waste products will not diffuse easily but must be transported in a specialized circulatory system. It will gather its water through its root system, which is also structured to support the tree in wind, but it may encounter flooding or drought. Therefore its needles and parts exposed to air must be sealed against too much water loss, and it must have means of adjusting its intake of water; it can easily drown or wilt. The water must be transported up to 60 feet into the air to be delivered to the uppermost needles, and the sugars produced by photosynthesis must be sent to the roots, which receive no light and cannot photosynthesize. The female part of its reproduction can hold the fat, nutrient-containing gamete (seed) to await fertilization by a smaller, mobile gamete (pollen) but the pollen must be capable of being transported by wind over great distances, and survive potentially drying out. Each problem that it faces must be dealt with by a different adaptation. One can carry out the same logic of reasoning for invertebrates (earthworms, mollusks, sea-dwelling arthropods such as lobsters or land-dwelling arthropods such as insects, and many lesser-known creatures) or for vertebrates (fish, amphibians,

Table 19.1. The major phyla of plants and animals

Phylum or family	Circulation	Respiration	Renal	Reproduction	Skin	Skeleton
Fish	2-chambered heart	Gills	Pronephros; excrete ammonia	Usually external; eggs -> fry -> fish	Scales, permeable	No true limbs
Amphibia	3-chambered heart; nearly complete separation into 4-chambered	Small lungs, skin	Mesonephros; excrete urea	External, unprotected eggs -> tadpoles -> adults	Moist, permeable or dry, semipermeable	4 limbs out to sides; some hop
Reptiles	3-chambered heart	Lungs	Mesonephros; excrete uric acid	Internal, hard shelled egg	Dry, impermeable, scales	4 limbs out to sides
Birds	4-chambered heart	Lungs; air sacs	Mesonephros; excrete uric acid	Internal, hard shelled egg	Dry, impermeable, feathers	4 limbs; wings and aligned legs
Mammals	4-chambered heart	Lungs	Metanephros; excrete urea	Internal, placental, no yolk	Dry, impermeable, hair or fur	4 limbs, aligned with body axis
Algae	Diffusion	Diffusion	Aquatic	Permeable	No skeleton	
Mosses	Diffusion	Diffusion, circulation	Wet areas, mobile gametes	Permeable	Modest vertical support	
Ferns	Tracheal system	Tracheal system	Minor semi-aquatic haploid phase, mobile gametes	Semipermeable	Tracheae, support structures	
Pines	Tracheal system, phloem & xylem	Tracheal system	Pollen and hidden seeds	Impermeable	Tracheae, support structures	
Flowering plants	Tracheal system, phloem & xylem	Tracheal system	Tube pollen and flowers	Impermeable	Tracheae, support structures	

reptiles, birds, mammals). These groups and their solutions are described in Table 19.1.

The fact that the major means that we have for grouping animals and plants relates to surviving on land might be an artificial intellectual construct. For instance, we might have classified animals according to whether they swam (fish, sharks, penguins, dolphins, and squid), flew (birds, insects, bats) or walked (frogs, lizards, and most mammals). However, these classifications do not explain many features of each animal and, today, we can show by DNA analysis that, for instance, penguins are birds and are not related to dolphins. Thus our groupings appear to have some validity. What this suggests is that access to land was an important step in the evolution of the organisms that we see on earth today. There were several independent solutions to each of the problems (an insect has Malpighian tubules, which are very different from vertebrate kidneys, while reptilian, bird, and mammalian kidneys each differ from each other in very pronounced ways) and those solutions that worked provided the ancestors of the animals and plants that we know today. Those animals and plants that got a foothold on the land found a vast and highly varied new territory in which to evolve, and each of those successful ancestors underwent an adaptive radiation into the land (see Chapter 24, page 335). We conclude that the new accessibility of land provided an outstanding opportunity to those animals and plants with the ingenuity to find ways to handle the difficulties of support, internal circulation, water retention and removal, and reproduction; that several solutions worked very well; and that we see today the descendents of those creatures that found the original solutions.

Oxygen in the atmosphere changes the earth in two major fashions: it provides a new, rich source of energy, and it makes the land habitable. Once the land is habitable, the first life on land transforms the earth in even more remarkable ways.

Once oxygen made the land habitable, the land was further transformed in another very important fashion. You have noticed, when the sun comes out after a summer rain, that earthworms that happened to be venturing onto a sidewalk or driveway when the sun reappeared often dry out and die. Or you may have tried to cross a driveway, street, or beach in your bare feet on a hot summer day. The ground is extremely hot, hot enough to burn you. You may have leaped off the driveway to the grass alongside. The grass is much cooler and refreshing. Even a thin layer of grass or moss is far cooler (Fig. 19.3)That is because living things contain water, and water can control heat and heat transfer far more comfortably than dry surfaces. Likewise, the first life that could establish itself on land, even a slimy coat of bacteria, would have made the surface of the earth far more tolerable than a dry, rocky surface. These creatures would have lived and died, and their remains would have accumulated and made the layer of water-containing organic material deeper and capable of supporting a more varied life. We see this kind of progression today, when life comes back to a burned-off land, or, more cogently, when living organisms begin to populate new land such as the lava from a recent volcanic eruption. At first the lava is very inhospitable to life, but once small plants such as mosses and lichens take hold, enough water is retained to sustain

Figure 19.3. If you were walking barefoot across a rocky terrain in full sun on a hot day, on what would you rather step? Even a very thin layer of moss can make a substantial difference in the surface temperature and in the ability of the substratum to transfer heat. Thus smaller organisms can survive on the moss where they cannot on the rock surface, and when they die they will contribute to the growing mass of soil building on the surface

slightly larger plants through the hottest and driest moments. Small animals can now live among these plants, and soon (over the course of a few years or a few tens of years) this previously harsh terrain is softened by an extensive assortment of life.

WHAT CAUSED THE CAMBRIAN EXPLOSION?

This is one interpretation of what it took for what is described as the Cambrian explosion—the apparently sudden proliferation of a wide variety of life, including most of the broad classifications of animals and plants that we know today, after a very long period in which life existed, but in very modest form. To recapitulate: in the approximately 4.5 billion year existence of the earth, we can detect evidence of life beginning approximately 3 billion years ago, or almost as soon as the earth was cool enough to tolerate organic molecules. Nevertheless, living organisms apparently did not become much more complex than photosynthetic bacteria and simple algae until less than 500 million years ago, and then by 350 million years ago, life had generated large numbers and multiple forms of plants and animals culminating, approximately 150 million years ago, in the early appearance of modern forms such as flowering plants and mammals. In terms of the history of the planet, the bursting forth of life occurred in a relative instant, and the question is, was life really relatively somnolent until this time, and what happened to kick evolution into high gear? One possible explanation is that land became accessible for life at this time. There are other possible explanations, including the development of specific types of genes (see page 283). There are no "correct" explanations for the Cambrian explosion; interpretations that we have are speculations based on the evidence that we have and logical interpretations of the evidence, but there is no criterion of falsifiability to solidify the hypothesis into a theory.

REFERENCES

Campbell, Neil A. and Reece, Jane B.(2004) Biology (7th Edition), Benjamin Cummings, Boston, MA

STUDY QUESTIONS

1. Why do we worry about ultraviolet light reaching the earth? What stops it?
2. Look at any plant or animal within your range. Can you identify how it resolves its problems of water gain and retention, support, and reproduction?
3. If you can find a rock that has a bit of moss on it, place it in the sun on a summer day. After about half an hour, measure the temperature at the surface where the rock is exposed and at the surface of the moss. What do you find?
4. On a sunny summer day, moisten a bit of peat moss and sprinkle it in a thin but complete layer on an asphalt driveway. Before the moss dries out, try to walk across the driveway where it is exposed, crossing onto the moss. What do you find?
5. If you live in an area where you can see lichens, look at them very closely and see how many other types of organisms live among the lichens.
6. Was there really a Cambrian explosion? Defend your argument.
7. If you feel that there was a Cambrian explosion, what do you think caused it? Defend your argument.

CHAPTER 20

THE GREAT AGES OF OUR PLANET

Geologists can easily distinguish various types of land by its characteristics, such as whether or not it was created by fire and melted (lava) or by the gradual settling and compaction of sand in a large body of water (sedimentary rock). Perhaps not so surprisingly, they also recognize characteristics of land that have been determined by life. The most obvious change is the sudden appearance of large amounts of organic carbon, producing black soils, beginning between 500 and 350 million years ago. See page 222 for the definition of "organic". Very simply, prior to that time there was not a large mass of living organisms leaving their traces on the earth. Beyond that startling change, other forms of land can be identified by the presence of huge amounts of the shells of shellfish (such as the limestone cliffs of the English side of the English Channel—the famous "white cliffs of Dover"), or remnants of ferns and other plants (the large coal beds of the world). Within these larger categories, the types of shells and other remnants are identifiable and distinguishable. For instance, in some soils and rocks we can clearly identify the fossil bones of various types of amphibia and insects, but no reptiles, birds, or mammals. In others, there are far more reptiles than amphibia, but no birds or mammals. In yet others, there are some but not many reptiles and amphibia, and many birds and mammals. These different layers of rock are found worldwide and in soils of different types, but are so consistent that the fossils served as the first means of comparing the soils and rocks of one country to those of another. For instance, they served as the basis for determining the probability of finding oil or other minerals. The characteristic types were given names that served to identify a specific era in the history of the earth. The names were derived from Latin and were either basically descriptive (Carboniferous = coal-bearing) or were chosen from the first or most prominent region from which the era was standardized (Devonian = from the region of Devonshire, England; Cambrian = from Wales [Cambria in Latin]; Jurassic = from the Jura region of Switzerland; Mississippian and Pennsylvanian, for formations found in these regions of the continental United States, etc. In a more fanciful fashion, sometimes names were given to commemorate ancient tribes believed to have inhabited the land [Ordovician, Silurian]). Using Steno's rule, that in sedimentary deposits the deepest deposit is the oldest, geologists assembled the geological strata, or eras, into a sequence, though, of course, the actual timing remained in question. The sequence indicated a striking series of events. In the oldest strata of the earth, there appeared to be no life. (This interpretation has been revised,

279

as is described on page 281.) Then, suddenly, at a point known as the Cambrian, life proliferated. This era was characterized by many animals and plants, but was most spectacularly identifiable by the presence of fossils of amphibia and primitive fish. After several variants on this theme, the Age of Amphibians was superseded by an era characterized by predominantly reptiles, some of gigantic size. Some but not all of the giants were the dinosaurs. In the most recent layers, the giant reptiles were gone, but mammals and birds had joined the fossil record. These eras were then given Greek names to define them: The Age of Amphibians was the Paleozoic ("old animals"); the Age of Reptiles was the Mesozoic ("middle animals") and the Age of Mammals, our current age, is the Cenozoic ("new animals"). Of course there were characteristic plants for each era—the ferns of the late Paleozoic are the coal fields of the Carboniferous (a subdivision of the Paleozoic), while flowering plants did not appear until the Cenozoic—but since we are animals, we tend to emphasize the animal names. A list of the primary geologic eras, their characteristics, and the translation of their names, is given in Table 19.1 and the eras themselves in Table 19.2.

The various types of strata had been recognized by Nicolas Steno at the end of the 17th Century, when he demonstrated that strata were similar over extensive areas of Europe and could therefore be easily compared. In the 18th Century Abraham Werner refined Steno's observations by dividing the categories of the earth's crust into four groups, appropriately if simply named Primary, Secondary, Tertiary, and Quaternary types of rocks, going from the oldest (deepest) toward the youngest. Alexandre Brogniart and Georges Cuvier in France and William Smith in England were using fossils to subdivide the major categories, recognizing as well that these subdivisions based on biology were valid for France, Italy, Germany, England, and Wales (the latter being an area from which many ancient rocks were taken.) Ultimately the generalization was extended to the New World, where, it could be seen, similar strata were easily identifiable. By 1850—before Darwin published *Origin of the Species*—the major subdivisions of the eras (periods) were defined, with the names of the periods (including "Mississippian" and "Pennsylvanian") reflecting the consensus that the geological patterns, including their fossils, represented changing conditions on a worldwide basis. The timing, however, was a matter of speculation. A river might dig through the rock where it is flowing fast and deposit the sediment slowly in a more slowly-moving stretch, but in a single major flood it might deposit a foot of sediment. A volcanic eruption might deposit anything from a dusting to ten feet of ash, or lava flows of equally broad range. Thus it is impossible to read dates from the order of layering alone. It is plausible to say that currently, the river extends the delta by an inch a year, so that a delta one mile long is approximately 65,000 years old (there are 63,360 inches in a mile). It is however equally plausible to argue that the delta was almost entirely created by Noah's flood approximately 6000 years ago and has variably expanded since that time. It was not possible to establish accurate time scales until measurements could be made of radioisotopes (Chapter 8), which are formed by mechanisms that are not affected by temperature, concentration, or presence of other chemicals. However, there were many ways to try, and many scientists set about to make the assessment. These methods are described in Chapter 9, page 114.

We now have an accurate assessment of these eras and periods, and our astonishment remains: each of the great eras represents a very different type of life on the planet, and the transition from one era to the next was, in geological terms, rather sudden. Why was life so different at each stage, and what caused the sudden and dramatic changes? In order to examine these questions, we need to understand at least in broad terms what the eras were, and how they differed.

THE PRECAMBRIAN

Until approximately 40 years ago, the Precambrian period (from 4 billion up to approximately 2.5 billion years ago) was considered to be a period without life, during which chemical reactions produced organic molecules (Chapter 15, page 222) and in general the earth, having cooled to a point where life was possible, awaited the appearance of the first living organisms, which burst forth in huge variety as the world entered the Cambrian era. There was much evidence for this occurrence, as indeed there was a sudden and startling increase in living organisms (Chapter 19, page 277). In fact, it was argued that the late appearance of life on this earth (570 million years of 4 billion years, or in the last 12% of the earth's history) was evidence in support of the complexity and enormous difficulty of life's appearing on a planet. Such arguments were presented to defend the proposition that the creation of life was an extremely difficult task and consequently extremely rare in the universe. More current interpretations are that life arose quite early and is not likely to be so rare in the universe. Thus the original argument has been turned on its head. How could such a change occur? As always in science, evidence trumps theory, and ultimately new evidence forced the re-interpretation.

First there was the logic. The logic consisted of two parts, one intellectual and one practical. From the intellectual standpoint, there was the problem that at the beginning of the Cambrian era, most of the major types of life—vertebrates, arthropods, mollusks, starfish, to address only the animals—were already present. Where did they come from? Did they not evolve from other, simpler, organisms? How could it be that so many different types of plants and animals appeared suddenly and in profusion? This intellectual question led to the realization of an extremely practical problem: that fossils might have existed, but not be found or recognized. Darwin and later scientists had recognized that the formation of a fossil was likely to be a chancy affair. An organism would have to die and sink into a muddy environment to protect it from bacterial destruction or crushing, in one scenario; it would have to have sufficient rigidity to hold its shape in the mud, or a skeleton or shell that would survive; the mud would ultimately have to solidify into rock that would be buried and remain undisturbed by earthquake, volcano, or erosion, until one of these processes brought the rock back to the surface again, to be found and identified by limited number of appropriately skilled people during the last two hundred years of our existence. A slug or an earthworm would probably decay before being fossilized, or otherwise produce an unidentifiable record.

What happened in the mid-twentieth century was the realization that the latter situation was in fact likely for early organisms. The Cambrian animals had large shells, clearly used for aggression or defense, thus bespeaking a highly competitive environment. But the first life appearing on earth was by definition not competing against other organisms, and therefore was unlikely to have been selected for bulky, metabolically expensive armor. Early life was not likely to have been shelled. Furthermore, if living organisms developed from non-living emulsions or coacervates (Chapter 17, page 251), there was no reason to assume that these organisms would necessarily be large enough to see with the naked eye, or macroscopic. Perhaps early life was microscopic and soft. It would therefore be necessary to search not for fossils but for signs of life.

There are many potential signs of life, including some of the following:

- Organic molecules can be formed in "left-handed" and "right-handed" forms. The difference is that molecules can have the same number of atoms arranged in the same way, except that their relationship is that of a left glove compared to a right glove. Chemical reactions produce equal numbers of both kinds, but living organisms vastly prefer one kind, such as the left-handed version of amino acids that are the basis of proteins. Organic molecules are found in various soils and rocks. If the left- and right-hand forms are equally common, they were most likely formed by strictly chemical processes. If one type predominates, they were most likely formed by living organisms.

- Although the differences between water (H_2O) and heavy water (also H_2O, but with each hydrogen containing an extra neutron) are extremely small, living organisms can discriminate between the two. Heavy water exists naturally in nature but in very low amounts. Water is often incorporated into molecules being made by living things. If these molecules contain less heavy water than occurs naturally, they probably were made by living organisms.

- Although many organic molecules can be formed by strictly chemical means, some, such as molecules like cholesterol, are not spontaneously formed by any non-living process known. Thus, if some of these molecules are identified in the sediment, the sediment probably contained living organisms.

- In all known plausible reactions other than photosynthesis, free oxygen is consumed rather than produced. If oxygen increases in the atmosphere, its origin is most likely from photosynthesis.

- Tiny droplets in emulsions or coacervates come in a range of sizes, whereas living organisms such as bacteria are uniform in size. If microscopic droplet type structures in rocks are very uniform in size, and particularly if they contain carbon, or if they form more oval or elongate shapes rather than spheres, they may have been produced by bacteria.

• Large constructions may be formed by microorganisms. For instance, certain types of bacteria found along seashores tend to accumulate in large piles, and by secreting calcium salts into the water, produce large stone-like structures called stromatoliths. Characteristic structures such as these can be identified in fossil-bearing rock. (See Chapter 18, page 257)

Once these issues and characteristics were fully understood, and radioisotope dating (Chapter 8, page 104) was recognized as sufficiently accurate to identify the ages of rocks dating to before the Cambrian era, a new search of these rocks was undertaken, by electron microscopy and other techniques. Electron microscopy can enlarge images to over 200,000 times, and thus provide detailed pictures of objects far smaller than can be seen using even the best light microscope. (Because of the limit of resolution of the human eye and the physical characteristics of visible light, the theoretical limit of resolution is approximately 2,000X or 1/10,000 of one millimeter. The theoretical limit is not easily obtained, and the best one can often do is to obtain an image of a dot or a small rod that represents a bacterium, but with no internal detail.) Various groups also analyzed the types of molecules found in carbon-bearing rocks, as well as the distribution of isotopes such as deuterium, the basis of heavy water. As the data from these types of analyses came in, two major themes emerged: first, there was considerable evidence for life well before the Cambrian era; and second, that a few previously known old rocks, such as some in Greenland, China, and Canada (the Burgess Shale in Canada, which was first recognized as fossil-bearing rock by Charles Walcott in 1909) unequivocally dated from an era prior to the Cambrian era.

As is discussed below, these fossils were quite remarkable in many ways, but they also raised many questions. First, if the various data were to be believed, life first appeared much earlier in the earth's history, so early in fact that it could be interpreted as a rather easy or probable event. In this case it might be very likely that some form of life might be found on an appropriate planet elsewhere in the universe. Thus interplanetary probes and landers such as the Mars Lander explore for chemical and microscopic signatures such as those mentioned above. The other issue was the conversion of the question, "Why did it take so long for life to arise?" to "If the earth is 4.5 billion years old, and took a few hundred million years to cool to a temperature at which life could exist, and life appeared as early as 4 billion years ago, what delayed the proliferation of many varieties of large, complex animals and plants for 3½ billion years?" We have many theories, based mostly on logic but with some evidence (drawn from extrapolating backwards from the relationships of animals and plants) to support the theories. As is explained in Chapters 17–20, the primary theories include:

1. The invention of homeotic genes, allowing animals (but not plants, which use different mechanisms) to develop complex structures with head and tail and back and belly, and appearance of oxygen in the atmosphere.

2. Available oxygen rendered possible oxidative metabolism, which was capable of yielding far more energy than fermentation and most importantly blocked the brutal UV that bombarded the planet and made it possible for living organisms

to move onto land. Thus, the theory goes, for almost four billion years life consisted of bacteria, algae, and similar forms, living in the seas and at most along shorelines that were usually below the tideline.

The end of the Precambrian

Then, approximately 600,000 years ago, the world began to change. And what a change it was! What happened was originally most clearly illustrated in the Burgess Shale. Shale is mud hardened into rock, often splitting into sheets because the mud accumulates at different times, for instance during big floods, and other material accumulates on the mud between the floods. The Burgess Shale, now in the Rocky Mountains of western Canada, was once a muddy flat subjected to such flooding. The floods and mudslides buried what shore life was living there, encasing the creatures in the mud that eventually hardened into the shale. If one cracks apart the shale, there are many strange forms found in it. The forms are clearly biological in origin, as we know of no chemical or physical process that could produce them. Although none truly resemble anything found on earth today, some are similar enough to animals known today, or to later fossils that are themselves similar to modern creatures, that we can classify them. For instance, there are shells that are related to the shells of the fossil trilobites, which are tolerably similar to today's horseshoe crabs. Thus there is no reason to assume that these fossils are not primitive shellfish. Others can be compared to sea urchins and starfish, or to some modern marine worms. Still others look like no creature seen or imagined by humans. In fact, scientists who tried to identify them christened one with the name Hallucigenia! One has three eyes. Another is a bunch of feathery fronds—gills— that do not seem to connect to anything. Another is, perhaps, a walking worm with spines along its back (Fig. 20.1). The shale formation is now reliably dated as being approximately 505 million years old.

What we have learned from this formation, as well as from similar formations in Greenland, Australia, and China, is three things:

1. There was considerable life well before the Cambrian.
2. Nevertheless, the varieties and numbers of creatures (the total living mass of the earth) began to increase rapidly in the later pre-Paleozoic period. Without living creatures, the element carbon is found mostly as the gas carbon dioxide and as its equivalent after reaction with other atoms, such as carbonates. (Chalk is calcium carbonate, and many semi-precious stones are other types of carbonates.) When living creatures die, their organic molecules ultimately deteriorate to pure carbon. There is a substantial increase in the amount of carbon in soils during this period, so much so that the transition line is quite distinct, with dark, carbon-bearing rocks overlying less carbon-rich rocks. (Fig. 20.2)
3. Many of the creatures of this period are heavily shelled, bear spines, or are heavily armored. As noted above, there will be no selection for bulky structures that take considerable investment in energy to build unless the structure offers some advantage to its bearer. Thus the appearance of shells bears witness to

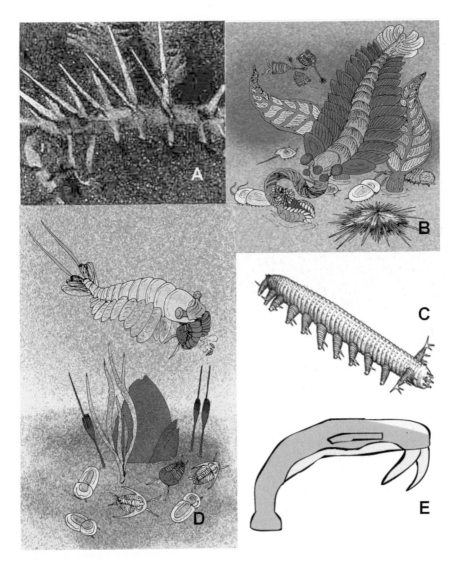

Figure 20.1. Precambrian animals. A. Hallucigenia. For a long while it was not certain which end of this animal was up and which was down, but it now is considered to a bit like a velvet worm (See Fig. 5.5) but with spines. B. An artist's reconstruction of a creature not known to be related to any modern form, Opabinia, eating a sea worm. The other creatures are apparently related to modern arthropods, sponges, and sea anemones. C. Another velvet worm-like creature, Ashyia. D. Artist's reconstruction of the rather large preCambrian arthropod-like animal Anomalocaris feeding on trilobites E. Artist's reconstruction of the mysterious soft-bodied creature named Amiskwia. It has no known descendents. Credits: Wikipedia.org

Figure 20.2. Cambrian-precambrian transition. The sudden appearance of black in the rocks indicates a substantial increase in carbon dioxide converted to organic molecules, as the amount of life suddenly escalated. The later disappearance of the carbon reflects the increase in shellfish in the rocks. Credits: http://wwwalt.uni-wuerzburg.de/palaeontologie/Stuff/ campics/n02.jpg–Wikipedia

predation and competition among animals. It also creates a situation in which fossils will be much easier to recognize.

4. Although we can describe a lineage from some of the creatures to creatures that we know today, many are completely foreign to us. They are not only extinct, but the entire group of animals that they represent is extinct. In fact, they do not seem to have survived much into the Paleozoic period. Thus, we can suggest, as did Steven Jay Gould, that this period was one in which many varieties of life appeared. It was, in his words, a period of enormous "biological experimentation". Of all these fantastic creatures, only a few proved to be sufficiently successful and adaptable to survive into the present. The ones that did are the ancestors of the handful of groups, or phyla, that we know today. Phyla, as defined in Chapter 5, are groups such as mollusks, arthropods, and chordates (the true vertebrates and their close relatives). In other words, the period was one of considerable competition, variation, selection, and survival or elimination—evolution in its most dramatic and rapid form. As is discussed in Chapters 17–19, although many of the phyla remain predominantly in the oceans, the families or major subdivisions of the phyla, such as fish, amphibians, reptiles, birds, and mammals among the chordates; crustacea and insects among

the arthropods; or mosses, ferns, evergreens, and broad-leafed flowering plants among the plants, represent adaptations to life on land.

THE PALEOZOIC

Beginning a bit less than 600 million years ago, the world that we can recognize began to take shape, with a vast increase in the number of living organisms on earth as well as representatives of today's organisms. During this period, which lasted a little less than 300 million years, there appeared fish, corals, primitive land plants, insects. The middle of the period saw the appearance of amphibia, flying insects, and the first plants that could truly stand erect on land (the vascular plants, including mosses and ferns). By the end of this period, slightly less than 300 million years ago, the ferns were giant (tree ferns) and so numerous that their remnants gave rise to the eponymous Carboniferous period, the fossils that created the great coal fields of the Americas and elsewhere. We have already addressed the issue of why the Cambrian was such a fertile period, and we will speculate in Chapter 23 why it finally closed, giving rise to the Mesozoic. We also need to ask what the world was like and how we can tell what it was like.

As will be discussed in Chapter 22, there is substantial evidence that the continents were not in their present locations and that the present configuration of mountain ranges, deserts, rain forests, and plains was probably not present. There are several means of determining that the climate was warm and moist. To take the simplest and most obvious, ferns today are found only in moist locations. They are not hardened to retain water like a cactus, and their means of reproduction requires that the gametes (the technical equivalent of eggs and sperm) find each other in water. Furthermore, ferns undergo what is known as a two-stage life cycle, in which one stage is a small, inconspicuous organism that would quickly shrivel up and die if left in a dry situation. Fossil ferns appear to have reproduced the same way. The presence of amphibia, which reproduce almost exclusively in water, and of ferns, which likewise require water for reproduction, indicates that wherever we find such fossils, the land was at least damp. Today we find ferns mostly in damp locations, tree ferns in tropical rainforests, and by far the largest number of amphibia in rainy areas. We have no reason to assume that they lived otherwise in earlier times. Also, with the exception of one frog, the crab-eating frog, which has very special adaptations, frogs, toads, salamanders, and newts cannot survive salt water and are consequently very rarely found on islands. Although the sea was once less salty, the presence of amphibians on all continents argues for land bridges connecting the continents during this time.

THE MESOZOIC

As is discussed in Chapters 22 and 23, mountain building creates many more environments, or niches, in which individual animal and plant species can survive, and it also creates regions of much more highly variable temperatures and levels of humidity. This resource produces great potential for any creature that can exploit the opportunity, but to exploit the opportunity means acquiring the several physiological features needed to survive on land. These include skeletons capable of supporting animals or plants erect on land; means of motility on land; means of protecting sperm (pollen), eggs, or both for reproduction protected from the drying effects of air; skins capable of minimizing water loss; respiratory structures adapted for air rather than water; and excretory systems (kidneys and rectums) capable of reabsorbing water. These adaptations were first completely achieved by reptiles and the plants that became the pines, spruces, and similar evergreens (described, significantly, as Gymnosperms, or plants with hidden seeds). Unlike the frogs, toads, and salamanders, the reptiles have dry skins; they have internal fertilization, meaning that the male implants sperm into the female rather than spreading it over eggs already in water; they have eggs with hard shells, impermeable to water; their legs and shoulders lift them farther off the ground than the legs and shoulders of frogs or salamanders; and they process urine and feces in such a manner that very little water is lost. Insects have similar adaptations. For animals and plants such as these, the land was a paradise waiting to be exploited, more so than the fabled El Dorado ("The gold-covered land" that the Spanish sought in South America). The reptiles moved into the land, expanding to fill every possible slot for life, in what can be described as "adaptive radiation" or development of forms to occupy every position from herbivores to carnivores, from lowly creeping creatures to turtles to rhinoceros-like creatures to the powerful and fast raptors beloved of children. (Unfortunately, judging from the skeletons—see Chapter 2, page 28, and Chapter 3—it certainly would NOT have been fun to meet a real raptor.) Some took to the air to become the first flying vertebrates, and others returned to the sea as porpoise-like animals (see Fig. 3.4.) This is what we really mean by adaptive radiation. Each particular role on the planet—large, fast-moving aquatic carnivore, small, mouse-like creature eating seeds and any dead animal or plant material, large leaf-eating herbivore—is called a niche, from the French for nest, and if no creature currently occupies the niche, something will evolve to do so. Thus in the absence of porpoises there were pleisiosaurs.

The age of the Middle Animals, the Mesozoic, had begun. It was quite a successful period, lasting for 180 million years, three times as long as the current modern era, the age of mammals, has lasted. Again, judging from the types of plants and animals that we can recognize, the climate was tolerably mild and moist. Almost every variation on animals and plants appeared among the reptiles and the plants, with some exceptions. There were no true flowering plants, fertilized by insects. Some of the reptiles must have been warm-blooded in the sense that they were usually much warmer than the environment, and a few had begun to develop feathers, presumably as insulation; but feathers and control of body temperature were probably not common. Finally, they laid eggs, and eggs were vulnerable to predation or to sudden changes in climate. Nevertheless, the Mesozoic persisted for almost 200 million years before rapidly collapsing approximately 65 million years ago. The cause of its demise is a matter of considerable curiosity and speculation, as is described in

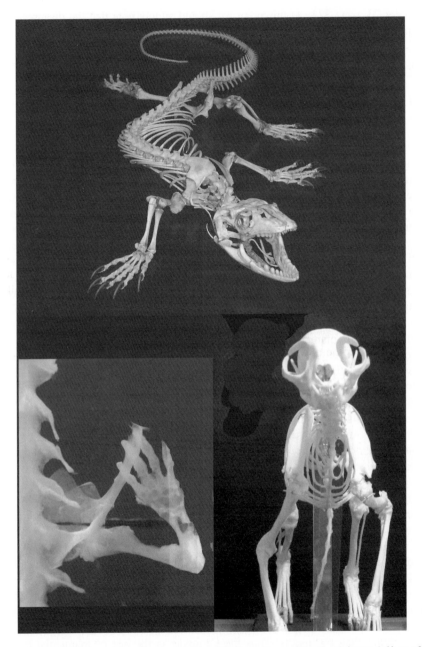

Figure 20.3. Reptilian vs mammalian leg position. The legs of a monitor lizard (upper left) or of a salamander (lower left) go to the sides of the body, providing stability but not much power for a forward motion. Thus, reptiles often run by moving their bodies from side to side. Those of a mammal (cat, lower right) align vertically relative to gravity, allowing an efficient forward-to-backward movement. Credits: Cat, salamander: Author's Photos; lizard:http:// bioweb.wku.edu/faculty/Huskey/default.html Skeleton articulated and photographed by Steve Huskey, Ph.D

Chapter 23, page 323. At this point we should simply note that the disappearance of most of the reptiles again created the opportunity for adaptive radiation.

THE CENOZOIC

The dinosaurs have not really disappeared. Modern-day birds are the descendents of one particular group of dinosaurs, judging from similarities in their skeletons, and the mammals are descendents of another large group of Mesozoan reptiles, again judging from the skeletons. One feature of the skeleton that is easy to see is the structure of the shoulder. Reptiles run with their limbs out to the sides, and they tend to waddle or creep along the ground. (The name "reptile" derives from a Latin word referring to the tendency of these animals to creep.) They can be very fast—you do not want to be in the position of trying to outrun a charging alligator—but mammals swing their legs alongside their bodies in what is physically a more efficient motion (Fig. 20.3) and lifts them higher off the ground. One group of reptiles, the therapsids, had developed this shoulder and hip structure, and mammals came from this group. This group also developed homeothermy, or the ability to maintain a constant warm body temperature, and the feathers-without-the-side-barbs that are modern hairs to insulate the body and limit both excess heat loss and excess heat absorption from the sun. Also, whereas some reptiles retain their eggs in the body of the mother until they hatch, mammals did away with the eggshell and the yolk altogether and invented an entirely different means of protecting the baby. They allow the baby to attach to the mother, through the placenta, as a sort of parasite, drawing its nutrition from the mother; and finally, after birth, to be nourished by a very protein-rich, highly nutritious form of sweat called milk, produced in the specialized mammary glands (for which mammals are named) or breasts.

Mammals existed during the Mesozoic but were not very common or very conspicuous. They were small, rodent-like creatures, presumably capable of sneaking around at night, since they were warm-blooded, perhaps stealing eggs. (Since all chemical reactions increase 2, 3, or more times for each 18° F rise in temperature, an animal that gets cold at night cannot move quickly. This is why flies found by windows in the fall seem to be very "sleepy".) But they did not radiate, or expand, until niches occupied by the reptiles became available. Once they were, at the beginning of the Cenozoic (the "recent or new animals"), they quickly radiated. Today we have large and small carnivores and herbivores, flying mammals (bats), and mammals completely confined to the sea (whales and porpoises). Similarly, springtime in any part of the world can convince you of the importance of the flowering plants. Even plants that do not appear to have flowers, such as birch or oak trees, in fact have less visible flowers adapted to being pollinated by the wind rather than by insects. Thus the Cenozoic is the age of mammals, insects, and flowering plants. Essentially all the broad-leafed plants that you see today are members of this group, technically named the "tube-seeded" plants from the manner in which a pollen grain grows to meet the seed—again a reproductive adaptation to life on land. The major eras and their characteristics are outlined in Table 20.1, and the translation of their names is listed in Table 20.2. It is worth having some familiarity with these names, if only because the names are frequently used in many contexts.

Table 20.1. Geologic eras

GEOLOGIC TIME

EON	ERA	PERIODS AND SYSTEMS	EPOCHS AND SERIES	BEGINNING OF INTERVAL*	BIOLOGICAL FORMS
Phanerozoic	Cenozoic	Quaternary	Holocene	0.01	
			Pleistocene	1.6	Earliest humans
		Tertiary	Pliocene	5	
			Miocene	24	Earliest hominids
			Oligocene	37	
			Eocene	58	Earliest grasses
			Paleocene	65	Earliest large mammals

Cretaceous-Tertiary boundary (65 million years ago): extinction of dinosaurs

	Mesozoic	Cretaceous	Upper	98	
			Lower	144	Earliest flowering plants; dinosaurs in ascendance
		Jurassic	208		Earliest birds & mammals
		Triassic	245		Age of Dinosaurs begins

Permian-Triassic boundary (245 million years ago): Major extinction

	Paleozoic	Permian		286	
		Carboniferous			
		Pennsylvanian		320	Earliest reptiles
		Mississippian		360	Earliest winged insects
		Devonian		408	Earliest vascular plants (as ferns & mosses) & amphibians
		Silurian		438	Earliest land plants & insects
		Ordovician		505	Earliest corals
		Cambrian		570	Earliest fish
Proterozoic	Precambrian			2500	Earliest colonial algae & soft-bodied invertebrates
Archean				4000	Life appears;earliest algae & primitive bacteria

---

* In millions of years before the present – from Merriamwebster.com

Table 20.2. Names and characteristics of geological eras

Era	Period or Epoch	Translation
Cenozoic		New or recent life
	Quaternary	Fourth period
	Holocene	Entirely recent
	Pleistocene	Most recent
	Miocene	Less recent
	Oligocene	Fewer recent (mammals)
	Eocene	Dawn [of the] new
	Paleocene	Older new
Mesozoic		Middle life or middle animals
	Cretaceous	Chalky era (period of White Cliffs of Dover)
	Jurassic	Like the limestone of the Jura Mountains (Switzerland)
	Triassic	Third period (of the Mesozoic)
Paleozoic		Old life or animals
	Permian	Like Perm, a former province in Russia
	Carboniferous	Carbon bearing (coal bearing)
	Pennsylvanian	Like the soils of Pennsylvania
	Mississippian	Like the soils of the Mississippi valley
	Devonian	Like the soils of Devon, England
	Silurian	Like the soils of the land of the Silures*
	Ordovician	Named after an ancient people of Wales
	Cambrian	Like the lands of Wales
Proterozoic		Before life
	Precambrian	Before the Cambrian
Archean		Ancient

* An ancient mythical people in England

REFERENCES

http://www.beyond.fr/history/geology.html (Description of geology, based on Provence, France, in English)

http://www.gpc.edu/~pgore/geology/geo102/burgess/burgess.htm (Description of the Burgess Shale, from Georgia Perimeter College)

http://www.burgess-shale.bc.ca/ (Burgess Shale site)

STUDY QUESTIONS

1. What was the practical value of giving names to different geologic eras?
2. How sharp are the boundaries of the different eras?
3. What reasons can you give for why there should be boundaries between eras?
4. What evidence is there for when life began on earth? How reliable is this evidence?
5. How do interplanetary probes "search for life"? What do they look for? Why?

6. How do preCambrian creatures differ from those of the Cambrian era?
7. What happened to most of the preCambrian creatures?
8. How do fossils indicate climate?
9. What do we mean by the term "adaptive radiation"?
10. What do we mean by the term "niche"?

CHAPTER 21

RETURN TO WATER AND TO LAND

If one looks at a porpoise or a whale, one realizes that its existence is a conundrum, or puzzle. It lives in the water, which can withdraw heat very rapidly, but it insists on being warm-blooded. Some of the smaller sea-going mammals devote half of their body mass to insulation just to keep warm (Fig. 21.1). It has to come to the surface quite frequently to breathe air. Giving birth is a problem, since the baby is born with no air in its lungs and might sink and so must be lifted to the surface (by "midwife" females) to take its first breath. Nursing or suckling also must be arranged so that the infant can breathe; the mother frequently lies on her back in the water. And yet they are so fish-like, the only difference being that their tails go up-and-down, like a loping mammal, rather than side-to-side like a fish.

These characteristics define a whale or a porpoise as a mammal, but mammals radiated because they were so successful on land. A similar story could be argued for bats: Birds don't have teeth because of the efficiency of maintaining balance in the air (Chapter 3) but bats, for the most part carnivores, do. Likewise, one advantage of developing a large egg quickly and then laying it is the same reason that airlines charge so much for excess baggage: It is very expensive to carry a large infant around in the air. But bats have true placentas and suckle their young. These questions lead to the more general question of how they arose, and what their ancestry was. The ancestry of the whales has been rather well documented. First, following the discovery of skeletons of what appeared to be the ancestors of whales, the basic scenario was described. Today the similarities of the DNAs confirm the outline, while correcting a few details. Rather than trace the story of analysis in historical sequence, we will pick out some of the details and clues.

First, consider the lifestyles of mammals comfortable in water. They are highly varied but telling. The capybara, a sort of giant, long-legged, South American guinea pig, grazes on land but is also fond of water plants and spends a large part of the night swimming and diving for these plants. It can stay underwater for several minutes. Otherwise it breeds and moves easily on land. Beavers are adapted to water primarily by building underwater entrances to air-filled houses, and having flat tails that like swim fins push it rapidly through water. An otter, a carnivore, does the same, though because of its body shape and flexibility it is more agile and adept in water than it is on land. A hippopotamus grazes on water plants and spends most of its time in water to help it keep cool in the hot, dry lands in which it lives. Next we move to the sea lions, which have ears and a bilobed tail fin that

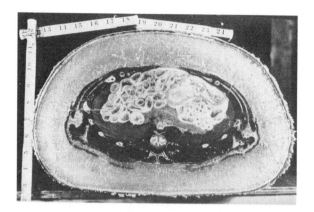

Figure 21.1. Cross section of seal. The total blubber (outer ring) represents 58% of the cross-section. Because water can extract heat much more rapidly and effectively than air can, while mammals are warm-blooded and must maintain temperatures inside their bodies of 98–99° F (37.5° C), marine mammals must maintain very thick insulation. The scale is in inches, Credits: Schmidt-Nielsen, K. Animal Physiology: Adaptation and Environment, Cambridge University Press, 1975;from P.F. Scholander, Biological Bulletin 1950, 99: 234

is clearly highly modified hind feet, and seals, which do not have external ears but have a tail that is much harder to recognize (unless one sees the skeleton) as related to hind feet. All of these animals nevertheless breed and live on land, and are illustrated in Fig. 21.2.

What happens in the evolution of something like a hippopotamus that forages farther and farther into the ocean and returns as little as possible to the shore? The first clues were collected from the skeletons of large creatures similar to hippopotamus and horses that clearly were not very graceful on land. They had large, bulky bodies and quite stumpy legs (Fig. 21.3). No flight of imagination would make these animals anything other than clumsy, lumbering creatures on land. And yet some of them had skulls and teeth that defined them as carnivores (page 35). To be successful predators, they had to hunt in water, where their bulk would not be so cumbersome. In fact, they also had strong tails, rather like alligators and crocodiles, whose shape they generally matched. A strong tail could propel them through the water even though their legs were not strong. Their large jaws and powerful teeth would likewise compensate for legs not powerful enough to use claws as major weapons. Such skeletons were found in the mountains of Pakistan and, more recently, in China. (Why they are found in mountainous regions is addressed in Chapter 22.)

In the next series of skeletons the legs are clearly useless (Fig. 21.3). Such legs could never lift this creature off the ground, and would not have much function in the water, unless, like the forelimb, they were modified as paddles. We have no information about its reproduction, but this creature is similar enough to a whale to be recognizable as one. In the last of the fossil whales, as well as some modern ones, there is no sign of an external hindlimb, but there is a small and apparently useless pelvis. Such a situation is still encountered in today's anacondas and boa

Figure 21.2. Progressive adaptation to life in the water A. hippopotamus. The first adjective to come to your mind in looking at a hippo on land is unlikely to be "graceful," meaning that you recognize that its small legs are not very effective in supporting its weight and making it agile. Its name translates to "river horse" and it is more comfortable grazing in water than on land. B. A hippo is an herbivore. To be a carnivore, one needs to be more quick and agile, as this sleek river otter is. Note that an otter is four-legged but that it has a strong tail to aid it to swim. C. Sea lions, also water-dwelling predators, have two flippers for front legs and their hind legs are fused into a tail flipper. However, they rest on land and breed on land. D. Manatees, which are slow-moving herbivores, have full tail fins and can no longer climb onto land. Credits: Manatee - © Photographer: Wayne Johnson | Agency: Dreamstime.com, Hippo - © Photographer: N joy Neish | Agency: Dreamstime.com, Otter - Image provided by Dreamstime.com" Sea lion - © Photographer: Ravter Bostjan | Agency: Dreamstime.com

constrictors, which have very small bones representing the vestiges of the pelvis of the legged reptiles from which they descended (Fig. 21.4). Today's whales, porpoises, and manatees no longer have any signs of pelvises, but the line of descent is clear. The DNA trail confirms the story (Fig. 21.5). In fact, the evidence is overwhelming that whales derived from land animals. The types of evidence are numerous, and include the multiple, independent sources that we have emphasized before (page 114):

- Fetal whales have body hair, lost before birth, and fetal baleen whales (which have a hard, horny net to filter out small swimming animals, rather than teeth) also have teeth.
- In a study of 72 different mammals, Miyamoto and Goodman found that, in whales, several proteins common to all mammals were biochemically most closely related to those of the pig/hippo group.

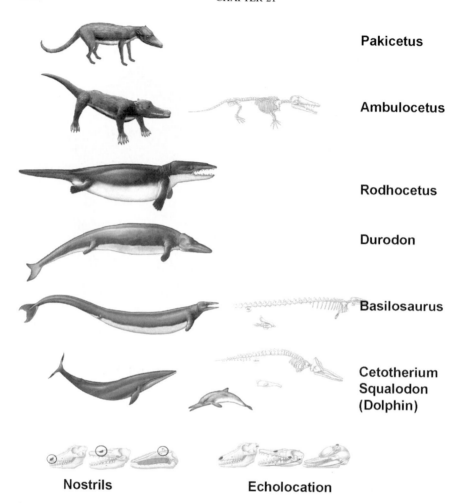

Pakicetus

Ambulocetus

Rodhocetus

Durodon

Basilosaurus

**Cetotherium
Squalodon
(Dolphin)**

Nostrils   **Echolocation**

Figure 21.3. Evolution of whales from 4-legged, swimming carnivores into completely aquatic mammals, according to the fossil record. The approximate ages of the fossils are Pakicetus, 50 million years; Ambulocetus, uncertain; Rodhocetus, 47 million years; Durodon and Basilosaurus, 38 million years; Cetotherium, approximately 20 million years; and Squalodon, 23 to 5 million years. In the lower panel is shown the gradual movement of the nostrils to the top of the head (blowhole) and the development of the bulbous structure that allows modern whales to hear and locate sound under water. Credits: Summary Adapted from National Geographic, November 2001, The Evolution of Whales, by Douglas H. Chadwick, Shawn Gould and Robert Clark Re-illustrated for public access distribution by Sharon Mooney ©2006,Wikipedia.org

- In many whale embryos external hind limb buds appear but, by ceasing to grow, disappear before the whale is born. They also have vestiges of external ears.
- In whale embryos, the nostrils start out at the front of the head but get pushed back to a blowhole by the growth of the nose beneath them.

Figure 21.4. Rudiments of a hind limb and pelvis in a modern sperm whale. Credits: Summary Adapted from National Geographic, November 2001, The Evolution of Whales, by Douglas H. Chadwick, Shawn Gould and Robert Clark Re-illustrated for public access distribution by Sharon Mooney ©2006,Wikipedia.org

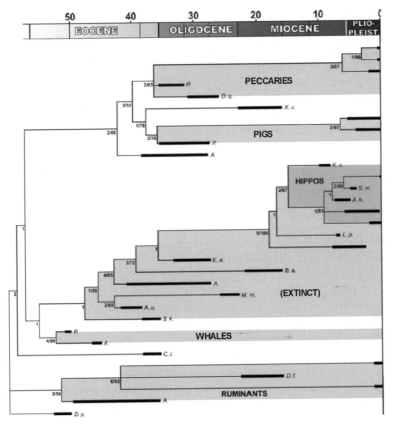

Figure 21.5. A modern interpretation of the evolution of whales. This sequence was derived primarily from interpretation of bone structures, but comparison of proteins of modern hippos, elephants, horses, and whales produces a similar sequence. Credits: The position of Hippopotamidae within CetartiodactylaJean-Renaud Boisserie, Fabrice Lihoreau, and Michel Brunet doi:10.1073/pnas.0409518102 2005;102;1537-1541; originally published online Jan 26, 2005; PNAS This information is current as of October 2006. Rights & Permissions To reproduce this article in part (figures, tables) or in entirety, see: www.pnas.org/misc/rightperm.html

- The record of fossils confirms in many respects their history. The first whales, which had legs, are found in local regions; their bones have a composition that indicates that they drank fresh water; and the fossils are found in soils including washout characteristic of fresh-water lakes. Furthermore their skeletons have not developed the adaptations of the ear or thorax that are required for deep diving, and their nostrils are toward the front of their heads. All the indications are that these beasts (Pakicetus) lived on the shores of freshwater lakes and occasionally dived for food. Amblocetus, on the other hand, is found along with marine fossils, and Rodhocetus is found with deepwater soils.
- The localization of the fossils is confirmatory. Pakicetus has been found along the shores of presumptive lakebeds in Pakistan. The range expands until Basilosaurus is found worldwide, from Asia to lands bordering the Gulf of Mexico and the Canadian Pacific Coast.

The story of the whales also raises the question of the meaning of vestigial organs. Do they have any other functions? Why do they disappear, and why do they not disappear instead of gradually becoming less useful? All of these questions, when phrased properly, lead to interesting explorations and greater understanding of how evolution works.

The first interesting answer relates to the fact that young embryos of many species look far more alike than the older embryos (page 30). For instance, all vertebrate embryos have notochords, a firm gelatinous rod that serves as a structure against which muscles can pull in the most primitive chordates (page 66) but in the higher chordates the skeleton serves this purpose, and the notochord ceases to grow, becoming a small, inconspicuous structure in the adult. In the embryo, however, the notochord does serve a purpose. In all chordates, it is the structure that defines the axes of the embryo, particularly what is left, center, and right; and what is backside (dorsal) and bellyside (ventral). If one experimentally interferes with the formation of the notochord, the embryo cannot build its organs in the right places, and it will die as a disorganized or poorly organized mass of tissues. Even though in the adult the notochord apparently does not function, the first chordate embryos used the notochord as the basis of their organization, and this very important role has not been assumed by any other structure or mechanism.

The second principle is that evolution does not occur in the absence of selection. For a structure to be gained or lost, there must be some interaction favoring its construction or its loss. In other words, one can select for more and more powerful claws, but the most important factor favoring the loss of an already useless part would be that the resources used to build it can provide an advantage if diverted to other functions. Thus small legs that do not really help a permanently aquatic animal might be a problem if they were likely to scrape against something and be cut, or if smaller predatory animals could bite them off and cause an injury. Otherwise, the only way in which they might be lost is the following: There is always a possibility of an error in copying DNA, or a mutation. Organisms have many means of identifying and correcting errors in DNA, but occasionally an error gets through. If the error causes a severe problem, it will be selected against (individuals bearing the error will not be as likely to leave progeny to the next generation), and therefore the gene will not change in the population. However, if the error has no particular impact, then there will be no selection against it. Thus it takes considerable biological energy to construct a hind limb. If the limb serves no purpose, then an error in its construction will not be selected against and will persist to the next generation. The laws of physics teach us that creating order requires energy, whereas decrease in order is spontaneous. It takes work to build a watch or a building, but the watch or the building can fail spontaneously and if there is no further use for the non-functioning watch or building, it will simply continue to disintegrate. As long as a limb or an eye serves a valuable function, individuals with poorly functioning limbs or eyes will be selected against. However, if for instance an animal that lives in caves learns to move and hunt using sounds or smells only, there is no further selection to preserve vision. A non-functional eye is only a source of potential injury, and there may be selection against it. Thus structures can gradually disappear, but will do so only gradually.

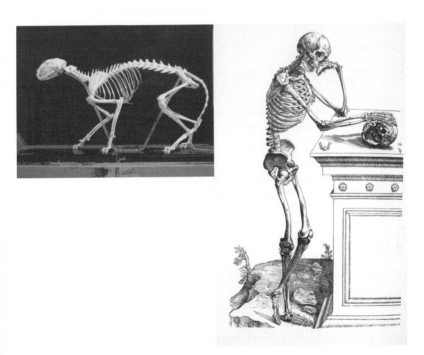

Figure 21.6. Backbones of tetrapods and humans. The backbone of a human is curved in an S-shape, the better to balance the torso. That of a tetrapod (cat) is arched to support the weight on four legs. Note also in the human skeleton how the weight is centered on the pelvis, the angle at which the spine joins the pelvis, the angle of the femurs (thigh bones) relative to the pelvis, and the angle and balance of the head on the spine. Note particularly the foramen magnum (the hole in the base of the skull on the table, where the spinal cord enters the skull)

Finally, evolution will always build on what is available and will not purposefully invent structures. The swimming motion of a whale or porpoise will be based on the abdomen-to-backside motion of the backbone of a running animal rather than the side-to-side motion of a fish. Likewise, the spine of an erect human has developed a few curves to help balance the body—the spine, upper body, and head are balanced on the pelvis much like balancing a broomstick on one's hand—but is basically a rather crude means of converting the suspension bridge-type construction of the spine of a four-footed mammal into an entirely different support system. The construction is the origin of many problems ranging from fatigue in standing to backaches and spinal injuries (Fig. 21.6).

Although starting with the amphibia the story of the vertebrates has essentially been a story of finding better ways to survive on land, the search for new niches has led many animals to return to the water and to the air. We can recognize from evolutionary relationships and from the structures themselves whether the return represents a single evolutionary sequence or several. From the structure of the bones, the return to the air of flying (now extinct) reptiles, birds and bats were different sequences (Fig. 12.2) as were the return to water of plesiosaurs and swimming mammals; but whales and porpoises have a common ancestry.

REFERENCES

Boisserie, Jean-Renaud ; Lihoreau, Fabrice ; and Brunet, Michel (2005). The position of Hippopotamidae within Cetartiodactyla. PNAS 2005;102;1537–1541

Campbell, Neil A. and Reece, Jane B.(2004) Biology (7th Edition), Benjamin Cummings, Boston, MA

Schmidt-Nielsen, K, 1975, Animal Physiology, Adaptation and Environment, Cambridge University Press, 1975, p 327 (From Scholander, P.F., Walters, V, Hock R, and Irving L 1950, Biological Bulletin 99: 225–236. (figure on p 224))

http://www.pbs.org/wgbh/evolution/library/03/4/l_034_05.html (Origin of whales from Public Broadcasting System series on evolution)

http://www.talkorigins.org/features/whales/ (Origin of whales from TalkOrigins)

http://en.wikipedia.org/wiki/Whale (Origin of whales from Wikipedia)

http://en.wikipedia.org/wiki/Evolution_of_cetaceans (Origin of whales from Wikipedia)

STUDY QUESTIONS

1. What characteristics define a whale as a mammal? Could you argue that it is a fish?
2. Some sharks have a placenta-like structure and bear their young alive. Are they mammals? Why or why not?
3. Why does a seal need such a heavy layer of insulation?
4. What other explanations can you come up with to explain the existence of what we argue are vestigial pelvic bones?
5. What other explanations can you devise to explain the existence of small limbs on the embryos of porpoises and whales?
6. What other explanations can you give for a mechanism whereby unused structures are lost during evolution?

CHAPTER 22

EVIDENCE FOR EXTINCTIONS—WHY DO WE GET THEM?

THEY DOMINATED THE EARTH, AND THEN WERE GONE—HOW DOES THE WORLD CHANGE COMPLETELY?

We use the term "dominant species" or "dominant type" in a rather loose and arrogant fashion. If we wanted to be accurate, we should refer to those types of organisms that make up the largest biomass (the total weight of all individuals of that type on the earth) and in this sense we would have to conclude, in a somewhat humiliating fashion, that—using different assumptions and calculations—the dominant species of our era are bacteria, nematode worms (bloodworms), or insects. This viewpoint however does not appeal to us, and we prefer to designate as dominant species obvious large organisms on the planet. Thus if you are suddenly dropped into an undisturbed temperate area on the globe, for instance northeastern United States, what would strike your eye would be oaks, maples, and many mammals ranging from rabbits to deer, bear, wolves, and coyotes. You might notice some birds, but reptiles and amphibia would not be particularly prominent and insects, though numerous, would not attract too much of your attention. Thus, from our viewpoint, the dominant species would be the big broad-leafed trees and the mammals. We would therefore fully agree that this is the age of mammals.

But now that we can read the fossil record fairly well, we understand that the world was not always so. Before the Cambrian era, there were many very strange apparently invertebrate animals, but almost all of them died by the beginning of the Cambrian. Then there was an era (the Paleozoic) filled with fish, amphibians, tree ferns, and some insects and relatively small plants, but no reptiles, birds, or mammals and no flowering plants or true trees. These too nearly vanished—one can hardly consider today's world to be overwhelmed by frogs and ferns—and we encounter an era (the Mesozoic) in which reptiles are everywhere. If you lived near a forest at that time, first, the forest would not have flowering trees, and there would be no true flowers; and, second, the large animals that browsed or hunted in the forest would have been reptiles. Anything that flew overhead was also a reptile, and instead of porpoises, otters, and seals, there would have been reptilian equivalents. Some of the dinosaurs—a subgroup of reptiles—were indeed truly massive, much larger than elephants (though still smaller than the largest mammal, a blue whale). The only mammals were small rodent-like creatures, perhaps adept at stealing eggs at night. Yet the dinosaurs vanished. Although the occasional alligator or anaconda is indeed a threatening creature, by and large we do not spend our lives in fear of these creatures. Why do they no longer rule the world? And is it possible that we could go the way of the dinosaurs, and that the age of mammals and flowering plants (the Cenozoic) could be superseded by an era of very different creatures? We must ask these questions.

THE VIOLENCE OF THE EARTH—CONTINENTAL DRIFT

To most of us, the land that we know is a pretty stable thing. People are born, live, and die in the shadow of a mountain or along the shores of a lake. Ancient paintings illustrate landscape features that we recognize today. References in the Bible, Koran, or other holy books identify mountains and rivers that are still present. The idea, therefore, that the earth was once very different is certainly not an intuitive one. Thus, when Antonio Snider-Pellegrini pointed out in 1858 (the year before Origin of the Species was published) the close fit of the eastern coasts of the Americas to the western coasts of Europe and Africa (Fig. 22.1) he was only emphasizing what others had seen, but adding also the note that the soils of the opposing sides had some relationship to each other. The curious mirror symmetry between the East and West Coasts of the Atlantic Ocean is even more similar if one follows not the current coastline but the continental shelves. (In several areas of the world, the

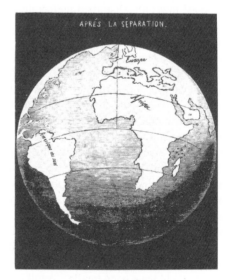

Figure 22.1. Snider-Pellegrini's identification of the close fit between the east and west coasts of the Atlantic Ocean. He was a little bit fanciful—note how he wrapped South America around the horn of Africa—but he was generally quite correct in his supposition. The title on the left is "Before the Separation" and on the right, "After the Separation". Credits: Snider-Pellegrini map - http://pubs.usgs.gov/gip/dynamic/historical.html

land surface continues its gradual descent for several miles into the ocean suddenly dropping off in a cliff-like structure to a depth hundreds or thousands of feet deep. This dropoff can sometimes be recognized as an abrupt change in the color of the water, if it occurs where the initial depth of the sea is sufficiently shallow to reflect light from the bottom, and is called the continental shelf. The mid 19th C was a period of formation of concepts of geology and the fit was noted as a curiosity. Snider-Pellegrini suggested that the lands had once been connected (Fig. 22.1). Once it was understood that the earth was quite ancient, and that the earth might have formed as a very hot body that had pulled off from the sun, the physics of the structure of the earth became more obvious. Granite rocks are heavy. Technically, the appropriate description is that they are dense or have a high specific gravity, meaning that a given volume (one cubic centimeter, or a cube approximately ½ inch on each side) weighs quite a bit. Rocks are dense, but they are not as dense as metals. Thus they will float on liquid metals. For instance, if you put a rock of specific gravity approximately 3 g/cc onto a pool of liquid mercury, specific gravity approximately 13 g/cc, the rock will float on the mercury in the same way that oil (specific gravity approximately 0.8 g/cc) will float on water (specific gravity 1 g/cc). If the earth was once hot enough so that metals such as iron were molten, then the rocks (continents) would float on them, much as pieces of bread float on a dish of soup. From here it is a very short step to assume that the continents broke apart during this period before they finally settled into place.

As late as the 1950's, this was the argument that was taught: The continents bordering the Atlantic had once been part of the same mass and had subsequently separated, but this separation had happened very early in the history of the earth and had no bearing on evolution. Still, there were anomalies that were confusing. Among them were the following:

1. The fossil record indicated tropical plants and animals in locations that are today far too cold to support them.
2. The natural magnetism of certain rocks has a north and south pole, like any magnet, but sometimes these rocks—part of a large mass, not just loose on the ground—point somewhere other than to the north pole.
3. The distribution of animals and plants on the planet is rather curious. There are large flightless birds in Africa (ostriches), Australia (emus), New Zealand (the now extinct moa and its cousin the kiwi), and South America (rheas). In the 1950's it was assumed that these were the result of convergent evolution, or similar circumstances producing similar selection and similar results among unrelated organisms (see Fig. 7.3). However, similar birds were not found in North America, Europe, or Asia. Likewise, although monkeys are found in South America, they are all of the tailed variety, and there are no tailless, or anthropoid, apes in South America. Marsupials (animals with pouches for their young) are found in Australia and South America. Some situations undoubtedly provoke convergent evolution, such as the similar shapes of seagoing creatures (see Fig. 3.4), but the similarities of ostriches, rheas, and emus derives from different forces, as was gradually understood during the 1960's and dramatically

confirmed in the 1990's and the 21st Century by analysis of DNA. The story
of this changed viewpoint represents a realization derived from the convergence
of findings from several disciplines, one of the most convincing arguments of
science (Chapter 9, page 114). Watching the change in attitude is one of the
marvelous stories that makes the history of science so similar to a mystery novel.

4. The distribution of volcanoes and earthquake zones on earth is not random.
   Both tend to occur in coastal mountain zones and otherwise in high and rough
   mountain settings. In fact, they are so characteristic of the Pacific coasts that
   almost the entire boundary of the Pacific Ocean is known as a "ring of fire"
   (Fig. 22.2). Since science is always about how things work, geologists were
   anxious to understand what drove volcanoes and earthquakes.

The story begins in the 1960's with a seemingly useless project of the type that
is often described by politicians as a "waste of taxpayers' money". The National
Science Foundation agreed to sponsor the full mapping of the floors of the ocean,
especially the Atlantic, to understand the trenches, mountain ranges, and volcanoes
of the Atlantic. It was known that there was one major by a mountain range,
down the center of the Atlantic, and that the volcanoes of the Atlantic (Iceland,
the Canary Islands, the Cabo Verde Islands, Tristan da Cuna) bordered the trench.

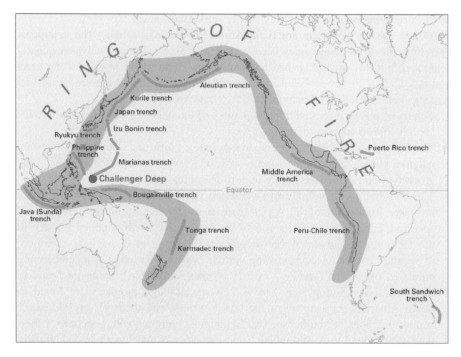

Figure 22.2. The Ring of Fire. The dark gray area is the Ring of Fire, the belt along which numerous
volcanoes are found. The darker lines indicate deep trenches, areas in which there is substantial
subduction. Credits: Ring of fire - http://pubs.usgs.gov/publications/text/fire.html

The boats of the NSF were to map the depths, the variations of the magnetic fields, and other physical measurements, while crisscrossing the ocean. The project was to learn more about the structure of the oceans, on the basis of pure science as well as for the practical purposes of aiding navigation and fishing industries. As a part of the project they undertook to map the magnetic fields of the Atlantic Ocean, the modest deviations from true north that affect magnets when the magnets are near iron-containing rock. What they found very much altered our view of the history of the earth as well as of the mechanics of evolution. In brief, the evidence indicated that the continents had actually drifted during the period in which plants and animals were evolving, and that they were continuing to drift. As the great French physiologist, Claude Bernard, had remarked, if the evidence contradicts the hypothesis, one must accept the evidence, even if the hypothesis is supported by the most influential scientists. In this case the hypothesis was that the position of the continents was set prior to the origin of life on earth. The evidence, however, indicated that the continents could drift and were still drifting. The evidence was very solid, and it gave an interpretation to the mechanics of earthquakes and volcanoes. The evidence was as follows:

- In the middle of the oceans there are usually ridges, such as the mid-Atlantic ridge running north-to-south in the middle of the Atlantic (Fig. 22.3). They are in effect underwater mountain ranges, occasionally bursting above the surface to form, in the case of the mid-Atlantic ridge, islands such as Iceland, the Canary Islands, Tristan da Cunha, and the Falkland (Malvinas) Islands. These islands are typically volcanic and subject also to earthquakes.

- Near the shores of continents are often deep trenches, such as along the Pacific coast of the Americas. These trenches are also areas of earthquake activity.

- Along the mid-ocean ridges and elsewhere, there are often underwater volcanoes, which erupt but, because of the weight and cooling properties of the water, never reach the ocean surface.

- If you heat a magnet to red heat or even to molten iron and allow it to cool, when it solidifies it takes on the orientation of the prevailing magnetic field. Thus, if you heated an iron rod and laid it in an east-west direction to cool, you would create a magnet for which one side of the rod was the north pole and the other side was the south pole. If you laid the hot iron rod in a north-south direction, the entire rod would be magnetized so that one end was north and the other end was south (Fig. 22.4).

- Molten lava contains iron and other magnetic metals. Thus, when lava solidifies, it creates a permanent magnetic record of the direction of the magnetic poles at the time that the lava cooled.

- The magnetic poles of the earth are known to drift slowly, so that they are near to but not identical with the celestial poles (the axis on which the earth turns). This drift is reflected in the magnetism of successive lava flows from a volcano.

The surprising result of this basic information was that, along the mid-Atlantic ridge, successive changes in magnetism indicated that the oldest lava was furthest from the ridge, and that the changes in magnetism were symmetrical on both sides

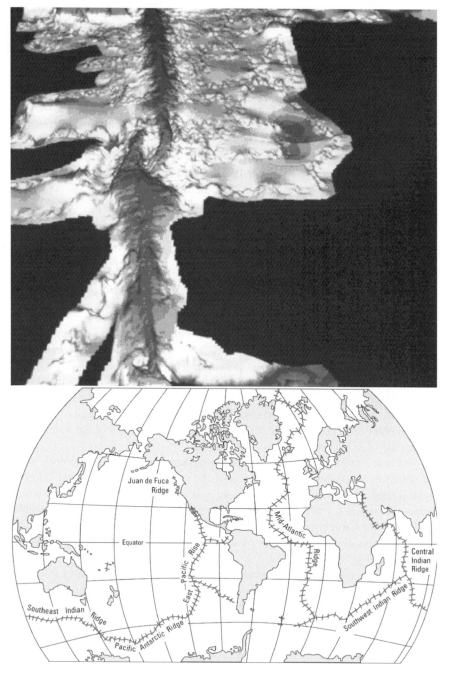

Figure 22.3. Upper: False-color image of the Mid-Atlantic Ridge. Lower: The major ocean ridges. Credits: Mid-Atlantic ridge - http://pubs.usgs.gov/publications/text/topomap.html, Last updated: 05.05.99, Major ridges - http://pubs.usgs.gov/gip/dynamic/baseball.html, Ocean ridges - http://pubs.us.gov/gip/dynamic/developing.html

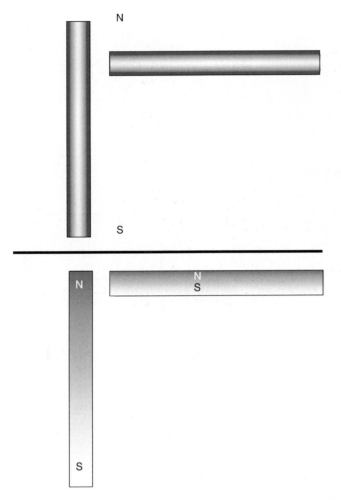

Figure 22.4. Alignment of magnets. If iron rods are heated to red hot and allowed to cool in a magnetic field, they will become magnets with the orientation of the field to which they were exposed

of the ridge (Fig. 22.5). This result indicated that the lava was spreading in both directions from the center of the ridge. Other geological evidence indicated that the lava was in fact the formation of new ocean floor, and that it was shoving the outer edges further away.

The next task is to measure the speed at which this expansion occurs. Again judging from evidence of magnetic fields on land, and from the ages of the rocks dredged from the ocean floor, it is possible to measure the age of the rocks on the east and west sides of the ridge as well as the distance from the ridge. In other words, it is possible to measure the rate of expansion, and to estimate when the continents were connected and since when they are being pushed apart by the

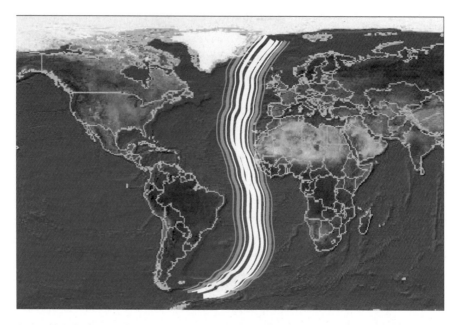

Figure 22.5. Symmetry of magnetic ridges. The explorations of the National Science Foundation traced the magnetic anomalies (indicated by lines of different colors and widths) in the Atlantic Ocean. The results produced two surprises: First, the farther the anomaly was from the Mid-Atlantic Ridge (thickest line in center), the older it was. Second, the anomalies were symmetrical on both sides of the ridge. In other words, if the anomaly in the second line to the west of the ridge pointed 3° west of today's magnetic north, the second line to the east of the ridge also pointed 3° west of magnetic north. The most reasonable interpretation was that there was outflow from the ridge, pushing both eastward and westward

expansion of the Atlantic floor. To the surprise of almost everyone, the calculations indicated that the continents were connected in biological time. Snider-Pellegrini was right, but the time scales were very different. Today, satellites are capable of measuring the movement of the continents and have confirmed that the Atlantic Ocean is expanding at a rate of a few inches per year.

Trenches, mountain ranges, and volcanoes exist in many parts of the world, and similar arguments have led to a very different image of the earth. The center of the earth is hot and molten, as is easily recognized from volcanoes; and the movement of this fluid phase is responsible for the gradual change in the magnetic fields of the earth. The continents—rocks that are less dense than the metal-rich liquid core—float like pieces of bread in soup on the core, drifting apart and occasionally colliding. The collisions create crumpling, like two trains colliding. From space, the folds of the Appalachian mountains look very much like a blanket that has been shoved up from the side, as indeed they are: they represent the buckling of the continent (Fig. 22.6). Where the movement is more rapid and more recent, the collisions lift the mountain ranges of the Alps (Africa colliding with Eurasia), the

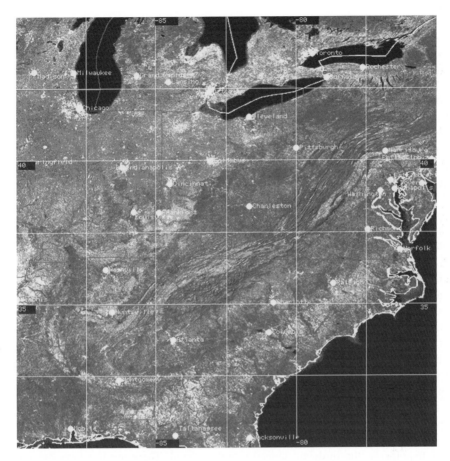

Figure 22.6. Buckling of Appalachians. Note, in this satellite image, that the Appalachian mountains are folded like a blanket. This folding comes from the pressure of plates pushing against the continent. Credits: http://eol.jsc.nasa.gov/sseop/clickmap/map078.htm

Himalayas, (India colliding with Asia), and the Rockies through the Andes (the westward movement of North and South America against the Pacific plates). In some areas of the world, the mechanics of the process is obvious to the trained eye. In northern India and Tibet, the soils and rock structures on the north and south sides of some valleys are very different, since the valley represents the line of collision. (It is a valley, created by the process of subduction, described immediately below.) Where the pieces of bread pull apart, they leave deep trenches, such as the deep depression that forms the Dead Sea, the Red Sea, and the Afar Depression of East Africa, or Death Valley in the United States.

Sometimes, again like the collision of two trains, one continent (or tectonic plate) slips beneath the other. This process is called subduction. Its slipping beneath creates a deep rift or valley, often dipping deeply enough to allow the escape of molten lava and the production of volcanoes. Particularly in the Pacific, volcanoes follow the coastlines in what is called the ring of fire (Fig. 22.2); and the subduction lifts mountain ranges beside the trenches. The continual movement causes earthquakes, much as continual tugging on a rusty bolt evokes occasional jerky slips of the bolt.

This interpretation gave a plausible interpretation of the coastal mountain ranges, which in this sense would be buckled as the push of the ocean floor crumpled the edge of a continent, as a throw rug would buckle if you pushed it from one edge (see below).

From the standpoint of evolution, the most important aspect of these findings was that, through radioactive dating (see Chapter 8, page 104), measurements of magnetism, and other techniques, the rate of this spread could be measured; and the measurements indicated, to most scientists' surprise, that the movement had occurred within biological time—that is, within the last 300 million years. (It is likely that there had been movements before this, but these are the ones that interest us). If indeed the continents were recently connected, some of the bizarre distributions of animals and plants can be understood. There are no placental mammals in Australia because Australia was connected with the other continents during the time that warm, furry animals were evolving but was separated before truly effective placentas had evolved. In competition with other animals, the true placental mammals succeeded very well, driving the marsupials to a very minor part of the animal world, but the Australian marsupials did not face this competition. The rhea of South America, the emu of Australia, and the ostrich of Africa were similar not because of convergent evolution but because a large flightless or nearly flightless ancestor roamed the land when South America, Africa, and Australia were connected (but were not connected to North America, Europe, or Asia). Their similarity reflects common ancestry, and their distribution the connectedness of the continents at the time of their appearance. Subsequent DNA testing has confirmed this hypothesis: the birds are indeed closely related. There are monkeys but not apes in South America because South America was still in contact with Africa when primates evolved but was separate before the apes evolved. The appearance of tropical fossils in currently cold lands may reflect a different position of the continent rather than a different temperature of the earth. Today the biology of distribution of organisms and the geological data confirm each other, and we have a far more meaningful understanding of the otherwise idiosyncratic distribution of organisms on earth.

The bulk of the geological evidence, including also the distribution of magnetic fields, comparison of soils, the activity of volcanoes and earthquake zones, and the existence of fossils inappropriate to current climatic conditions where they are found, leads to a hypothesis that is now very well defended and widely accepted: Approximately 250 million years ago, the continents were connected in what was essentially a single land mass, now given the name Pangaea. At this time, the period of the "old animals" (Paleozoic, see Chapter 20), there were fish, amphibia, and ferns, which were distributed worldwide. By the Mesozoic ("middle animal") period approximately 135 million years ago, the southern hemisphere continents and Africa

(Gondwana) were separating from the northern hemisphere continents (Laurasia). At this time, there were dinosaurs and early mammals, as well as the ancestors of birds. Dinosaur fossils are identical in eastern South America and western Africa. Fossils of the ancestors of the large flightless birds are found; they give rise to the ostriches, rheas, and emus. DNA evidence (Chapter 15) now has established that these birds are related and are not the result of convergent evolution (page 305). There are no ratite birds in the northern hemisphere (other than in Africa) because Laurasia and Gondwana were separated by this time (Fig. 22.7).

By the beginning of the Cenozoic ("recent animals"; modern period, 65 million years ago) South America had definitively separated from Africa, and Australia and Antarctica were likewise distant from the other continents. The Australian group had separated from the European complex before true placental mammals had become established, and thus it has no native true mammals, only marsupials (other than, of course, flying bats). The marsupials also reached South America. However, separation of South America was becoming definitive, and fewer mammalian groups reached it. Thus the rodent-like, arboreal (tree-living) monkeys that actively use their prehensile ("grabbing" or "holding") tails to climb arrived but the later tailless anthropoid apes did not.

The hypothesis of continental drift explains the peculiar distribution of animals and plants around the globe. (Differences in plant distributions are less marked, since seeds can be widely dispersed by wind, birds, or even sea currents. Darwin cultivated over 200 species of plants from islets consisting of one or two palm trees ripped from Caribbean shores by a hurricane and drifting in the Gulf Stream to England.) It also explains much of the violence of the earth. The violence of the earth is very similar to that of a traffic accident, in this instance one of a light-weight car colliding with a more massive one.

The violence of the earth. Volcanoes, glaciers, and meteors

Assume that the cars will not bounce, roll, or otherwise undergo any movement other than inexorably plowing into each other. In this case, two changes are likely to occur: they will crumple, and the lighter one will go above the top of the heavier one. This is what happens when the expanding sea floor plows into the side of a continent. The earth is divided into large pieces floating on the molten iron core, called plates, which include the major continents and the sea floor. These drifting plates bang into each other. The crumpling pushes the mountains up. The diving of the sea floor under the floating continent creates the trench. Since the push of one plate against another creates pressure, the pressure builds until one plate slips against the other, like pushing two boards against each other; the sudden slippage produces an earthquake. At the trench and where the rock cracks (fault lines) the heat of the inner earth pushes to the surface, melting rocks and producing volcanoes, hot springs, and lava (Fig. 22.8).

Now that we understand the mechanism, it is relatively easy to collect vast amounts of data, both dramatic and mundane, in support of the hypothesis of

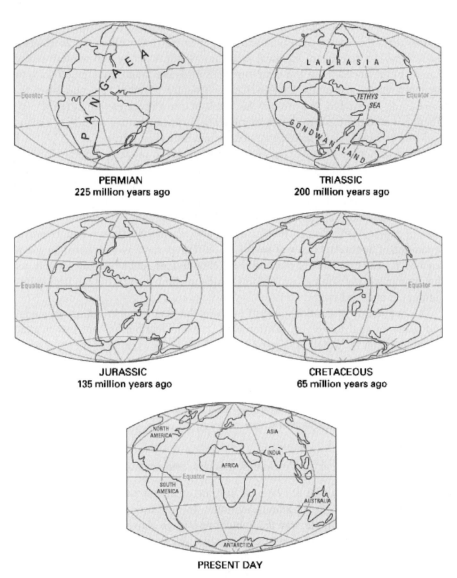

Figure 22.7. Movement of continents. The migration of the continents, based on evidence of magnetism, the fossil record, and calculations from plate tectonics. Credits: http://pubs.usgs.gov/gip/dynamic/historical.html

continental drift. Seen in overview, the central Appalachian mountains have the appearance of a crumpled blanket, as indeed they are, pushed into shape by the expanding plate (Fig. 22.6). Anomalies such as Staten Island, an island within a few hundred feet of the New Jersey shore, but with soil and rock compositions

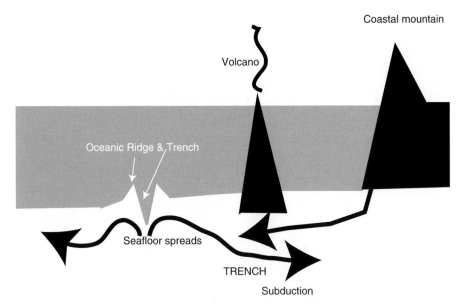

Figure 22.8. Subduction. (Left) Lava flows upward from a deep trench, flowing to either side of the trench and creating a ridge. (Right) The continent is pushing toward the left (west). (Middle) Where the two flows meet, the heavier material flows under the lighter material (subducts). Where it dives down creates a trench through which lava can flow, occasionally reaching the surface of the ocean and creating a volcano. The subducting material causes the overlying material to buckle, lifting a coastal mountain range

that do not at all resemble those of any neighboring region, are explained (with other evidence) by the argument that Staten Island first appeared far at sea and was pushed toward the coast by continental drift. In the Himalayas, there is a deep valley in which the hills and mountains on the north side differ considerably from those on the south side. The Indian subcontinent is pushing into the Asian continent. The more dense Indian side is subducting, or going under the edge of the Asian continent, shoving the Himalayas into the air. The region is a very active earthquake zone. In a similar fashion, the Alps are lifted by the gradual rotational (counter-clockwise) movement of Africa, pushing the eastern edge of Africa into Europe. The very deep valleys on earth, such as Death Valley in California and the Dead Sea valley in the middle east, are regions were two plates are separating, as if one pulled on two sides of a piece of bread. (The Dead Sea valley in fact extends from the Jordan River through the Dead Sea, through the Red Sea, and into Africa as the Afar Depression. The eastern edge of Africa is separating from the rest of the continent.) The current movement of the continents is summarized in Fig. 22.9.

Continental drift has consequences well beyond whether a land is tropical or temperate or whether or not animals and plants can migrate from one location to another. The mountain building and rifting create major differences in climates and therefore niches. Consider for instance that the Grand Canyon is such a barrier for

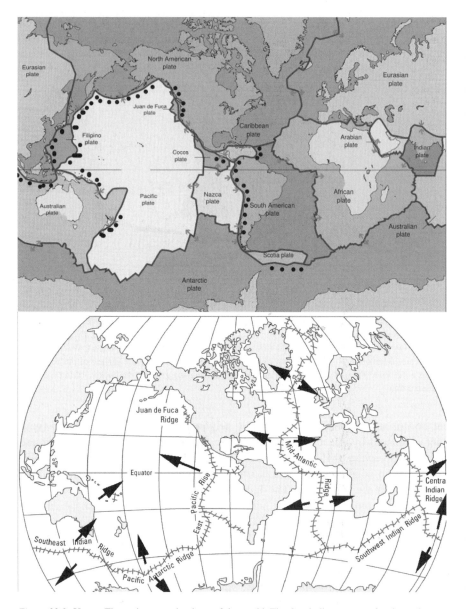

Figure 22.9. Upper: The major tectonic plates of the world. The dots indicate areas of active volcanoes, which are found along the edges of the plates. The volcanoes around the rim of the Pacific are called the Ring of Fire. Lower: Major current movements of the plates. (The movements are shown in finer detail, but more subtly, in the upper figure). Credits: Major tectonic plates - http://pubs.usgs.gov/publications/ text/slabs.html, Movement of plates - Plates_tect2_en.svg This is a file from the Wikimedia Commons

squirrels that the squirrels on the north side of the canyon are substantially different from those on the south side, or consider the difference in climate between a sea floor or seashore and a land, formerly seashore, now 7,000 feet above sea level. We will address these issues in the next chapter.

REFERENCES

McPhee, John (1982), Basin and Range, Farrar, Straus and Giroux, New York; Reissue edition
McPhee, John (1984), In Suspect Terrain, Farrar, Straus and Giroux; Reissue edition (January 1, 1984)
http://pubs.usgs.gov/gip/dynamic/world_map.html (Explanation of continental drift from US Geological Survey)

STUDY QUESTIONS

1. What reasons, besides intellectual curiosity, were there to explore the characteristics of the ocean bottom?
2. Would it have been possible before the mid-20th C to identify continental drift?
3. How many questions can be answered by the theory of continental drift?
4. If you can do so safely, heat a soft iron rod to red heat. Set it down and allow it to cool undisturbed, noting its position. Once it has fully cooled, by use of a magnet or iron filings determine if it is magnetic and, if so, where its north and south pole are. (Hint: if you hang it balanced on a string, its north pole will point toward north.) Repeat the experiment, but this time allow it to cool in proximity to as large a magnet as you can find. Are the results different?
5. What is the Ring of Fire? Is there only one? Why or why not?
6. What other limited distributions of animals and plants would you predict according to the hypothesis of continental drift? Why?

CHAPTER 23

THE VIOLENCE OF THE EARTH:
RAINSHADOWS AND VOLCANOES

The violence associated with continental drift has a substantial impact on living creatures. Individual earthquakes can be disruptive, but they generally are local and threaten a specific species only if the species is very limited in range and confined to a tiny area (but see also Chapter 29, page 326 on the Indian Ocean tsunami). However, in the longer range, mountains are built by these processes. The building of mountains creates new habitats for animals and plants and thus increases variation in environment, allowing a greater variety of species (see Chapter 24, page 336). As an example, note that one can find very cold-weather organisms, such as are found in the arctic and Antarctic, in the snow-covered mountains of the Andes at the equator. More importantly, mountains markedly alter the climate over a very large range.

RAINSHADOWS AND CLIMATE CHANGE

Changing geography, such as the building of mountains, changes climate in more ways than the obvious ones that mountains are much colder at the top than the bottom. They create barriers to the movement of many animals and plants, separating the populations on the two sides of the range. This can be an important force for speciation as is discussed on page 317, Chapter 22. One of the most important changes that mountain building brings about is climate change, by directing and limiting the pattern of winds. Some of the more curious among you may have noticed that the deserts in the west of North America lie just to the east of mountain ranges. The same is true of South America, southern Africa, and elsewhere. These deserts develop because of an effect called rain shadow. An excellent example is shown in Fig. 23.1, which depicts the north and south sides of the Alborz Mountains, photographed on the same day. In this part of the world, the prevailing winds blow south across the Caspian Sea, absorbing water and becoming moist as they go. When they reach the mountains, they are forced upward. The air expands because the weight of the air above it is less. If you have ever allowed gas to escape rapidly from a cylinder or a tire, you know that expanding gas cools rapidly. Also, if you have breathed out through your mouth on a cold day or seen "steam" collect on the inside of windows of a warm car on a cold day or on the outside of a glass of a

Figure 23.1.a The effect of mountains on rain. These pictures come from the Alborz mountains in Iran, which form the southern barrier of the Caspian Sea, blocking the prevailing southward winds off the sea. (a): (Upper left): The leeward (south) side of the mountains is dry and barren. The only plant life seen derives from water seeping through the mountains from springs. (b): (Upper right): The windward (north) side photographed on the same day, approximately 30 miles from the previous photograph. The mountains are cloud-covered and the forest is lush, indicating substantial rainfall. (c): (Lower left): Looking northward from an airplane, it is clear that the mountains form a barrier to the weather. The area to the north of the mountains is heavily covered by cloud (white area); to the south (bottom of picture) the mountains are dry. (d): (Lower right): Rainfall, and therefore forest, cease abruptly near the top of the mountains

cool drink on a warm day, you can understand that warm air can hold much more water than cold air, and that when warm air saturated with water cools, the water condenses. (This "explanation" is of course a hypothesis to interpret an observation. It was suggested by physicists starting in the 18th Century and tested extensively. You can easily devise an appropriate, controlled experiment to test the hypothesis. It should be suitable for class discussion.) Air cools as it rises on the windward side of mountains. The rain can be quite heavy, as in the illustration, and the forest can be lush and dense. On the other side of the mountain, not only does the air contain much less water, as it descends it is compressed by the weight of the air above it. Again, if you have filled a tire with air, you may have noticed that the tire becomes hot. When gas is compressed, it gets warmer. Thus the air descending the mountain becomes hotter and drier, absorbing water from the land rather than adding it in the form of rain. The line of demarcation can be quite sharp, marking

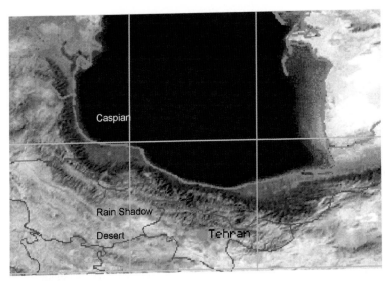

Figure 23.1.b The rainshadow as seen by satellite. The green valley bordering the Caspian gives way abruptly at the crest of the mountains to desert terrain to the south, Credits: http://eol.jsc.nasa.gov/sseop/clickmap/

the point at which the winds begin to sink. Because of these effects, the climate on both sides of the mountain is changed dramatically, and the effect can extend many miles, even hundreds of miles, beyond the mountains, well beyond the point at which the mountains are visible. On the windward side of the mountain, organisms accustomed to wetter climates will be found, while on the other side one would find animals and plants adapted to drier climates. The monsoons of India are generated by the wind coming off the Indian Ocean and rising over the Himalayas; on the north side of the mountains is the Gobi desert.

VOLCANOES

Volcanoes present another problem. An individual volcano can create, at least temporarily, world-wide climate change, by filling the atmosphere with enough soot to seriously limit light penetration to the earth. La Soufrière erupted in the Caribbean in 1812, followed by the Philippine volcano Mayon in 1814. The dust can stay in the upper atmosphere for several years. Thus in 1815, when Mount Tambora, a volcano in Indonesia ejected 400 km³ of dust (that is, a cube of packed dust a little over 3 miles on each side) into the upper atmosphere, the atmosphere was heavily laden, producing, in 1816, the "year without a summer". There were frosts every month of the year in New England, and crops in northern Europe, the American Northeast, and eastern Canada were all lost. Without trade of foodstuffs, there would have been massive starvation. As it was, there was famine in Europe, North America, and Russia. Not counted in the losses were all the insects that must have depended on specific leaves to eat and warm temperatures so that they could move around. Likewise any insects or other predators that lived on these insects would have been in severe trouble, as well as any annual plants that depended on insects to pollinate them. The world recovered from this situation since animals and plants from more southerly regions of North America and Europe could repopulate the region, but what if the explosions were more massive or more global? Much of eastern Siberia and India are covered with lava, indicating a far more violent phase of the earth's existence. Throughout the world there are lava caps—high promontories that did not erode because the area was covered with lava rather than, for instance, sand. And Tambora was far from the most violent explosion that the earth has seen. The eruption of Krakatoa, east of Java, in 1883 influenced climate for many years and killed 36,000 people. The explosion of Santorini (Thera) approximately 1650 BC produced a tsunami in nearby Crete, may have been a major blow to the Minoan civilization in that region of the Mediterranean, and may have been the source of the Biblical story of the parting of the Red Sea. On the remnant of the island of Santorini, volcanic ash can be 150 feet deep; the last Minoan town is buried under 6 feet of ash (Fig. 23.2). For living creatures on an island, to which replacement individuals of the same species have difficulty coming, such events very heavily influence the possibility of survival of the species.

Meteorites colliding with the earth can produce similar disarray. For instance, a meteorite that smashed into Arizona about 50,000 years ago (as dated by techniques described in Chapter 8) produced a half-mile wide crater (Fig. 23.3). Based on estimates of its speed, it is possible to calculate that the meteorite would have been approximately 80 ft in diameter. From evidence of other craters on the earth and on the moon, we can be certain that many other, far more massive, meteorites have struck the earth. In 1908 a meteorite exploded over Siberia, flattening or burning trees 20 miles away and decreasing the transparency of the atmosphere worldwide for several weeks. But the

Figure 23.2. Ash layers (left) and alternate accumulations of rock and ash (right) on Santorini Island (Thera) resulting from the explosions that destroyed its volcano approximately 3500 years ago. Some of the ash deposits were hundreds of feet deep. Credits: Ash, Santorini - © Tom Pfeiffer. www.decadevolcano.net/photos/keywords/ash.htm, Ash layers, Santorini - http://volcano.und.edu/ vwdocs/volc_images/europe_west_asia/santorini.html Photography copyrighted by Robert Decker. Decker, R., and Decker, B., 1989, Volcanoes: W.H. Freeman, New York, 285 p. This is a file from the Wikimedia Commons

Figure 23.3. Meteor crater in Arizona. Approximately 50,000 years ago, a meteorite roughly 80 feet in diameter hit, creating this half-mile wide hole. The white circle above the figure indicates the approximate size of the meteorite, Credits: http://neo.jpl.nasa.gov/images/meteorcrater.html

most intriguing story of all is the suggestion that the dinosaurs may have been destroyed by a meteorite. This last sentence is surely an exaggeration, but the effort to understand what happened to the dinosaurs is surely a fascinating detective story in science.

WHO KILLED THE DINOSAURS? THE CHICXULUB DETECTIVE STORY, OR CSI INVESTIGATES

The end of the dinosaur era seems to have been rather rapid. In evolutionary terms, this can mean hundreds of thousands of years, but even in these terms there is evidence of massive die-offs, for instance in the bends of former rivers in the American Great Plains, where huge numbers of dinosaur bones washed ashore and accumulated. There had been much speculation about various causes, ranging from the rise of mammals—not likely, since mammals were not-very-impressive small, rodent-like creatures at the time, and in any case the rise of numerous variations on the mammal theme is a typical adaptive radiation (page 35, Chapter 3) or expansion of mammal varieties to fill the niches left by the already-gone dinosaurs. Other explanations included temperature changes: if the earth got too cold, a large, cold-blooded reptile might never get warm enough to move easily; or availability of water might have been an issue. Whatever happened, there was a massive change. Something like 57% of all land species died very abruptly. The change was so sudden that there is a shift in the appearance of the soil, owing to a drastic reduction in carbon content (from decaying animal and plant life) at this period, and the transition is known as the K-T boundary (from K for Cretaceous or chalk-bearing, and the K used to distinguish this era from the earlier Carboniferous or carbon-bearing era; and T for Tertiary, or the beginning of the era of mammals) (Fig. 23.4). Thus the world had changed abruptly from one filled with reptiles and tree ferns to a more impoverished one and finally to one dominated by mammals.

The detective story begins with the effort of Walter Alvarez to get his father, Luis Alvarez, who had won a Nobel Prize for his work on hydrogen bubble chambers,

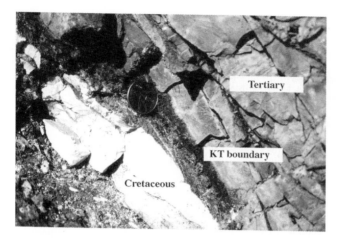

Figure 23.4. The K-T (Cretaceous-Tertiary) boundary is rather sharp and is marked by a narrow layer of increased iridium and a sharp decrease in the carbon content of the soil, reflecting a decrease in the number of organisms alive, Credits: http://www.icdp-online.de/sites/chicxulub/ ICDP-Chix/Figures/Fig1.jpg (no longer available on web)

to help him understand a question of geology. As is described in Chapter 8, one can date rocks using radioisotopes and composition, but there are occasional anomalies. For instance, the metal iridium, which can be used to assess certain dates, is relatively uniformly and rarely found in the earth's crust, though there is much more deep within the earth. In 1973 Walter Alvarez enlisted his father to help understand why, in a region of Italy, there was an exceptionally high deposit of iridium, hundreds of times higher than normal. What they determined through exploration was three observations, which they summarized in 1980:

- The iridium anomaly was found in a very thin layer of the earth, rather than being generally distributed;
- The same anomaly could be identified not only in Italy but far from Italy, in Spain and even in the Americas;
- And, the iridium layer could be dated to 65 million years ago—the same date as the K-T boundary, or the disappearance of the dinosaurs.

There is another point that they understood, which applies to this situation. Iridium is quite rare on the surface of the earth, but much more common in meteorites. They therefore made the following hypothesis:

A giant meteorite had struck the earth 65 million years ago, scattering iridium as dust throughout the world.

This meteorite had thrown up such a dust cloud that it seriously interfered with photosynthesis, disrupting the ecosystem of the planet and causing massive die-off of dinosaurs and many other organisms. The minerals in the dust would have been converted into acid, creating acid rains that would have further seriously damaged the ability of organisms to survive.

This was a very interesting hypothesis, but there was no other evidence for it. The question therefore became whether one could find independent evidence for such a meteor strike. The problem was that there were no craters on earth that could be linked to such an event. However, this was not a fatal argument or true falsification, since 2/3rds of the earth is under water and a crater might not be noticed; or it might have existed in now-eroded lands. Nevertheless a hypothesis is interesting when it suggests an experiment and the question was then what other evidence of a meteorite might one expect to find.

If a giant meteorite were to hit the earth, the shock would be immense. The sound of the Siberian meteorite exploding could be heard for hundreds of miles. Likewise, the explosion of Krakatoa produced shock waves and tsunamis that were registered on machines around the earth, and the sound of the explosion was heard 2000 miles away. Such powerful events shatter and melt rock, and therein lay our clues, in the form of tektites and shocked quartz. The Alvarez hypothesis was so compelling that a search began for such further supporting evidence of a meteorite. The first item sought was tektites (Fig. 23.5).

If you were to take ball of molten glass and to drop it into a tub of water or otherwise cool it very quickly, the outer glass would cool very quickly into a hard, rigid, shell, while the inside would cool more slowly and either have to conform to the shell or fit uncomfortably within it. Such a phenomenon is exploited in the construction of the glass insulators that hold the high voltage electrical lines to telephone poles. In this case the interior glass exerts considerable pressure on the shell. If the shell cracks, the whole insulator shatters. This makes it much easier to spot from the ground a defective insulator than if, for instance, the insulator were ordinary glass that could fail by developing a small, inconspicuous crack. In any event, when examined by appropriate microscopy, the shell should be distinguishable from the inner, more slowly cooling, mass. If a meteorite hits a rocky surface, it will be hot enough and exert enough force to melt the rock, such that sand could be converted into glass and other minerals into similar glasslike substances. These will be ejected, in molten form, from the impact site and will cool rapidly in the atmosphere, forming such a shell. Small stones formed in this fashion are called tektites, and they mark sites of ejection of molten materials, such as volcanoes and impact sites. Even more interesting, depending on how high they go into the atmosphere and how rapidly they cool, they may take on teardrop shapes. If the place in which they land is completely undisturbed, they may even lie in an orientation that suggests their origin. If the trajectory was relatively low and long, their pointed tails will face the direction from which they come. Also, the smaller, lighter ones will fly farther than the larger, heavier ones, and one can trace the source if there is a geographic distribution in size. Tektites are common and had not been much studied, but a second look demonstrated that many, particularly in the Caribbean, seemed to be about the right age of having been formed 65 million years ago. That turned much more attention to the Caribbean. That turned more attention to some peculiar deposits of sand well inland in Texas. The sand apparently had come from a large body of water, but there was no really good evidence for a lake or an ocean in that region. The distribution of sand however suggested a violent origin such as a tsunami or other sudden inrushing of water. But tsunamis come a few or tens of miles inland, not hundreds of miles inland.

There was another peculiarity with the Texas sand. It was shocked quartz. You are familiar with the situation in which a piece of glass, particularly a hard glass such as Pyrex®, breaks. A small crack forms, and this crack propagates rapidly, branching and spreading until there are cracks all over the glass. Quartz is very similar to glass, and even harder. The atoms in quartz tend to be aligned, so that polarized (aligned) light shining through quartz looks different according to the angle. When the alignment of the quartz is in the same direction as that of the light, light passes through. When the alignment is at right angles, the light is blocked, and at intermediate positions, one sees different colors. (You can see the light-blocking effect by taking two polarized sunglasses and rotating one relative to the other while

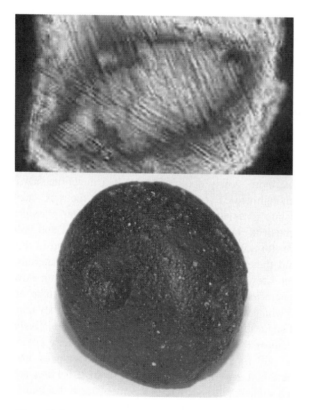

Figure 23.5. Upper: shocked quartz, as seen through crossed polarized light. The fractures in two directions are clearly visible. Material such as this is found deep into Texas. Lower: a tektite. The locations, sizes, and orientations of the tektites allow one to infer their origin. See Fig. 23.6. Credits: Shocked quartz - http://rst.gsfc.nasa.gov/Sect18/h_impact_shocked_quartz_03.jpg, Tektite - Wikipedia commons

looking at the sky. The polarizers will become opaque when they are "crossed" or at right angles to each other.) The value of this insight is that polarized light allows us to see the structure of quartz. Using this technique, we can see if the quartz has been shattered, or is "shocked quartz". It still holds together, but the crack lines are there (Fig. 23.5).

The sand in Texas is not only in the wrong place, it is shocked. What this suggests is that a powerful impact produced both the shock and a huge tsunami that flooded Texas. Again, the various dating mechanisms suggested a date of 65 million years ago. In the eastern Caribbean, tektites suggested a source to the west, and the flow lines of the presumed Texas tsunami pointed eastward, drawing attention to the Yucatan Peninsula in Mexico. But there was no crater in the Yucatan.

Or was there? From the surface, none is apparent, though there is a modest quarter-circle of a small ridge on the north coast. However, the Yucatan is basically an old coral reef, easily eroded, and sands can easily shift and fill holes in the sea floor. More sophisticated techniques were obviously necessary. These included forms of radar and sonar, techniques that send microwave and sound wave signals, respectively, and listen for the echoes. Harder material will send a more identifiable echo, and the timing of the echo will indicate its distance or depth. These techniques produced a very surprising result: Although it was filled with sand, there was a very clear circular outline along the north coast of the Yucatan Peninsula.

The crater is approximately 100 miles across, suggesting the impact of a meteor approximately 4 ½ miles across. It is approximately 65 million years old, and it has another interesting property: judging from the structure of the crater, the meteorite came in at a very shallow angle from the south-southeast (Fig. 23.6).

The size and angle lead to several interesting predictions. First, it could have generated a Texas-bound tsunami of sufficient size to carry the sand appropriately far into Texas. Second, the heat generated would have been sufficient to generate a firestorm over much of North America. From various lines of evidence the dying of the dinosaurs appeared to begin in North America and then to spread to Europe and Asia. The direction of the meteorite could justify this argument.

Thus this beautiful detective story meets our criteria in terms of **Evidence**, particularly multiple, independent, sources of evidence, and **Logic**, in that there is a reason for the dinosaurs to die. If the explosion of Tamboura or Krakatoa was bad, this would have been much, much worse. Sufficient dust would have been ejected to seriously undermine photosynthesis for at least a decade, leading to massive starvation and collapse of ecological cycles of dependency. The loss of larger numbers of land than marine animals would likewise follow, since the ocean would have at least been protected from fires. To summarize the evidence and logic:

• There is substantial evidence that a giant meteorite hit the earth 65 million years ago. The iridium layer, the tektites, the shocked quartz, the evidence of a tsunami, and a crater all point to an impact site on the coast of the Yucatan Peninsula in Mexico.

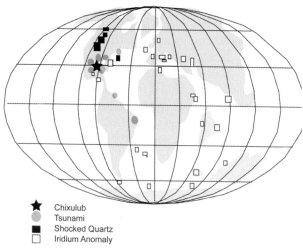

★ Chixulub
● Tsunami
■ Shocked Quartz
□ Iridium Anomaly

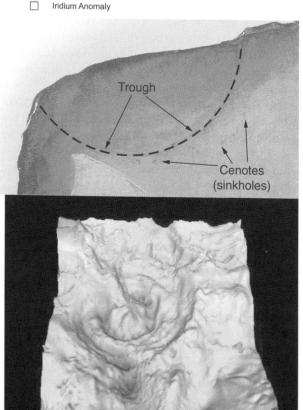

- There is substantial evidence for a massive die-off of reptiles and many other organisms 65 million years ago, so substantial that it marked the end of one era and the beginning of another.
- Logic—that is, calculations—indicate that a meteorite big enough to make the Chixulub crater could have created a firestorm and a dust cloud large enough to strongly decrease photosynthesis and could cool the earth enough to make life very difficult for cold-blooded creatures. Certainly the changes to the climate would have done great damage.

ARGUMENTS AGAINST THE CHICXULUB HYPOTHESIS

What we do not have is true falsifiability. Obviously, we have no desire to replicate the experiment, but is there any way that we can test the hypothesis? Not really, but there are those who do not consider what is now called the Chicxulub hypothesis to be correct. They base their arguments on four issues:

The calculations all are based on many assumptions, and there is some disagreement over the assumptions. For instance, dust in the atmosphere will cool the earth, but so-called greenhouse gases will warm it. (Greenhouse gases act like one-way mirrors: they allow sunlight to reach the earth but, when the sunlight warms objects on the surface of the earth and the heat rises into the atmosphere, the gas reflects the heat back to the earth, trapping the heat.) How much the two effects will cancel each other out is a matter of dispute.

An impact the size of this meteorite would certainly have repercussions around the world. The immense lava flows of Siberia and India were generated at approximately this time. Perhaps the meteorite triggered this activity, perhaps not; but in any event the volcanic activity could have generated the dusts and the greenhouse gases without invoking a meteorite.

Figure 23.6. Chixulub. (upper). Map of locations of evidence of a tsunami (along the American coasts); tektite fallout, which are in a gradient with larger tektites being found to the north and west of the Yucatan peninsula; and iridium anomalies, which are world-wide but more equatorial than elsewhere. The map illustrates the positions of the continents at the time of the Cretaceous mass extinction, with India not yet having collided with Asia. The approximate site of Chixulub is marked with an X. (middle) Evidence of the crater at Chixulub. There are traces of a semicircle on the north coast of the Yucatan peninsula, with sinkholes all along the semicircle. The sinkholes were caused by shattering of the coral structure with the impact. (lower) A computerized image of the variations in gravitational fields, a measurement of the density of the soil and rock and therefore a measurement of the total metal and compression of the soil, along the coast. Today's coastline runs from left to right approximately through the center of the impact zone. All the evidence is consistent with a collision of a meteorite coming in from the southeast and plowing toward the northwest. Credits: Chixulub map - http://www.ugr.es/~mlamolda/congresos/bioeventos/claeys.html, Chixulub trough - http://photojournal.jpl.nasa.gov/catalog/PIA03379, Chixulub gravity - http://www.lpl.arizona.edu/SIC/impact_cratering/Chicxulub/Drilling_Project.html and http://antwrp.gsfc.nasa.gov/apod/ap000226.html

There was evidence for a decline in dinosaur populations before the K-T transition. It is not clear what caused this decline, or whether the populations were already in severe trouble. Thus the impact might have been the death knell for a deteriorating pattern of life, or have had minimal impact on an already imminent collapse. We cannot tell if the "rapid" collapse occurred in weeks or over thousands of years.

The K-T event, while spectacular since it is relatively recent and involved gigantic animals, is only one of several population collapses, and it is far from the most massive. There is no solid evidence that meteorites triggered the other collapses.

The detective story of "Who killed the dinosaurs" illustrates very well the limitations of science. We have a wonderful story, complete with evidence and logic, that a meteorite wiped out the dinosaurs. Today it is enshrined in children's tales and movies. Just because it is appealing however does not make it so, and many questions remain and must be answered before we can really consider this hypothesis to be a theory. The most important consideration is that of the other massive

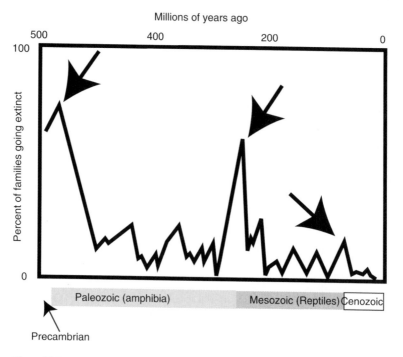

Figure 23.7. Mass extinctions. 75% of all existing families died out at the end of the Precambrian and the beginning of the Cambrian (large arrow at left). The disappearance of over 60% of all families marked the end of the Paleozoic in an event termed the Permian mass extinction (middle arrow). In contrast, the Cretaceous mass extinction, ending the reign of the dinosaurs and marking the transition from the Mesozoic to the Cenozoic, though spectacular from our viewpoint, was far less drastic than the earlier ones, perhaps because the total variety of creatures on earth has been steadily increasing. Credits: Author's drawing from textbook figure

declines in life on earth (Fig. 23.7). Although people are looking for evidence of meteorite impact at these times, the further back in time one goes, because of erosion, continental drift, and other processes that replace or alter exposed rock the harder it becomes to document so transient an event as the impact of a meteorite. We are left to conclude that the impact of the Chicxulub meteorite was too close to the crash of the dinosaurs for it to be entirely coincidental, but we cannot prove it. We are of course very interested in what causes these crashes, from both the intellectual and purely selfish viewpoints. The loss of 80% of the weirdest possible creatures cleared the way for the rise in recognizable life forms that we know as the Cambrian, the beginning of the Paleozoic; these creatures thrived for almost 300 million years before the amphibia died off (the Permian mass extinction, eliminating up to 60% of existing families) and left an opening for the reptiles, marking the end of the Paleozoic and the beginning of the Mesozoic (some people claim to have located a crater representing a meteorite "the size of Mount Everest," but this finding is disputed); the reptiles dominated the earth for approximately 120 million years before the K-T extinction swept away at least 20% of all existing families of organisms; and in their swath, the mammals rose to characterize the Cenozoic. If we look to the future, we have to ask: Will we die off? What might cause us to collapse? What might replace us?

REFERENCES

http://en.wikipedia.org/wiki/Image:Trinlake2.JPG (Wikipedia: the K-T boundary)

http://en.wikipedia.org/wiki/Chicxulub_Crater (Wikipedia, on the Chicxulub Crater)

http://www.lpl.arizona.edu/SIC/impact_cratering/Chicxulub/Chicx_title.html (Chicxulub, from the Lunar and Planetary Laboratory, University of Arizona at Tucson)

http://antwrp.gsfc.nasa.gov/apod/ap000226.html (Images of data from Chicxulub, from NASA)

STUDY QUESTIONS

1. Closely follow weather reports for a week or so, and notice any features of the weather in your or other regions that appear to be dependent on the geography of the region. Note mountains, large bodies of water, and the presence of large cities.
2. Look up reports of the last large volcanic explosions. Did they influence the weather in any way?
3. What other explanations can you give for the sudden discontinuity at the K-T boundary? (There are several possible explanations.)
4. What other explanations are possible to explain the close of the Mesozoic?
5. If the Chixulub hypothesis is true, why has it not been possible to confirm the argument for the other massive die-offs or extinctions that have occurred?

PART 6

THE ORIGIN OF SPECIES

CHAPTER 24

COMPETITION AMONG SPECIES

BIZARRE LIFESTYLES

Now that we have a full range of creatures on earth, what pressures drive evolution? In other words, what forces cause selection? Some are obvious: Prey animals need to avoid predators, and predators need to catch prey. Other mechanisms are less obvious. Why do we have creatures that live in the most horrible environments or eat the most restricted and to our minds unpalatable foods, and why do some creatures spend so much time in courtship, or sport such bizarre appurtenances of their sex that their movement is actually restricted, subjecting them to severe danger from predators? We will discuss these issues in this section on how species form, Chapters 24 through 27. Let us begin by considering some of the more bizarre lifestyles.

The argument that God made creatures to fulfill all needs on earth, from scavengers of carrion and offal to parasites of parasites[10] is at some level satisfying but, to take just the human situation, most of us tend to understand that life in a freezing land where there may be no sun for half the year, or in the midst of a hot (over 100° F), humid, rainforest filled with disease-bearing insects, is not as pleasant as it might be, and we might ask why people live in such conditions. For that matter, we can ask why there are creatures living in springs so hot that the water would quickly burn our skins; in water three to ten times as salty as the ocean; in supercooled water (water below its freezing point, which because of technical reasons has not turned to ice, but can do so if it is shaken or a bit of dust falls into it); in deserts so dry that the creatures can move about only one or two days per year. Or one can ask by some cicadas live as larvae for 17 years and emerge as adults to live and mate in the space of a week or so, or mayflies up to three years as a larva and less than one day as an adult; or why a plant should take ten years to bloom, with the bloom lasting less than one day; or why some animals exist in remarkably restricted existences, such as being a parasite on a certain part

---

[10] "Big bugs have little bugs
On their backs to bite'm
Little bugs have lesser bugs
And so on ad infinitum."
Believe it or not, this little ditty was popular, in Latin, among students in medieval times.

of a single species of animal (the lice that inhabit human head hair, body hair, and pubic hair are different species) or subsisting on food that is either noxious or of extremely poor nutritional quality. As usual, a question beginning "Why" does not point us in a very fruitful direction, but we can rephrase it to wonder what selection pressures function such that there would be an advantage for organisms choosing the most ascetic of lifestyles. Here we must truly deal with the issue of the biological niche.

THE BIOLOGICAL NICHE

"Niche," as you recall, is the French word meaning "nest," and it refers to the particular segment of worldwide activity occupied by a given species. The definition of the term "segment" is particularly vague, as it could be anything from "eater of mice" to "catcher of insects flying at dawn" to "parasite in the intestines of termites" to "grass that can grow at temperatures below 35° F". What is important is that the species fulfill a unique function and role, for two species that fulfill the same role will compete with each other, and in the long run only one contender will survive. A creature may occupy several niches simultaneously—where it nests, what it eats in the summer, what it eats in the winter—but it must defend its position in each niche. The idea that these roles exist and can be defined is well supported by the previous existence of marsupial dog-like creatures called Tasmanian wolves in Australia (Fig. 24.1) and the existence of marsupials, likewise in Australia where there were no other true mammals, filling the roles of rabbits, mice, squirrels, and cows. We see similar expansions of roles in many isolated settings, particularly islands, where lizards can be the primary grazers, or a completely surprising species can fill a role normally associated with another species. For instance, in the Galapagos Islands, one of the finches acts like a woodpecker, a bird that can bang a hole into a tree to dig out an insect living in the wood. The finch does not have the reinforced head and neck or the long tongue of a woodpecker, and it digs insects out of less woody cactus by using a thorn to dig the hole.

More interesting is the question of why there should be so many variations of finches in the Galapagos. It has a lot to do with the question of competing for niches. Remember that a niche can be anything—we will discuss below niches not related to food—but you can understand the concept by imagining that humans and squirrels live on an island in which the only edible material is nuts, but there are many kinds of nuts available. They range from giant nuts like the seeds of mangos or avocados to tiny, dust-like seeds like grass seeds, including in the range peanuts, walnuts, cashews, etc. If the humans and the squirrels both find that they cannot open the avocado seeds and cannot get enough nutrition from the grass seeds to make them worth-while, both will seek the sunflower seed to walnut type of seed, and between the two may consume all of the available seeds, so that both finally starve; or eventually one species will be able to drive the other away from the food, so that the latter will starve and only one species will survive. In this case, it is not necessarily the human. There have been many instances in which humans were unable to protect grain stores against rats, or even molds, and the human population has perished.

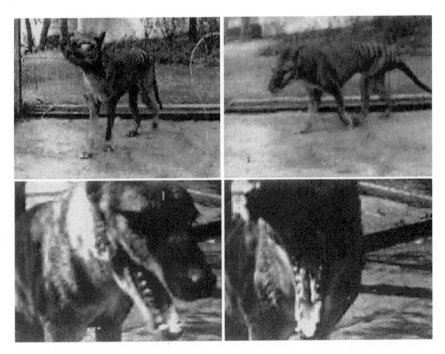

Figure 24.1. Tasmanian wolf. This animal was not a dog but a marsupial related to kangaroos and opossums. It loped like a German shepherd and in almost every way resembled in lifestyle a hunting dog. You can tell that it is not a dog because its hind feet are more flat, like those of a kangaroo; it has very narrow hips, since its young are extremely tiny when they are born; and its teeth, though clearly those of a carnivore, lack the large canines that distinguish the dog family. In Australia there are rabbit-like, mouse-like, groundhog-like, and badger-like marsupials, Credits: http://cas.bellarmine.edu/tietjen/images/tasmanian_wolf.htm (public domain)

If humans manage to learn to crack open the avocado seeds, which the squirrels cannot touch; or if the squirrels manage to tolerate the grass seeds, coexistence will become possible. More likely, among all the squirrels, a few, perhaps the smaller ones, in desperation will try the smaller seeds, and some of these will survive while the others, competing with humans for the larger seeds which the humans have now learned to protect with nets, will starve. The generation of surviving squirrels will be the smaller ones, carrying genes for smaller size, and the species will survive by evolving. Eventually, both species will continue to survive on the island, the humans eating the larger seeds and the squirrels the smaller ones. This is what we mean when we make the statement that only one species can occupy a given niche, and that the other species must evolve to define a new niche or it will become extinct. However, this example is rather extreme and meant to make a point. Most typically competition will occur between two closely-related species that, after being separated, expand their ranges to encounter each other. In this circumstance both species will have to evolve to avoid competition with each other.

If one species feeds in the morning and the other in the evening, these adaptations may be sufficient to avoid the competition. Likewise, if one species adapts to eat an otherwise unpalatable animal or plant, this too may be sufficient.

FOUNDER EFFECT

Sometimes speciation may be accidental. Pigeons show a wide variety of colors. Suppose that, during a hurricane, a pair of pigeons or even a single pregnant female were to be blown to an island where she could survive. This female, however, was a white pigeon. The colony that she founded, unlike the parent colony, might consist entirely of white pigeons and, evolving on the island, could end up very different from the parent colony. Such transferals have been seen many times. In addition to the presumed origin of the Galapagos and Cabo Verde birds, in the 1970's a few parrots were blown by a hurricane from Guatemala to Florida, where they have now taken up residence and their numbers are expanding. It is very doubtful that they represent the entire range of genetic possibilities of the Guatemala parrots.

ALLOPATRIC AND SYMPATRIC SPECIATION

The above examples illustrate species evolution caused by competition with very different organisms and accidental isolation. The former might cause change in a species but not necessarily speciation, or creation of a new species. This often occurs when dissimilar populations of a single species compete with each other. This hypothesis argues that new species often arise when one common population separates into two isolated populations. Each population continues in its path of adaptation until one conquers the barrier and the two populations meet again. If the two populations now differ so much that the hybrids are less successful, any individual that chooses a partner from the other population will stand less chance of leaving young to the next generation, and selection will operate to eliminate such cross-breeding. Once the barrier to cross-breeding is complete, the single species is now two, and each will continue its evolution independent of the other. Such a process appears to have occurred in the Galapagos Islands, in which several different species of finch are distinguished essentially by the size of their beaks, which defines what they can eat. Some finches eat insects. Others eat large, medium, or small seeds. Such a process, repeated isolation, adaptation, re-encounter, and reproductive isolation can generate very large numbers of species, such as the thirteen or fourteen types of finches in the Galapagos and the many varieties of animals and plants that appear when a totally virgin land is first occupied, such as the appearance of many types of carnivorous and herbivorous marsupials in Australia. Under these conditions it is called **adaptive radiation**. We do not need a totally virgin land. The rise of the reptiles was presaged by the disappearance of the Amphibia from their niches, as was the rise of mammals following the disappearance of the reptiles. This type of speciation, allopatric speciation, seems to be the most common, and the interpretation of ring species (page 59, Chapter 5) is that, without geographic isolation, gulls or leopard frogs or house mice can never become truly separate species. There is however evidence that species can become divided into two populations even within the same geographical region by changing specialization. A flea that infests both birds and mammals can develop populations that vastly prefer birds and populations that vastly prefer mammals. If the two populations do not cross during breeding, they are effectively isolated from each other. Although it is less obvious in the Galapagos, most birds change food with the seasons, eating insects when they are plentiful and seeds when insects are not plentiful. Two birds that prefer the same seed could differentiate into two niches if the seasons at which they switched from seeds to birds and back were different. One bird might eat only the green seeds, and another only the mature seeds. Or one bird might forage only the seeds that fell to the ground, and the other might take them from the tree. Or one bird might fly and eat seeds only above 10 feet off the ground, and the other only from the lower branches. Any mechanism that lowers the competition between two species will allow both to survive or, from the standpoint of the species less favored, will allow it to escape extinction.

This competition involves more than food. Many other factors can enter into the competition. Nesting sites provide greater or lesser protection from predators or weather, and in this sense a nest on the most peripheral part of a branch differs from one in the crotch of a branch, and both differ from one in a hole in the tree; or a nest in a spruce tree differs from one in a maple tree. A hunter that can find food when the ground is snow-covered or can survive periods of snow cover by hibernating or not eating can extend its territory into regions where its competitors cannot go. An animal that can tunnel into the ground can live differently than an animal that can merely dig a hole.

COMPETITION AMONG SPECIES

This is the crux of the argument: Animals and plants compete for every resource on the planet and the rules of the game are simple if brutal: Win the competition; find another game; or die. It may or may not be God's Plan to find a creature to eat poison ivy or to put a brine shrimp in the Great Salt Lake, but the mechanics of the process are straightforward: the attractiveness of the bizarre, precarious, or rare niche is that it offers freedom from competition. The organism that can deal with that situation can live without being driven out by another organism, and the species can evolve to become more and more specialized to deal with the unusual circumstances, rendering it even more capable of fending off future challenges. Its digestive tract may evolve to deal with foods of the most questionable nutritional value or to detoxify particularly vile toxins. Some well-known examples include termites, which contain in their digestive tracts single-celled animals that can digest wood; several animals including the larva of the monarch butterfly, which incorporate deadly toxins from the plants they eat, so that they are not harmed but they remain toxic to other animals; others that reabsorb water from the driest of foods and even animals and plants that can absorb water from the humidity in the air. Others can expel salts from their bodies in the saltiest of environments or extract necessary salts from the most unpromising of environments. Desert rodents can extract and make enough water from seeds to live their entire lives without a single drink, and camels can store enough water resources to cross the desert. Other animals and plants have evolved modifications of proteins so that they can survive extremes of heat, cold, acidity, and alkalinity. Some frogs can complete their life cycles in a temporary pond that lasts less than one month, while others can do so in the water that collects in a leaf in the rain forest. The carnivorous plants make up for the poor nitrogen content of acid bogs by trapping and digesting insects. There are creatures that live only in the ears of a specific species of mammal or bird or on the roots of a specific species of plant. Some fish and invertebrates live, feed, and hide among deadly stinging corals. Others are found only in the mist of great waterfalls, or in the water accumulated by certain aerial plants in a rainforest, or in the nests of specific species of ants or bees. Other creatures nest or lay their eggs in the most unusual places. There are creatures that live in permanent darkness, miles beneath the surface of the ocean, subsisting on detritus that sinks from the surface or on chemicals produced by undersea volcanoes, and lichens and bacteria that survive on whatever they can consume in Antarctica. One can assume that, if a given environment is not completely hostile to life, some animal or plant has found a way to live there. The general principle is that, if a species can specialize to avoid competition with other species, it has the potential of surviving. The alternative is too horrible to contemplate. The mechanics of adaptation to apparently dreadful conditions is escape from annihilating competition.

REFERENCES

Campbell, Neil A. and Reece, Jane B.(2004) Biology (7th Edition), Benjamin Cummings, Boston, MA

STUDY QUESTIONS

1. Carefully observe any wild animal or plant in your neighborhood. Can you define the characteristics of its niche? Can you tell what competes with it for the niche? What are the limitations that it encounters, for instance, changes of food with seasons?
2. What processes might carry isolated members of a species to become founder colonies? Hint: Consider migrations caused or allowed by humans, as well as mechanisms unrelated to weather. Bacteria and viruses may also be considered founders. Search the archives of newspapers and television for reports that you suspect to illustrate founders, and prepare to discuss this possibility.
3. Do you think that mammals could have spread throughout the world if the dinosaurs had not disappeared? Why or why not?
4. What is the worst condition that you can imagine under which animals or plants can survive? What lives under these conditions? What allows it to live there, while other organisms cannot?

CHAPTER 25

SEXUAL SELECTION

Any teenager will attest to the fact that courting takes considerable energy and time, whether the courtship consists of earning enough money and investing the time to have an appealing car, practicing dance steps, working to become a prominent athlete, investing in personal care products or clothing, joining specific activities or working for acceptance in specific institutions, or any other of the myriad investments designed to improve one's chances of success. Many societies follow through with this principle, maintaining elaborate and extremely athletic activities, such as folk dances, in which young people participate and for which the finest or champion participants are highly admired.

What is not often appreciated is the importance of such activities in all sexual species on earth. We can initiate this discussion by noting that animals undertake the following activities that do not obviously promote a long and healthy life: (1) They fight among themselves, most commonly males fighting over females; (2) They carry out long and arduous courtships, which not only interfere with food gathering, they can be exhausting (In some animals, such as the sea elephant, one male may collect a harem and then spend the entire summer defending the harem against other males, not even stopping to feed.); (3) In the course of courting, they expose themselves to considerable danger, by bright coloration, prominent public display (a bullfrog will attempt to position itself in the middle of a pond to sing; herons and other wading birds are aware of this and will seek it as prey), or large, complex structures—a peacock's tail—that ultimately must be considered a hindrance to its movement, camouflage, or escape; (4) many animals have extremely elaborate and exotic genitalia, so complex that sometimes the copulating partners get stuck to each other and cannot separate. The principle of elaborate courtship appears to apply even to the presumptively dumbest and least reflective of animals, such as a cockroach. Surely these creatures are not writing ballads extolling the beauty or charm of their mate, and they obviously care little beyond the sexual act, for the partners separate and go their separate ways once consummation has been achieved. Although it is not obvious, plants display similar selectivity in the identification of the appropriate pollen by the appropriate seed. Though for obvious reasons dances or other movements are not part of the courtship, the showy flowers that modern plants produce to attract insects or other pollinators are very expensive in terms of the energy needed to build them and the energy that could have been captured if the flower had been a leaf, the risk that they pose in decreasing the stability of the plant to wind, and the amount of water that they can lose.

The question, then, is why almost all sexual organisms invest such energy to attract and choose partners. Organisms such as sea urchins, starfish, and mollusks, which do not move around much, are content simply to disperse eggs and sperm into the water. There is a certain level of selectivity, as specific proteins on the surface of the sperm and egg must interact to assure that the right sperm fertilizes the right egg, but this selection is at the level of the species, not the individual. We have no evidence that sperm A finds egg A' specifically attractive or vice-versa; it simply is a matter of which sperm and egg are simultaneously available at a given location. The question therefore is this: although it might not be nearly as much fun, would there not be more profit if coupling were indifferent or random and the money were invested in college education or a house, rather than a sexy sports car or an elegant gown? For this as for all questions in evolution, the argument is that there is no mechanism for propagating within a species any modification that does not provide tangible benefit, meaning improved chance of leaving offspring to the next generation. At best, a modification of neutral impact will persist at the very low frequency at which it appeared (if it appeared in one individual of 10,000, it will remain a characteristic of one out of 10,000). If it is deleterious in any respect, it will be driven from the population[11]. Thus if we find a characteristic widely shared in a large population, we must assume that there has been selection for the trait. In other words, the first bearers of the trait were more successful in leaving young to the next generation than those who did not have it. If we find similar characteristics in many diverse groups of organisms, the strength of the argument is redoubled: the characteristic must have decided value in evolution.

By means of hypothesis, observation, experimentation, and analysis we have come to recognize several virtues defining the value of courtship. These include first, the value of sexual recombination to survival of the species; second, the importance of distinguishing between appropriate partners (partners of the same species) and inappropriate partners (similar-appearing creatures of another species); third, synchronizing the state of readiness of the partners; and fourth and extremely important, using courtship to identify and select the healthiest and most desirable partners.

VALUE OF COURTSHIP

There are many asexual creatures in this world, starting with bacteria, amoebae and other small organisms that reproduce simply by dividing—no muss, no fuss. Even some small animals such as flatworms and sea anemones do this quite regularly.

---

[11] This is not strictly true. For instance, a small population of birds may be blown to an island in a storm. If this population consisted of one pregnant white pigeon from a large population of mostly grey pigeons, the island could continue as a white, not grey, population. But this is an exception, as discussed in Chapter 24. The argument here assumes that there is no isolation of a small, non-random segment of the population.

Many plants propagate by sending out runners, dropping branches to take root (Fig. 25.1), or producing seeds or plantlets without benefit of sexual recombination, and there are races of predominantly sexual organisms such as lizards in which females lay fertile eggs without troubling themselves to find a male. By and large, however, the vast majority of the living world is sexual. The sexuality may take highly ingenious forms, with several sexes for a single species (yeast), sex decided by temperature at which the egg is incubated (some reptiles) or by the environment

Figure 25.1 Plant runners. (a) A banyan tree. This relative of the popular houseplant called the Benjamin Fig native to the South Pacific but has been imported into Florida and Hawaii. The branches drop runners that, when they touch ground, develop into new trunks. One tree in Lahaina, Hawaii, has hundreds of trunks and now covers more than a city block. (b) A strawberry plant. Note that this single runner has started six new plantlets (white arrows)

or presence of a potential partner of the opposite sex (some mollusks, worms, and fish), readily convertible sexes (some fish), species in which the sexual choices are male or hermaphrodite (combined male and female), and species in which the male is reduced to small, parasitic bump on the female (some fish, insects, and other animals). There are organisms (ferns) in which the sexual phase, the equivalent of us, is a tiny, microscopic structure, while an asexual phase, which if it existed in mammals might be roughly equivalent to our eggs and sperm forming entire organisms on their own, is a large, dramatic plant, even reaching the size of trees. There are others (aphids) that spend the entire summer reproducing asexually, female begetting female in a series of immaculate conceptions—no time wasted or risk taken in courtship. To be anthropomorphic about it, she invests no money or time in makeup, lipstick, or fancy dresses, does not go to dances, nightclubs, or bars, and there are no males performing dangerous stunts to show off or getting into fights over her. This might not be so much fun but, in terms of efficiency, the energy invested will produce far more young. But, even in the case of the aphid, when fall comes she produces a winged generation of both males and females, and these undertake the usual forms of courtship and mating to produce overwintering eggs. So, the question remains, why do the vast majority of animals and plants undertake sexual reproduction, with all its attendant hazards and costs? Nevertheless, to common and justifiable impression, the world is sexual: "Male and female, created He them." We do not question the role of sex for all life: two-year olds will insist that you do not mis-classify them—"I'm a boy, not a girl!"—and three- and four-year olds will want to know, when encountering any animal, if it is a boy or a girl animal. Why should the world be so constructed?

What we can do is to look at what sexual reproduction achieves that is different from asexual reproduction, generate a hypothesis as to the benefit of this difference, and, hopefully, design an experiment to test the hypothesis. The assumption that was common in earlier textbooks was that the mixing of genes produced healthier children and greater variety. The argument that sexual recombination produced healthier children derived from the phenomenon of heterosis, the observation that hybrid progeny (children) were often stronger and healthier than their purebred parents. This difference probably derives from the fact that purebreds often have two copies rather than one of mildly defective genes, and thus minor defects, while in their children the defective genes are covered or compensated by good copies (from the other parent) of the genes. However, in nature most organisms are not purebred, and it appears that the heterosis effect is an artifact of human tendency to create purebreds for human purposes. Thus the issue is more an inbred weakness of the purebred rather than an increase in health of the hybrid. We therefore need better arguments.

The argument of variety is rather theoretical: It is based on the assumption that, while species are frequently well adapted to their environments, eventually the environment will change in some manner, and that some of the variant forms produced by sexual reproduction will prove better adapted, more "fit," than their parents. While this is possible, it does not explain the retention of sexuality in organisms such as horseshoe crabs (pages 37 and 163), living in relatively constant environments, which have consequently evolved very little throughout evolutionary history.

A new hypothesis was based on keen observations of the characteristics of the winners and losers of sexual selection, which led to experiments to test the validity of interpretations of choice, and finally to experiments to test the effectiveness of the choice. The hypothesis was that sexual selection chose the healthiest specimens. Recognizing that sexual characteristics require high metabolic or other demands, researchers tested two related sub-hypotheses: that females chose the showiest or most spectacular males, and that these males were the healthiest or most likely to successfully father and, if necessary, raise a brood. (This argument is presented on the assumption that males display or otherwise court females, and that females make their choice among competing males. This is often but not always the case; there are instances in which females court males, or in which the courtship is mutual.)

The first sub-hypothesis was easy to test. While there were many attempts to test the hypothesis, perhaps the easiest to picture is the spectacular display of a peacock's tail (Fig. 25.2). There is no doubt that this is a sexual display: the male proudly opens his tail feathers and struts in front of the female, posturing so as to show his tail to best advantage. The female observes these preening displays of several males and ultimately moves to and accepts the overtures of the lucky male. Researchers counted the numbers of feathers and spots on the tails of the competing males and found that the females almost invariably selected the male with the most eyespots on his tail. They then plucked a few feathers from the tail of the successful male, whereupon he immediately dropped in popularity with the young ladies. This observation has been amply confirmed among many species: females select the male with the most symmetrical tail feathers or with the brightest or showiest color; where males compete in battle or other physical struggle, they select the strongest, the winner, or the male that does the most elaborate dance. Some spiders, insects, and ospreys bring their dates "nuptial offerings"—food or other indication of what good providers they will be. The females select the suitor with the best gift. Where it is possible to interfere with the native physical characteristics of the males, dulling the colors of a champion, or improving the colors of low-ranking males, or similar adjustments, predictably alter the ranking of the male. There is not much thought or nuance in these decisions. The females identify one characteristic, called a releaser, and respond to it, ignoring any other parameter. For some animals this response can experimentally be led to the ridiculous. In stickleback fish, for instance, the belly of the male turns red during courtship. Males will drive each other away, and females will come to red-bellied males. However, males will attack, and females will come to, disks with the lower half painted red, and if one paints the belly of a female red, this hapless creature will be harassed by males and approached by females. Observation of giraffes, and of the neck bones (and therefore neck musculature) of brontosaurus, suggest that seeking of food was not the advantage achieved by their long necks. Rather, male giraffes spar during courtship by pushing each other back with their heads and necks. The long neck is more likely to be an appurtenance of sexual display, courtship, or rivalry. Similarly, since it appears that the long-necked

Figure 25.2 Elaborate sexual displays. Upper: A male peacock displays before a not-very-interested peahen. Lower: The male guppy (left) is brightly colored and has elaborate fins, whereas the female is uncolored and with normal-shaped fins, Credits: Peacock - Taken by Darkros and released to the public domain. (Wikipedia), Guppy - Date 2005-10-11 Author Silvana Gericke (http://people.freenet.de/silvanagericke/startsaqua.htm) (Wikipedia)

dinosaurs did not normally carry their heads erect, it is possible that they reared to full height only to show off, presumably the males toward the females.

If females always prefer the most spectacular males, what is the value of this choice? This question led to further examination of the hypothesis. Certainly in tests of physical strength, the more powerful male will be able to defend the best nesting site or more successfully drive off predators, but even color or agility in dancing tells something important. Sickly animals are not very peppy, they are a bit ragged, and their colors are not very good—sallow is the term we use for an unhealthy complexion. Researchers investigated bright and dull male birds and fish and quickly realized that the duller, less energetic males bore the heaviest load of parasites such as fleas and ticks. Bright, metabolically expensive, risky, or arduous activities were indicators of good health and freedom from disease, by the following logic: "I have so much energy available that I can squander it on this elaborate display that will not improve, and may even risk, my life." For the females, this is a very attractive pick-up line. Even for humans this type of logic plays some role, as manifested in strenuous folk dances, sometimes requiring great agility and speed, as well as the ability to perform with swords, fire, or other dangerous implements. Also, it is striking how often human concepts of ugliness correlate with the symptoms of common or previously common diseases.

THE VALUE OF SEX

Thus the observational and experimental evidence supports the hypothesis that selection of mate is selection of the best provider, whether this is represented as the ability to secure and defend the preferred nesting site or the least burdened by parasitic or other chronic disease. But what about the fact of sexuality itself? Would it not be easier simply to avoid all the fuss and muss of courtship, and simply produce young without a mate? It is possible, as seen in many animals, to skip all the expense and time involved in courtship and get straight to reproduction. Picture a world in which there were no teenage boys, and teenage girls at, say, the age of 18, simply became pregnant by themselves with female clones of themselves. It might not be much fun, but consider how much less effort it would be: no expensive cars, clothes, makeup, hairdos; no time invested in trying to meet an attractive partner, initiate a conversation, coyly ask for a date; no expense or time wasted on dates; no bravado or risky stunts to impress someone; no anxiety or hours spent on fretting about marriage, or planning and conducting a marriage; no need to arrange privacy for coupling—the list can go on and on. Sexual reproduction is a big investment. Again, using the evolutionary argument, such energy is not invested unless there is a big (biological) reward.

Here, unlike the earlier argument about heterosis raised above, there is a logical hypothesis that can be tested. Taking a cue from the experimenters who recognized the influence of parasitism, the idea developed from the 1970's that parasitism was itself the basis for sexuality. The logic of the hypothesis is the predicament in which predator and prey find themselves. This predicament may be described as follows: if the predator gets too efficient, it will capture all its prey and ultimately starve to death. If the prey become too good at escaping the predator, they will quickly overbreed and strip the land of all food, finally starving to death. Thus neither can become too good at what they do. This problem, called the Red Queen Hypothesis, is discussed more fully on page 366 and, in the case of the lynx and the snowshoe hare, on page 364 in Chapter 27. For sexuality, let us consider what genetic recombination can do for guppies, a small tropical fish from Central and South America. They live in small ponds, which often contain fish parasites. Suppose that the parasite becomes very good at what it does, and attacks all the guppies in the pond. They all become very sick or die, the parasite has no more hosts, and everyone loses. However, parasites are often very specific in their choice of hosts. One defense that the fish has is that there is perhaps a variation that, for one reason or another, is not very appealing to the parasite. This variant guppy will survive and flourish, while its more susceptible brothers and sisters suffer from the parasite. Ultimately, the pond will be filled with the resistant fish. However, the parasite is now in trouble, unless, of course, it too can generate varieties of parasites. One variety may be able to attack the flourishing variety of guppy. Thus, eventually, the parasite comes back (evolves) to infest this type of guppy. The guppy species' primary defense is to generate a variety resistant to this new variety of parasite, and the round goes on.

Clones or purebred lines cannot easily generate many varieties, but heterozygous populations can by sexual intermixture of genes. Thus, the logic says, sexuality provides the range of variation necessary to maintain these host-parasite interactions. This suggests an experiment: if we release guppies into ponds teeming with parasites or free of parasites, will we see a greater range of variation in the pond with the parasites? The experiment was done in Trinidad, where volcanic springs have recently in geological history produced ponds that have not yet been reached by either guppy or parasite. The results were as predicted. Fish thrived in ponds with parasites only when they maintained a large genetic variety, subject at any time to considerable mixing. (It was not possible to assess the genetic variability of the parasite; this was assumed.) Thus the argument is as follows: for the *population*, if the population maintains variety, it has the capability of resisting invasions of new parasites. Although the selection at the minute-to-minute level works on the individual, on the larger scale it is the selection for the highly variable population that can adapt to numerous, rapidly changing, onslaughts of parasites, as opposed to the genetically pure, non-variant population, that counts. Perhaps the genetically pure population can do extremely well in some circumstances, but it is vulnerable to massive destruction. This is a big problem for our agriculture, in which we have sought specific high-yield strains and raise them in monoculture. An epidemic disease can wipe out the entire crop. If there were many varieties of, for instance, wheat or corn, the population as a whole would resist much better, but then its yield might even in good years be much less than the monoculture. We try to get the best of both worlds, by holding in reserve varieties of seeds that are resistant to various diseases, but it is difficult. As a final note and a consideration for our understanding of experimental science, it took us a very long time to recognize this very important and ultimately obvious advantage of sexual reproduction precisely because, in the laboratory setting, we strive to maintain the healthiest cultures of animals and therefore attempt to eliminate from our cultures parasitic and other diseases.

REFERENCES

Campbell, Neil A. and Reece, Jane B.(2004) Biology (7th Edition), Benjamin Cummings, Boston, MA
Gould, Stephen Jay, 2002, The structure of evolutionary theory. Harvard University Press, Cambridge MA.
Wilson, Edward O. 1995, Naturalist, Warner Books, New York.
Wilson, Edward O, 1996, In search of nature, Island Press/Shearwater Books, Washington, D.C.

STUDY QUESTIONS

1. For your own life, try to estimate what percent of your budget and your time are invested in activities reasonably considered to be courtship.
2. In the springtime, observe the courtship of any animals that you can readily see, such as pigeons or other birds, or aquarium fish. (Domestic animals are less easy to study, since they are usually selected to breed without much fuss or to-do.) To what extent do the courting partners do things that simply define who they are? To what extent do the partners undertake activities that appear to be displays of strength, endurance, or health?
3. Can you observe courtship patterns or display of traits that you can reasonably interpret to be hazardous to the displayer?
4. Can you identify behaviors teenage boys use to impress girls that might otherwise be considered hazardous?
5. Although farmers tend to raise at any one time a specific purebred variety of wheat, the U. S. Department of Agriculture maintains in reserve many varieties, each of which is resistant to a different type of disease or climatic condition. Is this equivalent to allowing free sexual recombination of the wheat? Is it more or less efficient?

CHAPTER 26

COEVOLUTION

One of the most peculiar and interesting aspects of evolution is the adaptation of one species to another. These adaptations are so common and so obvious that they are often presented as proof of God's wisdom and purpose. We have already mentioned in passing two of these pairings that caught Darwin's attention: the ability of the coconut crab to open coconuts (page 89) and the co-adaptation of a trumpet flower and a hawkmoth (page 91). There are many other adaptations, some of which can be identified by casual observation. Darwin in later years invested considerable effort in a study of the extreme variation in the varieties of orchids, noting that many orchids strongly resembled female insects. Some were so effective in this mimicry that males tried to copulate with the flowers, in the process pollinating them (Fig. 26.1). In a more local setting, if you go to any large outdoor display of flowers and you look closely, you will find insects that you perceive to be wasps or hornets, but which on closer inspection prove to be harmless flies that merely look like wasps (Fig. 26.2). If you have a sassafras tree in your vicinity, you may notice some leaves that are rolled up. If you unroll the leaf, suddenly a large green head with threatening eyes and an orange, evil-smelling tongue will jump out at you. If you are not too startled (birds have been seen to rear back in fright) you will see that it is only a relatively large green caterpillar pretending to be a snake (Fig. 26.3). If you live in the Southeast of the United States, you may have heard the rhyme by which you can distinguish between the deadly coral snake and the harmless but very similar-appearing milk snake: "Red, black, and yellow, dangerous fellow; red, yellow, and black, he's all right, Jack". The larva of the monarch butterfly feeds on milkweed and incorporates enough of its toxin to make a bird sick. A bird that eats one of these beautiful and showy butterflies will not touch another for years, and there is a completely harmless butterfly that resembles it so closely that animals and humans usually cannot tell the difference. Many species of wasps or other threatening creatures tend to look a lot alike. Even some hawkmoths have body coloring that make them look like wasps as they hover near a flower. These are all examples of aposematic mimicry, in which a harmless creature imitates a more dangerous one and thus garners the protection of mistaken identity. Other organisms cooperate, to their mutual benefit. Lichens, common in adverse climates, are known as individual species but they are a combination of an alga, which can conduct photosynthesis and thus produce biological energy, and a fungus, which provides a stable, water-retaining structure in which the alga can

Figure 26.1. Some orchids, such as these *Phalanopsis*, resemble and smell like female insects so that males of the species actually try to copulate with the flower. In doing so, they get the pollen on themselves and, when they visit the next flower (they don't learn from their mistakes), deposit the pollen onto that flower, Credits: © Photographer: Mario Koehler | Agency: Dreamstime.com

survive. The yucca plant, a large and spectacular succulent of the western desert, produces more long-lasting flowers than most desert plants. The female of one species of moth deliberately collects pollen from one plant, flies to another, and packs the pollen onto the pistil (the female reproductive organ) of the flower of the other plant. She then lays her eggs in the flower. The caterpillars hatch and eat some but not all of the seeds, and the bargain is complete: some seeds have been fertilized and mature, and the moth has likewise assured the survival of the next generation. Some marine animals are toxic because they allow deadly stinging anemones to attach to their shells. Other organisms use mimicry and other forms of adaptations for many purposes. Walking sticks and leaf insects closely resemble the plants on which they live, as do many other animals (Fig.11.1). An anglerfish

Figure 26.2. This fearsome-looking insect is not a wasp but a harmless fly. Even its front legs are shaped and held so that they appear to be the antennae (feelers) of the wasp, since flies have very unobtrusive antennae, Credits: http://www.cirrusimage.com/fly_syrphid_Temnostoma.htm Date06/15/2005 Author Bruce Marlin (wikipedia)

sports a feeler-like appendage that looks very much like a small, innocent but tasty fish, dangling directly over its large but otherwise inconspicuous mouth (Fig. 26.4). For many fruiting plants, the fruit is bitter and unpalatable or even poisonous (like nightshade or tomato) until it is ripe, whereupon it turns color and becomes sweet. Even then, the seed is either indigestible or unpalatable so that it is spit out or, better yet, passed intact through the digestive tract to be deposited in a little pile of fertilizer. Poison ivy, clinging to the trunks of trees, turns a spectacular red just as its berries ripen, attracting the attention of migrating birds, which eagerly consume the energy-rich berries, while spitting out or excreting the unpalatable seeds.

All of these examples describe a form of cooperation among two or more completely unrelated species, but they fall into different categories. In some instances, two species cooperate to provide an efficient assurance of survival or reproduction for both species. Many types of organisms hide by imitating plants, animals, or objects in their environment. In other instances, one species finds protection in its resemblance to an unrelated noxious species. Poisonous or dangerous species are frequently quite showy—they REALLY want their "don't mess with me" message to get out—and often look alike, so that a potential predator need learn only once that a particular showy pattern spells danger. We can classify the forms of cooperation into different groups.

Mutualism describes the general situation in which two unrelated species reside together in a cooperative arrangement. Mutualism can cover situations from the most casual to the most intimate. For instance, suckerfish or remoras cling to the bodies of sharks, eating scraps that these rather messy eaters drop, and otherwise

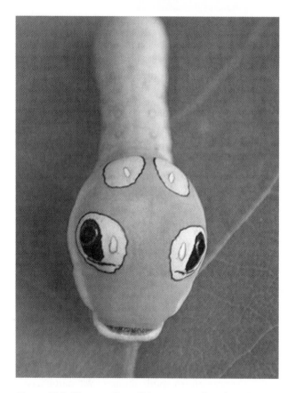

Figure 26.3. Tiger swallowtail larvae normally spin a silk pad so that a leaf tends to curl around them. If a curious bird pokes its head into the leaf, the caterpillar puffs up its anterior end and usually everts a forked orange gland that looks like a snake's tongue and smells awful. The "eyes" are entirely decoration; the small true head can be seen at the anterior end. Birds have been seen to give a shriek of horror and quickly fly away, Credits: © Photographer: Jill Lang | Agency: Dreamstime.com, Swallowtail larva 2 - http://www.ctbutterfly.org/tiger-swallowtail-larvae.jpg

Figure 26.4. This anglerfish has projecting from its head a remarkably fish-like attachment on a wand. It slowly waves this attachment back and forth. When another fish, looking for a nice snack, comes too close, the anglerfish lunges and grabs it in its very big mouth. The fish is also very well camouflaged. Credits: Humpback anglerfish (Melanocetus johnsonii), a species of black seadevil (Melanocetidae). From Brauer, A, 1906. Die Tiefsee-Fische. I. Systematischer Teil. In C. Chun. Wissenschaftl. Ergebnisse der deutschen Tiefsee-Expedition 'Valdivia' 1898–99. Jena 15: 1–432

removing parasites from the gills of the sharks. In this instance, both are free-living but benefit from the cooperative arrangement. The same is true for the cowbirds that hang around and sit on cattle, eating the parasitic flies that continually attack the cattle. But mutualism can go further, involving two species that cannot survive without each other, like lichens or the wood-digesting protozoa that live in the stomachs of termites. It is easily demonstrated that many rodents get many of their vitamins from the bacteria that reside in their intestines and will die if the bacteria are destroyed by antibiotics. Humans too derive many nutrients from the bacteria in their intestines, and cattle (ruminants, or animals that ''chew their cud") can live on grass because bacteria break down all the cellulose and turn it to sugar. The cud is a mass of grass and bacteria, held in a sort of pre-stomach, that the cattle can bring back up and re-chew, physically breaking it up so that the bacteria can get to it. Some protozoa benefit from commensal respiring or photosynthesizing bacteria, and today we feel that mitochondria and chloroplasts are the evolutionary descendents of an originally commensal arrangement. However, mitochondria and chloroplasts can no longer live independently.

Modifications of body shape and color are usually forms of *mimicry*, but the mimicry can be direct, protective or camouflage, or aposematic. Direct mimicry is used by one species to fool another, as does the orchid in mimicking an insect. Protective or camouflage mimicry serves the obvious purpose of making potential prey animals less visible to predators, and it can work extremely well. Not only does the skin bear the color and bumps of the environment, it is very important that the outline of the eyes, which have a similar form among all vertebrates, and the outline of the body be made less easy to discern. Thus markings and protuberances have the effect of confusing the overall discrimination of shape (Fig. 26.5). Even bright colors can work to this effect. The brilliant reflective blue colors of the morpho butterfly, for instance, are very confusing in its native environment. The light in a tropical rainforest is dim and dappled, interrupted by shafts of brilliant light. The undersides of the wings of the butterfly are brown and dappled, like the trunks and soil of the forest. When it flies, it briefly opens its wings, and one sees a bright flash of blue light, then it closes its wings and coasts, now becoming almost invisible. Suddenly a new flash of light appears elsewhere. It is almost impossible to track the flight of the butterfly. In a less spectacular fashion, the very common underwing moths perform the same trick. Their upper wings are quite cryptic, and their underwings are bright red or yellow and black. A predator pursues this brightly colored edible moth, but as soon as the moth lands it covers its hindwings, and the moth that the predator pursued has vanished. Finally, dangerous creatures take real advantage in flaunting their threat in having aposematic or warning coloration, well understood by most predators, and many animals profit by pretending that they, too, are dangerous. At a time when New York subways were considered to be dangerous, one defense considered to be reasonably effective was to maintain a posture and facial expression that communicated, "I just might be an undercover policeman, and armed." This is aposematic mimicry.

These types of interrelationships serve many purposes but they all address essentially the same issue. One means of assuring a niche is to develop a special

Figure 26.5. A pair of lionfish. Disguising one's outline is very important to success, in these cases to hunt but in other cases to hide. Normally the fish adjust their colors to blend in with the environment, but you would not be able to see the fish in a black-and-white figure. That, of course, is the point, Credits: Wikipedia.org

relationship with another organism, such that both species benefit and neither is destroyed by the other. This relationship can result in a considerable reduction in the cost of living, such as in the situation of the lichen. The alga would need to build elaborate cell walls or other means of retaining or trapping water in a harsh drying environment, but the fungus already has that capacity and can protect the alga, while the alga can trap sunlight to yield digestible products that the fungus can use. As astonishingly elaborate as these relationships are, they are assumed to have evolved by the continual selection of an initially incidental interaction or resemblance. One example would be the resemblance of the viceroy butterfly to the monarch (Fig. 26.6). Strictly speaking, this is not mutualism, since the monarch does not benefit from the relationship. Rather, it is commensal, meaning that only one partner benefits.

The closest relatives of the viceroy do not resemble monarchs, other than the overall structure of butterfly wings and the basic layout of butterfly wing patterns. The relatives are purplish-black butterflies. One can imagine a past situation in which among the ancestors of the viceroy there were individuals with variations on the color. Some had a mutation so that the purplish-black color was not so intense, and was a bit more red. Insects, as far as we can tell, see red as black but can distinguish colors a bit into the ultraviolet, so that the colors that we see may have been more or less obvious to predatory insects or birds. All that we need is for the variant to raise sufficient doubt in the mind of a bird or predatory insect that it hesitates and abandons the pursuit a few percent of the time. Selection will favor the variant and, in future generations, the selection will continue, always favoring the variant that is harder to distinguish from the toxic monarch. Today a skilled entomologist can easily distinguish the two. The viceroy is smaller, has an extra thin stripe on its hind wings, and glides with its wings flat rather than in a wide V as does the monarch. On the other hand, if you were going to quickly grab a fruit or a sandwich from a cafeteria line, and one of the fruits or sandwiches looked off-color, potentially spoiled capable of making you sick, what would you

Figure 26.6. The monarch (left) is poisonous but the viceroy (right), which in life is slightly smaller than the monarch, is a very good mimic. To a naturalist, the extra stripe on the hind wing gives it away. Credits: Viceroy - Date 2005-08-22 Author PiccoloNamek Permission English Wikipedia

do? Ultimately in many situations the advantage to both sides is such that selection of both partners moves to the same goal. For instance, although the yucca loses many seeds to the caterpillars of the yucca moth, the guaranteed pollination of some ensures more successful reproduction than randomly attracting pollinating insects in a desert situation in which water lost by flowers is very expensive for the plant.

This form of selection works so well that it is generally considered that the most virulent disease-causing organisms are those that have only recently begun to attack their hosts. Many organisms that cause severe disease in one host produce extremely mild or no symptoms in other hosts. In these cases it can often be demonstrated by DNA sequencing that the host with the mildest form of the disease is the original host. A bacterium or virus that promptly kills its host has not only lost its resource but may not have a means of finding a new host. A far better, if less dramatic, means of survival would be simply to go along for the ride, not weakening the host in any way and, if possible, supporting the survival of the host. Thus both parasite and host evolve toward a less noxious form of interaction. We will examine this consideration in Chapter 27.

REFERENCES

Darwin, Charles, 2004 (1859) The origin of the species, Introduction and notes by George Levine, Barnes and Noble Classics, New York.

Gould, Stephen Jay, 2002, The structure of evolutionary theory. Harvard University Press, Cambridge MA.

STUDY QUESTIONS

1. Research the following question: If you were to eliminate all the bacteria from your digestive tract, would your health improve or deteriorate? Explain.
2. Research the same question, but for cattle rather than humans.
3. Research the same question, but for termites rather than for humans.
4. What is your first reaction to seeing a very brightly-colored, as opposed to camouflaged, bug?
5. Do you recall ever having gotten sick shortly after having eaten something unusual, and having felt a revulsion to that food afterward? How long did that memory and behavior last? Do you think that the incident relates to the question of aposematic mimicry?

CHAPTER 27

THE IMPORTANCE OF DISEASE

A species can change very rapidly if it is forced through a bottleneck (a situation in which only a few members of a species survive). For instance, imagine a species of ants that exists only on a small Caribbean island. Normally ants propagate by producing, at specific times of the year, flying males (drones) and females (queens) that mate in flight. The queen then lands, digs a burrow, breaks off her wings, and lives off her atrophying wing muscles while she raises her first brood. The first brood of worker ants then begins to take care of her while she continues to lay eggs and build the new colony. Suppose that, just at the time of the mating flight, which is synchronous among all the colonies of that species, a hurricane came along and blew all of the flying ants out to sea. However, in one colony, there were mutant defective ants whose wings did not develop properly and they could not fly. If they could mate on the ground, they would be the only survivors. Their children would carry the mutation and therefore also be unable to produce flying drones and queens. The ants on this island would thereafter be completely flightless. In one incident, the species would have changed radically. In this case we would say that there was a 100% selection for flightlessness.

We will consider what this term means, but remember also that the bottleneck process can work in another way. Suppose that there is a species, like pigeons, in which the color varies widely. In this hurricane, a pregnant female or a pair of white pigeons is blown to a new island. They can easily live on the island, and they continue to breed. However, of the wide range of colors known in the species, only white birds are found on this island. This is called founder effect, and is also an important aspect of evolution, as is seen in the Galapagos Islands (Chapter 7).

In laboratory settings, and sometimes in the field, we can measure the extent to which a single mutation favors or disfavors its bearer. We call this selection and, strictly speaking, it refers to the likelihood that the bearer will leave progeny to the next generation. The point is that the estimation of selection pressure can in the simplest case be an easy mathematical calculation. The calculation is as follows: Let us assume a type of flower, in which one homozygous form (page 185, Chapter 13) (RR) is red, the heterozygous form (Rr) is pink, and the other homozygous form (rr) is white. (I am avoiding using "dominant" and "recessive" because in this case it is not true dominance. However, it is easier to follow.) Let us now cross two pink flowers:

(1) $Rr \times Rr \rightarrow RR, rR, Rr, rr$

In other words, we should get 25% red, 50% pink, and 25% white flowers. What if we do not? In the worst case, we might get 33% red and 67% pink, but no white flowers at all. In this case we might conclude that the white flower cannot survive or, in technical terms, that it is a recessive lethal. We could check this in various ways, for instance by looking for imperfect seeds or seeds that did not sprout, or simply noticing that Rr × Rr crosses produced about 25% fewer seeds than RR × Rr crosses. In this case we would state that the rr genotype or r phenotype is lethal, and that there is 100% selection against it. If we found some, but fewer than 25%, white flowers, we could calculate the **selection pressure** simply as the ratio of those found to those expected. If, with large numbers of plants, we found 20% white flowers, then the pressure would be 80% (100 × 20/25) or 20% against white flowers. The calculation is actually a bit more complex, but this is the general idea.

We can also calculate what would happen, all other things being equal, if this pressure were maintained from one generation to the next. We could modify the calculation to take into account the fact that there were fewer white flowers in the next generation, and continue pursuing the calculation over numerous generations. In the simplest and crudest calculation, it would be as follows:

Presume that two purebreds meet.

(2) $RR(red) \times rr(white) --> Rr(all\ red)$

The frequency of the gene "r" is 2 copies out of the 4 possible genes, or 2/4, or 0.5.

(3) $Rr \times Rr --> RR, Rr, Rr, rr$ (3 red to 1 white).

The frequency is still 4/8 or 0.5 However, let us assume that 20% of the whites die, perhaps without being seen. The expected frequency for r is 0.5, but what we find is the equivalent of RR=1; Rr=2; rr=0.8. The frequency with which we find r in the population is no longer 0.5 but (2+0.8+0.8)/(2+2+2+0.8+0.8) or the totals for r divided by the overall totals. The total for "r" is not 50% of the population, but 47% of the population. We then repeat the cross, plugging in the reduced frequency of "r" and again subtracting 20% of the resultant "rr".

Using this kind of calculation we could determine, for instance, whether or not a population could be converted in time from one type to another by the elimination of the less favored type or the greater success of the more favored type. The result depends on both the selection pressure and the size of the initial population. For instance, if one variation is favored over another by only 1% (for every 100 of the more favored variation that survive to leave young to the next generation, only 99 of the less favored variation survive to leave young) then the time needed to change the species to the more favored variation will be impossibly long. If the selection pressure is 10% (100 to 90 survival), then with an initial population of 600, it will take 68 generations before the entire population looks like the more favored variety, but if the new variation appears in a population of 6,000,000, then at the end of 100 generations it will still be seen in only 0.2% of the population or one in every 500—in other words, a rare variant. If the selection pressure is 50%

(one variant has a 2:1 advantage over the other) then if this new variant appears in a population of 600, it will completely replace the old variant in 16 generations. If the initial population is 6,000,000, then it will take 39 generations. What this says is that, first, most of this kind of change takes place in small, rather than large, populations (Chapter 29) and, second, that strong selection pressures can quickly force the conversion of a population from at least one characteristic to another (Fig. 27.1).

The thing about diseases is that they can quickly produce extremely strong selection pressures. If a disease breaks out that kills 95% of the population, and 5% survive because they carry a genetic trait that makes them resistant to the disease, then the population will quickly convert from being a sensitive to a resistant population. Since there are so few survivors, perhaps by accident or perhaps because of a connection of an otherwise irrelevant trait to the resistance, the survivors may well look or in other ways be different from the original population, and so the species will have changed.

Is this a realistic consideration for evolution? We have many examples in current events to indicate that such draconian selection does occur, and we have further evidence to suggest that it has happened before in history and that disease can play a major role in evolution. In fact, the whole purpose of sex may be to avoid disease. We will address each of these issues in turn.

Not the most obvious, but the easiest to understand, is a laboratory procedure conducted every day that is the basis of most of the most astonishing announcements from molecular biology. This is how to isolate a specific gene or to implant it into a certain cell. One can break up the DNA into gene-sized pieces and try to get the pieces into cells or bacteria, but how do you tell which cells have actually gotten a piece? The trick is to package the pieces with another piece of DNA, which has the information for resisting an antibiotic, usually a means of digesting the antibiotic (Fig. 27.2). Then the package is offered to cells that are normally killed by the antibiotic, and the cells are then grown in the presence of the antibiotic. All the cells that have not received the package are killed by the antibiotic, and only those that have absorbed the package—sometimes as few as 1 in 1,000,000—survive. We use this trick in research to quickly get rid of everything that we are not interested in but, within the universe of this Petri dish, it is evolution. Of the perhaps 10 or 100,000,000 bacteria in the dish, only a few, perhaps 10 to 1000, survive, and these are genetically different from all the rest. If the Petri dish had been the earth, the species would have changed.

Humans inadvertently conduct this type of selection on a regular basis. The poly-resistant bacteria (bacteria resistant to most antibiotics, "superbugs") that make the headlines come from this process. Antibiotics are common in the environment. They get there through several means: Many are used in obvious medical situations, though failure of patients to use the full killing dose often results in the selection of the most resistant organisms. Although it is discouraged today, small supplements of antibiotics added to the feed of animals raised for meat often stimulate faster growth of the livestock at very low cost. It is possible that the antibiotics kill

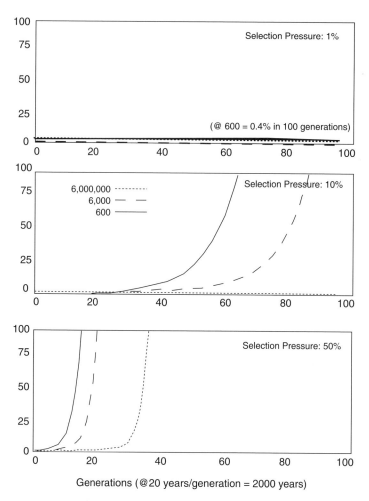

Figure 27.1. Selection pressures in theoretical populations. In each figure, the plots indicate the same selection pressures applied to populations of 600 individuals, 6,000 individuals, and 6,000,000 individuals. The selection is maintained over 100 generations, which would be approximately 2000 years for humans. The assumption is that a new recessive mutation appears in one individual in the population and is selected for at the pressure indicated. For a characteristic to go from essentially 0 frequency at the beginning to 100% frequency means that the entire population has been converted to the new characteristic. The selection pressures have been applied as follows: 50% selection (for every one individual of the normal type surviving to reproduce, two of the mutant survive to reproduce); 10% (for every ten normals, eleven mutants survive); and 1% (for every 100 normals, 101 mutants survive). With very low selection pressures, there is very little change. With very high selection pressures, even large size populations can be converted very quickly. With moderate (10%) selection, the freely-breeding population is converted only if it is relatively small. Calculations such as these, but of course considerably more complex, have led to the conclusion that relatively small human populations were under relatively severe selection pressure, leading to very rapid evolution

Figure 27.2. A Petri dish as a high selection pressure universe. Upper: 1,000,000 are placed in a Petri dish containing penicillin. Lower: Two days later, almost all of the bacteria have died, but 4 out of the 1,000,000 have survived and are now growing in the dish. Since each colony is a descendent of one bacterium, it is a clone and can easily grow to the numbers originally placed in the dish. If this dish were the world, the species would have changed

or limit the growth of the most noxious bacteria, thereby making the animals a bit healthier and faster growing, but the net result has been to expose all sorts of microorganisms to sublethal doses of antibiotics, thus selecting for those most resistant to the antibiotics. This practice has been a major means of generating bacterial resistance to antibiotics in our society. A third means of selecting for resistance has been misguided medical practice. For instance, during the Vietnam War, prostitutes in the Philippines were given prophylactic (preventative) doses of antibiotics to lower the spread of venereal disease. The theory was that the doses given, less than full curative treatment, would lower the number of bacteria that the prostitutes would carry and therefore protect them and their clients. Unfortunately prophylactic treatments work exactly in this fashion, allowing the survival of only the resistant bacteria. Thus syphilis and gonorrhea bacteria resistant to the most common antibiotics were selected for and their populations expanded.

Disease can frequently devastate a population. It is estimated that the native South American population decreased by 90% in the first 50 years after Europeans made contact—a bit by war, but mostly because the native population had no resistance to the diseases carried by the Europeans (see page 45 and this chapter, page 367). Fungal diseases carried by travelers have in recent times wiped out the American elm and the American chestnut, both lovely and highly desirable trees, and of course a fungus carried from South or Central America to Ireland in the space of two years destroyed virtually the entire potato crop. Such tragedies happen routinely. In the beginning of the 21st Century, diseases are destroying Central American frog populations, large numbers of crows have succumbed to West Nile Virus, and Asian bird flu is threatening many bird populations. Usually such a sequence begins when, through expansion of range of either the host or the parasite (virus, bacterium, or fungus) the parasite encounters a vulnerable host.

There are two likely consequences of such an encounter. In the worst scenario, the parasite will achieve 100% kill. In this case both the host and the disease-causing organism are out of luck, since the host is now extinct and the parasite, at least in territory of the host, has no further prey to attack. However, the niche (lifestyle, see Chapter 24, page 336) is now unoccupied, and a new species may expand or evolve to replace the extinct one. For instance, if the victim was a bird that caught flying insects, another species of bird might expand into the territory or adapt to exploit this food resource. The second likely consequence, which is seen quite frequently, is that the entire species will not be exterminated, owing to an existence within the gene pool of the species of some genes that confer resistance to the disease. In this case the survivors will be resistant to the disease and, when it is all over, the species will have changed or evolved. In most cases the resistance will not be absolute, and the parasite and host will continue in such a manner that the parasite does some but not catastrophic damage to the host species. Often such a situation works to the advantage of both species (see red queen hypothesis, page 366). For

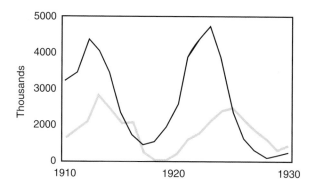

Figure 27.3. The population of lynxes depends on the population of hares and vice-versa. In this study by MacLulish in 1937, based on numbers of animals trapped in Canada, the lynx population (gray) increases approximately one year after the increase in the hare population (black). When the number of lynxes gets sufficiently high, they decimate the hare population, leading to starvation among the lynxes. The reduced number of lynxes then allows the hare population to build. The predator and prey cycle together, but neither wipes out the other. The scale for lynxes is 0 to 5000, since there are far more hares than lynxes

instance, the population of the lynx (a predator) tracks that of the snowshoe hare (Fig. 27.3). The populations oscillate, but then neither rises to unsupportable levels. This is very different from the misguided effort of the National Park Service to render the Grand Canyon more attractive to visitors. Considering that visitors liked to see deer and that the deer were attacked by coyotes, members of the service shot the coyotes. Sure enough, and as planned, over the next two years the numbers of deer rose, until the population had tripled. Unfortunately, there was an unforeseen complication. The many deer ate all available vegetation by December and almost all of them died before spring. It took many years before the population regained its original level. The lesson of biology was, better to choose your nemesis and live with it than to live without controls.

Besides the evidence of the damage that can be done to populations by the sudden outbreak of disease, there is substantial evidence that disease is a very strong driving force in evolution. The evidence ranges from the simple observation that what we, as humans, consider as ugliness in other humans is, far more often than not, an appearance caused by or very similar to an appearance caused by severe disease. It expands to the very large question of why animals and plants go to such lengths to maintain sexual reproduction. The first clue is to observe exactly what the courtship is all about. In most instances, males preen before females, and the females select their partners from among a few suitors. What determines their choice? Since birds undertake highly visible courtships involving aerial acrobatics or displays, bird courtship has been studied in some detail, but the rules apply to other animals as well. Usually the female notices one particular characteristic of the courting male. This may be any of a number of things: bright colors, symmetrical tail feathers, a strong, powerful, or magnificently varied song, or an elaborately built bower, filled with colorful objects. Male ospreys catch the largest fish that they can and display it, in flight, before females, to demonstrate what good providers they can be.

The preference of the females is toward the most of the characteristic, and the impact ranges from extravagant to ridiculous. Peacocks drag along ridiculously large tails, because when they are courting the female will choose the male whose tail has the most eyes. This can be easily demonstrated in zoos. One can take the male Don Juan or Rudolf Valentino, pluck a few of his tail feathers, and attach them to a less successful male. The less successful male will move up in the attractiveness rank, and the former champ will be far less successful in his courtship. Female swallows judge the symmetry of the courting male's tail. Again, if one captures a perfect, handsome, male and removes a few tail feathers, he immediately drops in the female's estimation.

This selectiveness on the part of the females can easily trend to the ridiculous. One can artificially exaggerate a characteristic that appears to be the object of selection, for instance by making display feathers even larger and, quite often, the female will be attracted to the artificially larger and more spectacular feature to the point of absurdity. From observations such as these, we conclude that some of the most bizarre structures and behaviors that we encounter in the animal world, such as giant antlers or the habit of American buffalo (bisons) to charge at each other over the length of a football field, slamming their heads into each other in a collision that can be heard for miles, are the evolved result of sexual selection or its counterpart, competition among males for preeminence in the herd or social structure. Part of the answer may lie in the description of these modifications as "extravagant". These features become desirable precisely because they are useless. The peacock that can invest so much food energy in building the biggest and best tail, or the buffalo that can literally knock his opponent senseless, is obviously healthy, with excess energy available. Diseases or parasites sap energy, making the animals weaker and less capable of very demanding acrobatics or other feats of strength or agility, less capable of building elaborate tail feathers, and less capable of synthesizing complex but brilliant pigments. These results have been documented. A parasite-ridden bird may not synthesize perfect feathers, or may get into more scrapes than a parasite-free bird; unhealthy cardinals are less brilliantly colored than healthy ones. What is happening when the female selects the showiest or most spectacular male is that she is selecting the most disease-free partner. Thus sexual selection works to minimize disease in the population.

THE RED QUEEN HYPOTHESIS

There is a second aspect of sexual selection that again argues that the function of sex, and the reason that most animals and plants, invest enormous energy in sex, is the limitation of disease. This aspect can be best answered by asking the question, why do animals never evolve perfect defenses against their enemies? An approach to answering this question is to consider the Red Queen Hypothesis. In *Alice in Wonderland*, the Red Queen challenges Alice to a race, which they conduct, only to end up at the place where they started. This image in biology applies to the question of why the fox never manages to catch all the hares, or the hares never evolve to escape the fox. Like all these questions, we can address it on several levels. The first, of course, is the obvious: Both the fox and the hare are capable of evolving. If the hare proves better at escaping, the force of selection for the fox will be improvement in the fox's ability to catch the hare; if the fox becomes too good, it will select for hares that can escape the fox. Diagrammatically, it might look like this (with "become" or the arrows being shorthand for an evolutionary hypothesis):

Foxes catch more hares–>only fastest hares survive, hares become faster–>foxes starve, only best hunters survive, foxes become better hunters–>foxes catch more hares–>only fastest hares survive, hares become faster–>foxes starve, only best hunters survive, foxes become better hunters–>foxes catch more hares–>etc.

This is the Red Queen model: everyone runs and runs to stay in the same place. Phrased another way, and a bit more profoundly, neither the fox nor the hare can afford to be perfect. If the fox is perfect, it will eat all of its prey and starve to death. If the hare is perfect, there will be no predation, and it will quickly overpopulate the land, eat all available food, and starve to death. Thus fox and hare are condemned to live together in a mutual race in which neither is allowed to win.

How does this apply to disease and to sexual selection? First, it is a bit more difficult to follow, but the relationship of hare to fox is the same as the relationship of hare to tularemia or any other disease of hares: if the disease organism kills off the entire species, it has no further hosts and itself will die out, and if the hare is completely disease-free, it risks overbreeding and consuming all its food. Second, sexual selection generates the variety necessary to allow the target of the disease organism to keep its position in the Red Queen race. Consider a situation of fish in a pond that are subject to heavy infestation by parasites. The parasites attack the fish by attaching to a specific protein in the gills. Perhaps there are five varieties of this protein in different individuals of this species of fish. This gives 25 (5 * 5) different combinations of the protein in individual fish. If the parasite is particularly effective in interacting with protein type A, all fish bearing the A type will be heavily infested and may die. However, fish bearing proteins type B, C, D, and E will be less infested. When type A dies, parasites that can identify only type A will also die, but among the parasites, which are also sexual, there are some parasites that can interact with fish bearing protein type B. Thus the parasite will evolve to attack these fish, and the cycle will repeat, until ultimately we will return to the few survivors carrying type A. If the fish were not sexual, but reproduced as clones like the aphids mentioned above, the varieties would not be available and one round of parasitism would kill off the population. Sexual recombination provides the variation necessary to survive parasites and disease.

Finally, disease may have played an important part in the development of modern society. Jared Diamond, in a monumental work (*Guns, Germs, and Steel*) argues that western societies spread across and dominated other societies because of a happenstance of geography. In one of the few regions of the world in which both animals and plants capable of being domesticated lived, humans in the Middle East domesticated both grains and large animals. The animals gave them the power to build large edifices and the speed and endurance to travel long distances rapidly— both giving strong military advantages—but also other advantages. For during the period during which the cattle, horses, and other animals were domesticated, probably many people contracted diseases from the animals and probably died. The survivors were those resistant to these diseases. When, however, the possessors of the animals explored further afield, such as Europeans reaching the New World, they carried the germs of diseases that to them were mild but that were devastating to the indigenous peoples. In some areas of Central and South America, as many as 90% of the native population died within 10 years of their first contact with Europeans. As Diamond notes, no wonder the natives were terrified of these strange, pale interlopers.

REFERENCES

Ameisen, Jean-Claude (2003), La Sculpture du vivant : Le suicide cellulaire ou la mort créatrice, Seuil, Paris (available only in French)

Diamond,Jared (1999, reprinted 2005) Guns, Germs, and Steel. The Fates of Human Societies, W.W.
    Norton, Co., New York.
Mann, Charles G., 2005, 1491, Alfred A. Knopf, New York.

STUDY QUESTIONS

1. Look up information about the effect of the arrival of explorers on the native populations in any land that you choose, and evaluate the impact of disease on the two populations.
2. Many people, though requested to complete the full prescription of an antibiotic, stop taking the antibiotic as soon as they feel better. What is likely to happen to the bacteria in this situation?
3. Look up the biological history of the potato famine of Ireland. What happened? Why?
4. Do the current spread of diseases such as West Nile Virus, Asian bird flu, or Mad Cow Disease, threaten severe selection to humans? How about to the animals that they infect? Argue your position with real data as gathered from appropriate reports.
5. Will the human population change because of AIDS?
6. Can you identify any alternative explanation that would address the high prevalence of sexuality among both animals and plants?

CHAPTER 28

THE AIDS MURDER MYSTERY—WHAT
CONSTITUTES PROOF?

Acquired Immune Deficiency Syndrome (AIDS) is a plague that, like the Black Death, threatens to have sufficient impact to change our genetic pool. It also is a disease that can illustrate the basis of scientific logic. When AIDS was first identified in the early 1980's, there ensued a rapid search to identify its cause as a first step in attempting to prevent or cure the disease. This search followed the primary principle of collecting evidence that could be fit into a consistent, logical argument, and attempts to falsify the hypothesis as a means of confirming that the hypothesis was correct. Part of this effort included an attempt to determine the origin of AIDS, which led to the hypothesis that AIDS was generated by human activity. This hypothesis, though proposed by amateurs—that is, people who were not professional scientists—was sufficiently plausible that it attracted the attention of the scientific community, and it certainly attracted public attention and threatened to increase social tensions at several levels. It was therefore subjected to quite close inspection, providing an excellent example of how scientific logic works. Because of the controversy, it is important for you to understand that the vast majority of the scientific community believes that the hypothesis has been adequately falsified, and that the evidence does not support an iatrogenic (doctor-caused) source of AIDS. The best means of addressing this issue is to describe what AIDS is and how it progressed around the world. Then we will examine the hypothesis that AIDS was introduced to humans through a vaccination program, and the arguments that to most people's minds satisfactorily refuted this hypothesis. Throughout the chapter, note both the information available concerning the disease and the structure of the arguments addressing the hypotheses.

WHAT AIDS IS

AIDS is a disease caused by the Human Immunodeficiency Virus (HIV). The virus attaches to and enters cells that make antibodies, killing them but more importantly causing the infected cells to induce nearby cells (bystander cells) to commit suicide. The virus is related to the virus that causes mononucleosis, which is a self-limiting disease, but in the case of HIV infection, although the immune system fights vigorously by producing more cells and antibodies, in the absence of treatment over

369

a period of a few years the loss of cells slightly exceeds the ability of the immune system to replace them, and the number of antibody-producing cells declines to a level at which there are not enough antibodies to prevent fungal infections or other infectious diseases. At this point the infection has become AIDS. Since part of the body's means of avoiding cancer is for specific cells of the immune system to watch for and destroy new cancer cells (immune surveillance) the loss of these cells leads to increased susceptibility to cancer, including rare forms of cancer, and infection of brain cells can produce neurological problems. AIDS patients do not die from the virus but from the infections, cancers, and other problems caused by the loss of immunity. When AIDS first appeared, people lived 5–7 years after the first signs of the disease but today a complex treatment designed both to interfere with the reproduction of the virus and to support the ability of the immune system to keep producing cells has considerably extended the lives of those lucky enough to have access to the treatment. The virus is not eliminated, and those infected remain capable of passing the disease to others and must continue the treatment indefinitely.

Part of the reason for the virulence of HIV is that it is a retrovirus, meaning that the intact virus contains not DNA but RNA, from which DNA is synthesized in the cell, and this DNA in turn makes the RNA for new virus. In the course of evolution, the DNA-to-DNA means of replication has been perfected, so that there are very few errors and most of those errors can be corrected, but the RNA-to-DNA-to-RNA route is far less perfect. "Less perfect" means more errors, or more mutations. The virus can mutate very quickly, and this is a means of defeating the ability of the immune system to destroy it. The immune system might make an antibody capable of trapping the virus, but a new variant of the virus appears that the immune system does not recognize.

AIDS is a sexually transmitted disease, and it is easily transmitted by heterosexual or homosexual activity, and by transfusion of infected blood. Although transmission by infected blood occurred during the first few years of the disease, better screening systems have drastically reduced the possibility of transmission by this route. The virus can be found in other body fluids such as saliva or tears, but it is very difficult to transmit it by these routes. It is not transmitted by mosquitoes, by sneezes, or through the digestive tract.

HOW AIDS SPREAD AROUND THE WORLD

The best hypothesis today is built on historical tracking of the disease as well as genetic evidence that can track the movement of the virus through the geographical, social, and historical distribution of new mutations or variants. These data lead to the conclusion that the virus entered the human population in west central Africa through close contact between humans and chimpanzees and between humans and mangabey monkeys. It was established in humans by the movement of infected workers between cities and villages until it was present in a substantial proportion of the population. As is frequent in situations in which there is a lot of poverty, prostitution is common. In this case, American homosexual men vacationed on the African Atlantic coast, where they sought male prostitutes, from whom they contracted AIDS. They also vacationed in Haiti, following the same behavior patterns and introducing AIDS to Haiti. The disease became established in Haiti, from which it was brought in quantity into the United States. In that carefree period, promiscuity among homosexuals was very widespread, and the disease propagated quite rapidly in this population. Some of this population were also drug users, and sharing of contaminated needles also spread the disease among addicts. Although AIDS was predominantly a heterosexual disease in Africa, in the United States it at first appeared to be a disease of homosexuals and addicts. We now recognize that this impression was due to the circumstances of its introduction into the United States.

HYPOTHESIS: AIDS WAS INTRODUCED TO HUMANS THROUGH VACCINATION AGAINST POLIO

In 1991, Louis Pascal from Australia and in 1992, Tom Curtis from the U.S. noticed a broad geographical and temporal correlation between the locale where AIDS was considered to have first appeared and efforts to test the first polio vaccines. Pascal attempted to present his analysis in several scientific forums but was ignored and his papers were rejected by prominent journals. Subsequently, Curtis published his hypothesis in the magazine *Rolling Stone,* and this article was picked up by several other news agencies, including the Amsterdam News. Finally, Edward Hooper in 2000 expanded the argument and published a highly readable and provocative book, entitled "The River". Thus the idea of a link between polio vaccine and AIDS became very widespread. What follows is a brief synopsis of the argument. To understand it, you need to know a bit about the history of the polio vaccine.

Until the 1950's poliomyelitis was an epidemic disease, crippling and killing tens of thousands of people, mostly children, every year. The disease was caused by a virus that, for several reasons, propagated most effectively in temperate climates such as those of the United States and Europe. Swimming pools and movie theaters were often shut during the summertime to control the spread of the disease. Although the disease was known to be caused by viruses, vaccines such as the vaccine against smallpox could not be made because the virus could not be readily grown in the laboratory. The virus could infect only higher primates (apes, monkeys, and humans). To study the virus, the only option was to infect a monkey and, before it died, attempt to influence the course of the disease. Needless to say, this was not a very effective approach. Finally, however, John Enders, a virologist from Harvard, demonstrated that the virus could be grown in cells isolated from the kidneys of green monkeys. These cells could be grown in culture and, while they would not survive indefinitely, they could proliferate enough to make substantial quantities of tissue. The cells in the Petri dishes could be infected with the virus, and the virus in turn would grow well, producing large quantities of virus that, though still dangerous, could be safely handled in secure laboratories. For his efforts, Enders was awarded a Nobel Prize in 1954, because this was the needed breakthrough.

Availability of virus had been the limiting factor in developing a polio vaccine, because the theory of how to produce a vaccine was already fairly advanced. To immunize someone against smallpox, the individual was in fact infected with another virus, cowpox or vaccinia virus (from which we get the name vaccination). Cowpox was known to be a highly mutated form of smallpox, which in humans produced a mild fever but nothing as devastating as smallpox. The body handled it like other moderate infections, producing antibodies to it and ultimately destroying it. Because the cowpox virus was a variant of the smallpox virus, the antibodies produced against cowpox were equally effective against smallpox, and the vaccinated individual was now immunized against smallpox.

This was one way to produce a vaccine: to get a much weakened form of the polio virus, so that the body could make antibodies to it that would be equally effective against the normal, virulent, form of the virus. Means were available, using radiation and a few different chemicals, to cause mutations in the virus. Though the mutations would be random, it would be possible to test in monkeys if the different mutated viruses could cause serious disease. There was a certain risk involved, because the virus would still be alive and it was theoretically possible, though unlikely, that it could reconstitute itself to the virulent form so that the vaccination would cause rather than prevent the disease. Cowpox differed from smallpox in many ways, making its back-mutation almost impossible, but it was unlikely that the weakened or attenuated polio virus could be made so different from the virulent form.

Therefore a second technique was also considered. In this technique, the virus would be killed, using formaldehyde, merthiolate, and other toxic agents. The trick here would be to kill it so gently that the virus would maintain its basic form so that antibodies produced against killed virus would still work against live virus. For instance, if one killed the virus with strong acids, hydrogen peroxide, or boiling, the destroyed virus would be much distorted—rather in the fashion that a shriveled piece of fruit or hamburger, left in the sun on a picnic table, soon looks very little like the original piece. If this version of the virus were to be used as an antigen (the material that provokes the production of antibodies), the body would likely make antibodies, but there was no guarantee that the antibodies would recognize or be effective against the real virus. The killed virus technique was also limited and carried a risk: Do too much damage to the virus, and you do not get an effective vaccine; do too little, and the virus being used to immunize might still be alive and dangerous.

Therefore the U.S. government decided to sponsor efforts to produce both kinds of vaccines, an attenuated virus vaccine and a killed-virus vaccine, and in effect instituted a race to see which groups could a safe and effective vaccine. Each type offered considerable promise but some risk, and all depended on reliable substantial quantities of virus, which could now be obtained from monkey kidney cells. Among the groups receiving support were those of three major laboratories: Jonas Salk, who was attempting to produce a killed-virus vaccine; and Albert Sabin and Hilary Koprowski, who were attempting to produce an attenuated-virus vaccine.

Finally, the government sponsored the release of the killed-virus (Salk) vaccine and one attenuated-virus (Sabin) vaccine. After several years of experience, in which the Sabin vaccine proved to be safe and more effective than the Salk vaccine, and including one incident in which a manufacturer's attempt to simplify the Salk technique resulted in the release of a batch of vaccine containing live virulent virus, the Department of Health recommended emphasis on the Sabin vaccine. Meanwhile, through the auspices of the United Nations, Koprowski tested his vaccine in several countries, including Poland and what was then the Belgian Congo (subsequently Zaire, then Congo).

What does this have to do with AIDS? Well, when the vaccines were developed, it was not known that apparently healthy cells in culture could carry viruses that reproduced, though no faster than the cells and thus did not kill the cells. These occult or latent viruses were detected many years later, using newly-developed techniques. Under specific circumstances, the latent viruses could begin to reproduce more rapidly and eventually kill the cell. The hypothesis proposed by Pascal, Curtis, and Hooper was the following: The virus that causes AIDS is closely related to viruses (Simian Immunodeficiency Viruses, or SIVs) that infect various monkeys, some to lethal effect but some less seriously. Koprowski was unlucky enough to use cells that came from a monkey or monkeys that were infected with one of these viruses. When the laboratory grew the polio virus in these cells, they unwittingly also grew this monkey virus, and it was present in the vaccine. When doctors immunized children in the Congo against polio, they introduced the SIV into humans, where it subsequently mutated and turned into HIV. AIDS, the hypothesis goes, appeared as a result of the vaccinations but the Belgian Congo was soon torn by anti-colonial rebellion and civil war. Under these conditions record-keeping disintegrated and no one realized that unusual numbers of children were dying. Nevertheless, the virus had been introduced into humans, to spread surreptitiously until it became evident during the 1980's.

REFUTATION OF THE HYPOTHESIS

This hypothesis was thoughtful and perceptive, but to scientists there were inconsistencies with the evidence and the logic. For instance, if the virus contaminated monkey cells, why did it not show up with the Sabin vaccine, potentially the Salk vaccine, or with the Koprowski vaccine used in Poland? Did HIV really descend from the SIV that infected green monkeys (the source of the cells) or did it descend from viruses that infected chimpanzees? Was it really likely that children dying of a wasting disease would not be noticed, even in a war-torn country? When AIDS was first identified in the US and Europe, the time from recognition to death was approximately seven years, and less in Africa. The time between the mid 1950's and the 1980's seemed excessively long. Finally, the Koprowski vaccine was given orally, but AIDS is very poorly transmitted through the digestive tract.

This is a long series of "if's": If only one monkey was contaminated, and if HIV derived from green monkey SIV, and if dying children would not be detected, and if it was possible for AIDS to pass through several generations before being detected, and if the first HIV could get through the digestive tract... To statisticians, the likelihood of two probabilities occurring together is the product of the probabilities. If the chance of getting heads when one flips a dime is 50%, and the chance of getting heads when one flips a quarter is 50%, then the chance of getting two heads when one flips the two coins is 50% × 50%, or 0.5 × 0.5, or 0.25 (25%). Similarly, if a single die has six sides, then the chance of rolling a 1 is 1/6 or 16.7%. With two dice, the chance of rolling two 1's is 1/6 × 1/6, or 2.7%. Thus the probability of accumulating all these unlikely events is rather low. But it is not infinitesimal. Therefore the hypothesis is valid as a hypothesis, and deserves to be tested.

Testing means attempting to falsify the hypothesis. It is obvious that no one will attempt to replicate the experiment by deliberately injecting HIV into humans, so the best that we can do is to test the premises and implications of the hypothesis. For instance, is it true that HIV derives from green monkey SIV?

As molecular genetics has become more powerful, it has become possible to analyze the genetic sequences of HIV and of the SIVs, and to compare them and determine both a probable evolutionary lineage and the probable time that HIV diverged from its ancestor (See page 118, Chapter 9). As it turns out, there are at least two sources of HIV, and neither is from the green monkey SIV. One source is a virus of chimpanzees, and the other is from mangabey monkeys (Fig. 28.1).

Is it possible then that the cells for the Koprowski vaccine were from chimpanzees or mangabeys rather than green monkeys? Koprowski had used chimpanzees in his experimental work but stated that only green monkey cells were used for production of vaccine. The laboratory records were less complete than one might have hoped, but did not contradict his asseveration. However, the chimpanzee virus related to HIV was found in chimpanzees from western Africa, whereas the chimpanzees used in the experiments came from eastern Africa, thus making it less likely that the laboratory chimpanzees were the source of the virus. There was more evidence that tended to falsify the hypothesis. Using the assumption that new mutations arise at a relatively steady average rate (See Chapter 9) it is possible to estimate the time at which HIV diverged from either form of SIV. There is some imprecision to the calculation, but all data point to a divergence between the mid-1940's to as late as the mid 1950's. What this means is that HIV was a separate entity well before the vaccine was first used, and that it was therefore not created by the introduction of SIV into humans.

The final evidence came from careful scientific procedure, meaning the attitude that data are sacrosanct. On the scales of judgment, data always outweigh hypothesis, prejudice, desire, or any emotion. Ultimately, samples from Koprowski's original preparations, which had been carefully stored for over 30 years, were relocated and tested, this time by very sensitive molecular techniques. They proved to be completely negative for HIV. The entire hypothesis was debated in open session of the Royal Academy of London, where by all accounts the audience strongly agreed that hypothesis of iatrogenic introduction of HIV had been proven false.

Like all scientific hypotheses, the conclusion that this hypothesis is false is tentative, always capable of being reversed should new data come forth. Even in

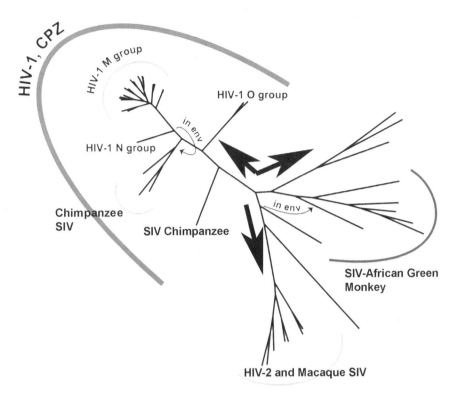

Figure 28.1. This diagram traces the relationship, as determined by base sequences, among simian immunodeficiency viruses and human immunodeficiency virus. Contrary to the original impression, HIV is much more closely related to the viruses of chimpanzees and mangabey monkeys than it is to the virus of the green monkey. Credits: Wikipedia from Z Naturforsch [C]. 1988 May-Jun;43(5–6):449–54. Serological and structural comparison of immunodeficiency viruses from man, African green monkey, rhesus monkey and sooty mangabey. Jurkiewicz E, Hunsmann G, Hayami M, Ohta Y, Schmitz H, Schneider J.Deutsches Primaten-Zentrum, Gottingen, Bundesrepublik Deutschland. (modified)

the present state, there is still wiggle room. As the proponents of the hypothesis still argue, the Koprowski samples that were left may (in spite of records that indicate to the contrary) have been from a different batch than that used for the Congo vaccines, and the Poland vaccines from a different batch than that used for the Congo vaccines, but the hypothesis is less and less tenable, and today other hypotheses seem much more probable.

WHY DID AIDS APPEAR NOW, AND NOT AT ANOTHER TIME?

An alternative hypothesis is that the ancestor viruses have been common in Africa since primates have existed in Africa. The viruses have always been capable of mutating and have done so in the past. In various parts of Africa, monkeys are kept as pets, often by prostitutes who spend considerable time alone, and they

are sometimes eaten. However, prior to decolonization and the social changes that
followed, social conditions were quite different. Rural villages were rather small
and isolated, limiting sexual promiscuity and migration of people. If AIDS did
appear, the individual infected passed the disease on to very few others before
becoming obviously sick, and the disease became self-limiting and died out. With
the construction of good roads and the industrialization of cities, men came from
the countryside to work in the city, frequenting prostitutes, and returning to their
villages frequently. They thus became the vectors for the disease, and it spread
to a large enough population to become self-sustaining. It was social rather than
biological factors that created the environment for AIDS to spread.

AIDS is an epidemic disease, one that threatens to kill a large segment of the
human population and possibly alter the percentage of specific genes in the surviving
population. It is not yet pandemic, the word pandemic meaning that the disease has
the potential to infect substantially all of the population. Diseases can substantially
alter the destiny of a species. The encounter of Europeans with natives of the New
World, for instance, frequently meant that up to 90% of the native population died
of European diseases. The ultimate impact is similar to that of a founder effect,
discussed on page 338, Chapter 24. There are several suggestions that diseases

Figure 28.2. Plague statues from Košice, Slovakia (left) and Kutná Hora, the Czech Republic (right).
These were erected throughout Europe to commemorate the end of a wave of plague. Credits: Plague
statue Slovakia - This is a file from the Wikimedia Commons. T Immaculata Statue at Hlavná ulica
(Main Street) in Košice, Slovakia. Author: Marian Gladis, Plague statue, Czech Rep - Plague column
in Kutná Hora, Czech Republic Built in 1713–1715. Photo taken by Miaow Miaow in June 2005. *GNU
Free Documentation License,*

have altered human population. For instance, some descendants of certain European countries display greater resistance to HIV than do other populations. Some authors have noted that these regions had previously been heavily hit by Plague, which likewise attacks through the immune system. These authors have hypothesized that genetic differences that allowed some individuals to survive Plague are likewise differences that allow them to resist AIDS. In other words, Plague selected a Plague-resistant population which is likewise, and probably through similar mechanisms, more resistant to HIV (Fig. 28.2).

WHY DID EUROPEANS SPREAD MORE DISEASE THAN THEY ACQUIRED?

The question of why European diseases such as colds, tuberculosis, and smallpox were so much more devastating to native Americans than native American diseases (perhaps syphilis and gonorrhea, though this is disputed) were to Europeans has been considered as well. Jared Diamond has suggested that domestication of animals brought many diseases to European populations, and that these diseases had killed the non-resistant members, thus selecting European populations who were resistant to the diseases. Native Americans, who had few domesticated animals, had not undergone this selection.

The importance of disease in evolution has been re-examined in considerable detail. As we have seen in Chapter 27, infectious disease, including viruses, bacteria, molds, and parasites, has been considered sufficiently important to account for elaborate and seemingly very costly sexual displays ranging from a peacock's feathers and the bright colors of male breeding birds to the athleticism of social folk dances. In fact, avoidance of disease has been hypothesized as an explanation for the existence of sex itself—the fact that most organisms, at considerable risk and investment of energy, contrive to share their genes with a partner. This logic ultimately leads to hypothesized answers to such questions as why many of the new deadly viruses come from Africa, and why many of the flus come from Asia. The hypothesis for the first question is fairly straight-forward: Viruses tend to jump among closely-related species, becoming more lethal in the newly-adopted species; Humans and primates have been in Africa longer than in any other continent; and there are more primates, and particularly anthropoid primates (great or tailless apes, see Chapter 29) in Africa. The second question is best considered in the context of how viruses change, discussed in Chapter 14, page 210.

Many of these arguments can be tested in animal and plant populations, though they obviously cannot be tested in humans. Nevertheless you will continue to recognize the ELF logic that underlies all of these investigations. One formulates a question such as whether polio vaccine could have been contaminated by an unrecognized live virus; one attempts to collect the evidence to test the hypothesis; one applies logic to the analysis, in the sense of whether the argument makes sense, whether the postulated mechanism is possible, and the timing and the quantity is correct; and one attempts to falsify the hypothesis by proposing alternative explanations, structuring specific logical expectations and predictions if the hypothesis is true or false and then attempting to identify a situation sufficiently constricted and well controlled that there are only two possible interpretations—true or false—of the result. The bulk of this chapter has focused on the development of this argument in the context of AIDS. An appropriate exercise would be for you to formulate a hypothesis concerning any of the topics addressed in these last few paragraphs and to construct a means for testing the hypothesis. Specific questions are offered in the thought questions, but there are many others. Remember that the hypothesis has to be sufficiently explicit that it leads to a testable question.

REFERENCES

Curtis Tom 1992 THE ORIGIN OF AIDS. A STARTLING NEW THEORY ATTEMPTS TO ANSWER THE QUESTION 'WAS IT AN ACT OF GOD OR AN ACT OF MAN?' Rolling Stone, Issue 626, 19 March 1992, pp. 54–59, 61, 106, 108. also available as http://www.uow.edu.au/arts/sts/bmartin/dissent/documents/AIDS/Curtis92.html

Health: 'Scientists started Aids epidemic' *BBC News Online - Thursday, August 26, 1999* http://www.aegis.com/news/bbc/1999/BB990810.html

AIDS Wars: http://www.cycnet.com/ englishcorner/essays/aids1008–5.htm

Polio vaccines and the origin of AIDS: some key writings http://www.uow.edu.au/ arts/sts/ bmartin/dissent/documents/AIDS/

Hooper, Edward 1999 *The River: A Journey Back to the Source of HIV and AIDS* (Harmondsworth: Penguin; Boston: Little, Brown, 1999; revised edition, Penguin, 2000).

Bagasra, Omar 1999, *HIV and Molecular Immunity: Prospects for the AIDS Vaccine* Eaton Publishing, Natick, MA: Biotechniques Books

Martin, Brian, 2001 "The Politics of a Scientific Meeting: the Origin-of-AIDS Debate at the Royal Society ", *Politics and the Life Sciences,* Vol. 20, No. 2, September 2001, pp. 119–130 [published 2005], also available at http://www.uow.edu.au/arts/sts/bmartin/pubs/05pls.pdf

Royal Society Discussion Meeting (and subsequent events) Origins of HIV and the AIDS Epidemic, London, 11–12 September 2000 Papers, press releases, media stories and responses http://www.uow.edu.au/arts/sts/bmartin/dissent/documents/AIDS/rs/index.html

Quest for the Origin of AIDS: Controversial book spurs search for how the worldwide scourge of HIV began http://www.stanford.edu/class/stat30/web1/aids1.html

STUDY QUESTIONS

1. What is the most likely evidence for the origin of AIDS? What supports the argument? What evidence is against the argument?
2. What are live-virus and killed-virus vaccines?
3. Has the hypothesis of contamination of the Koprowski samples been completely eliminated? Why or why not?
4. Why did AIDS appear when it did, and not before?
5. Is it likely that other diseases can appear in the same fashion that AIDS did? Why or why not?
6. Is it likely that a disease like AIDS will come out of South America? Why or why not?

PART 7

THE EVOLUTION OF HUMANS

CHAPTER 29

THE EVOLUTION OF HUMANS

WHAT IS A HUMAN?

A very meaningful question is to ask who or what is human. This question may seem relatively obvious or useless today, although various societies, promoting slavery, racism, or seizure of resources, have found ways to exclude one or another group from membership in the category of human. To a biologist, in spite of our very large variation in color, shape, and other characteristics, all nearly hairless apes capable of talking, forming societies, and of building elaborate tools belong to the species *Homo sapiens sapiens* by the criterion that we can interbreed freely and that all of our young, even those derived from the most diverse parents imaginable, are healthy, grow well, and are fertile. However, the question becomes more complicated as we move farther afield or delve into history. We may differ as to the rights and protections that an ape such as a chimpanzee should be accorded, but we have little difficulty in understanding that it is not human. But what about the fossils that we encounter?[12] We can evaluate them by many criteria: Did they walk fully upright? How large were their brains? What types of tools did they use? Can we determine what size colony they preferred? Did they build structures in which to dwell? Did they wear clothing? Did they use fire? Did they domesticate animals or plants? Did they have musical instruments? Did they have a sophisticated language? And, most importantly, did they leave any kind of artifact (artwork, statues, symbolic tools) that suggest that they thought about who they were and how they related to the earth? Did they care for wounded, deformed, or weak members of their society? Did they bury their dead or leave any indication that they had a sense of afterlife or a religion?

These questions have meaning when we consider our ancestors and most especially the ancestor that immediately preceded and to some extent overlapped with us. Current evidence indicates that the Neanderthal people—we will use that term—did not contribute to the population of humans that now covers the earth. Their DNA, insofar as it has been successfully analyzed, is too different from ours. And yet they met all or most of the criteria mentioned above. (There is still dispute, as described later in this chapter, as to whether or not they could have had a clear language.) In brief: they made tools, they buried their dead, perhaps with some

---

[12] See Wikipedia on Neanderthal, and Smithsonian (http://www.si.edu) on Shanidar cave (Neanderthal)

ceremony, they could control fire, they cared for their wounded, they decorated their bodies, and they apparently had musical instruments. And yet they did not survive.

Therefore the question becomes, what is the species we call *Homo sapiens sapiens*? Where did we come from, and how did we come to populate the world, as opposed to any other species similar to us or not? For this kind of analysis we look primarily to the fossil record, with some cross-referencing of our ability to interpret the record in our genes. For our purposes, we will use the following terms: *anthropoid* or human-ish: tailless apes that can stand erect on occasion; *hominid* (human-type): truly erect creatures with brain size larger than apes; *human*: truly erect large-brained creatures with sophisticated tool-making capability, the ability to control fire, and signs of culture. Our story begins approximately 4 million years ago (if one starts with the earliest creatures that resembled humans) 1,600,000 years ago (if one starts with creatures that were sufficiently like us to be considered within the genus Homo) or 160,000 years ago, if one considers those similar enough to us to be considered modern *Homo sapiens* with fragments of skeletons found in eastern Africa (Fig. 29.1). Because the skeletons are very fragmentary, much of what we understand about their lifestyles is inferential. In general, skulls or parts of skulls are more frequently preserved than other bones. Apes have sharp, tearing canine-like teeth, while modern humans have a mixture of grinding and more gently tearing teeth. Thus we can examine tooth structure. The size of the mandibles is also meaningful, since apes have more massive jaws. Although the intelligence of individuals does not correlate with brain size, in general populations of animals with larger brains are smarter than populations with smaller brains, and brain size has expanded very rapidly in human evolution. The vertebral column of humans follows an S-shaped curve, to balance the torso on the pelvis, and humans have

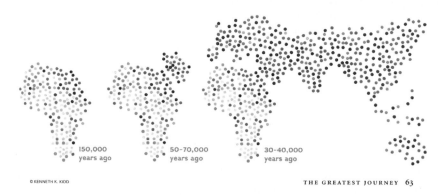

© KENNETH K. KIDD   THE GREATEST JOURNEY 63

Figure 29.1a. Human migration, as determined by genetic diversity and traces of human activities. *Upper:* Genetic diversity. The colors indicate different primary genetic markers. *Lower:* Migration patterns as traced by earliest evidence of human activity. In general, blood types, genetic evidence, and general appearance (skin tone, hair color, hair form, extent of beard and hirsuteness) are consistent with these values. Credits: Shreeve, James, 2006, The Greatest Journey, National Geographic, March 2006

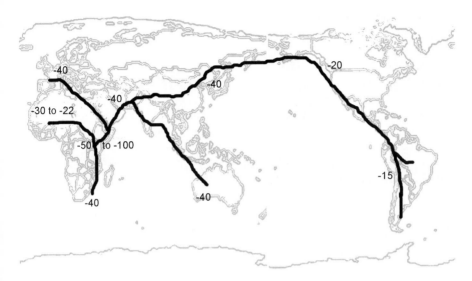

Figure 29.1b. (Continued)

Figure 29.2. Feet of a chimpanzee (left) and human (right). Note that the big toe of the chimp is more adapted for grasping, and that the angle at which the leg meets the foot is slightly acute

flat walking feet unlike the prehensile (grasping) feet of many apes (Fig. 29.2). It has been of some interest to know when humans became fully erect, especially since some hypotheses consider that the increase in intelligence followed the free use of hands by our ancestors; but unfortunately vertebrae and feet are only rarely found. In fact, it was typical of earlier dioramas, or model displays of early human existence, that these individuals were shown in tall grass so that the shape of their feet would remain ambiguous. There are, however, clues that can provide indirect evidence. In truly erect humans, the skull must be balanced on the spine, with weight toward the back approximating the weight in the front, and the line of the eyes is at right angles to the spine. An ape's head sits at more of an angle to the skull. Thus the angle of attachment of the spine to the skull can be interpreted (Fig. 29.3). In a similar manner, the femur (thigh bone) of a four-legged animal

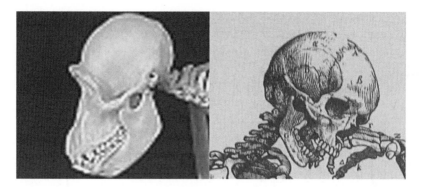

Figure 29.3. Skull and upper spine of chimpanzee (left) and human (right). Note the relative balance of the jaws vs the back of the skull, the position of the foramen magnum (where the spine meets the skull), and the angle of the spine relative to the skull. Using such criteria, it is possible to interpret the posture of fossils

rests naturally in the pelvis at approximately right angles to the spine, whereas in humans the femurs are almost aligned with the spine (Fig. 29.4). In the pelvis of an ape, which tends to "knuckle-walk" (move without putting its full weight on its front limbs, but using them for balance and to shift weight) the alignment of the pelvis is at an angle. In humans the pelvis is rotated relative to that of other animals to act as a basin for supporting the viscera. If a femur or a pelvis can be located, we can infer the posture of its owner. More recently, higher-resolution and reconstruction techniques have led to further inferences. For instance, an ape that climbs and swings from branches with its arms needs room for strong shoulder

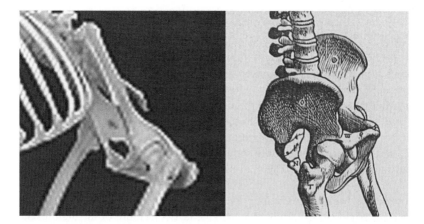

Figure 29.4. Angle of pelvis of chimpanzee (left) and human (right). Note that the angle of the pelvis to the spine, and of the legs to the pelvis, indicates that the chimp is clearly not erect

and forearm muscles. It therefore has a narrower, barrel-shaped chest than does a human, adapted for longer-endurance running and not particularly powerful arms.

THE IMPORTANCE OF LANGUAGE

Although all animals communicate with other members of their species, and some social animals cooperate to hunt, forage, or migrate (listen to migratory geese passing overhead) true language is much more powerful and much rarer. While porpoises seem to have a quite complex communication system, human language may be truly unique, and it is clearly much more efficient, creating the possibility of a species improving its probability of success by a means other than mutation. Consider, for instance, a pack of chimpanzees or wolves trying to encircle an animal that they hope to trap. Grunts or varying sounds and gestures may work well, but "Bill and Fred, go hide behind those bushes. Mike and I will drive the antelope toward you. When it gets to that rock, you should be able to spear it," is far more effective.

Recent history has given us an example of how selection for intelligence—here, the ability to speak, address subtle concepts, and remember—might have functioned. Just after Christmas 2004, a severe earthquake off the coast of Indonesia generated a huge tsunami ("tidal wave") in the Indian Ocean that devastated the coastline of Indonesia, India, Sri Lanka, Thailand, and other countries and resulted in well over 170,000 deaths. During the cleanup, rescuers assumed that the Moken people of the Andaman Islands, some low-lying islands directly in the path of the tsunami, had been lost. They were very startled to find that, although the villages were completely destroyed, most of the people had survived. What had happened was the following: First, the people of the islands derive from migrants who originated in southern Africa and who presumably populated the southern Indian Ocean coastlines, Melanesia, and Australia, where the descendants became the people known as the Melanesians and the Australian aborigines. They lead a simple life as fishermen in small villages, presumably similar to early human societies: they do not concern themselves with time, have no means of expressing how old individuals are in years, and have no words reflecting future and past. For instance, they have no word for "want"; if they desire something such as food, they simply "take" it. In their village life, they recount ancestral tales.

The water in a tsunami must come from somewhere and, because of the physics of wave motion in water, it pulls the water in from in front of the wave (Fig. 29.5). Thus a tsunami is preceded by what appears to be an extraordinarily low tide, most likely occurring at an unexpected time. For a tsunami of this size, generating waves up to 60 or 80 feet in height, the water withdrawal was enormous, and preceded the arrival of the wave by up to one hour. Many of the people who died in the tsunami had been intrigued by the surprising tide and had gone out to the suddenly-exposed beachfront to gather shellfish and stranded fish. Not so the people of the Andamans. When they saw the water retreat, they ran for the hills, and their fishermen who were at sea headed for the deepest water they could reach. Burmese fishermen

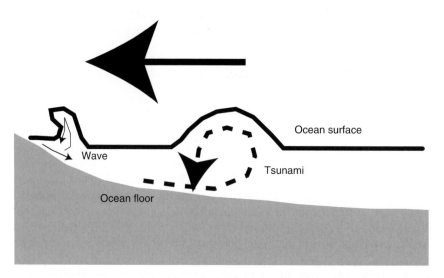

Figure 29.5. Mechanism of a tsunami. A wave is not a mass of water moving forward. Rather, it pulls in water from in front of it, and the water cycles through the wave and back under it. When a wave approaches the shore, the drag along the bottom slows the bottom relative to the top, and the wave builds height and finally falls over. Similarly, a tsunami draws the massive amount of water that it will take from in front of the wave, causing an unusually low tide or retreat of water from the coast, before it arrives

continued collecting squid and were lost, since the height of the tsunami is built in shallow water. (Watch how waves come in from the ocean: they get taller and finally break as their bottoms scrape the sand.)

Why did the Andaman people run? The tales that they told of their ancestors included stories of when the sea suddenly ran away and then came running back. It mattered little that they felt that the sea, responding to the spirits of their ancestors, was angry and came to eat the villages; the description in their legends was accurate, including the withdrawal of the sea and a first, smaller wave followed by a second, larger one. Geological records indicate that tsunamis had most recently occurred 40 and 200 years before, but the collective memory, explicitly described (as opposed to grunts) informed the people and allowed them to survive. The Andaman people lived very close to nature and observed it very well. Their comment about the Burmese fishermen was that the Burmese "were just looking at the squid, and were not looking at the water." Picture an earlier time, when small groups of the first modern humans were spread widely in the region, and the total population was small. In this situation certainly and most likely in many similar situations, the only survivors would have been that group that was able to profit from the memory of the earlier experience. There are two elements here: good memory (which is also shared by some animals) and above all, ability to communicate that memory from one generation to another (possessed by very few if any other animals) would have been an extremely influential factor in the evolution of the human race. It is probably quite meaningful that one human gene that has evolved quite rapidly and is very different from that of chimpanzees is a gene that is very important for the understanding and use of speech.

It is reported that some animals also fled to the hills. We are not aware that these animals were capable of communicating from one generation to another, for almost all species, adults do not interact with younger animals beyond the earliest phases of child-rearing. It is possible, however, that they could detect sounds lower or higher than humans can detect, or of lower intensity, and thus heard the incoming tsunami. They might have also detected other changes, such as in light or in magnetic fields; or they may simply have panicked in the face of a rapidly changing and unknown situation, not being as curious or self-assured as modern humans. At this point we do not know.

THE ORIGIN OF THE HUMAN SPECIES

Let us consider what we know about the origins of human beings. There are several lines of evidence. The oldest and most well-known evidence is that of fossil skeletons, or rather pieces of skeletons. These have been known, and in increasing numbers, since the mid 19th C. Other evidence includes traces of human activity in various parts of the world, coupled with radiometric and geologic dating; examination of the physical appearance of different groups of humans; consideration of the different versions of human language; and, today, several means of analyzing and comparing DNA from humans around the world. As has been discussed before, the evidence converges on common dates, and the convergence is a major basis for confidence in the hypothesis.

The physiology of humans and that of apes is extremely similar, as is the biochemistry. In fact, over 98% of human DNA is identical to that of the chimpanzee and slightly more to that of the pygmy chimpanzee or bonobo. To understand human evolution, we must first agree on how humans and apes differ and, hopefully, identify means by which those differences can be detected. The obvious differences include the size of the brain in humans, which is approximately three times that of chimpanzees; the relative hairlessness of humans, which cannot be measured in the fossil record (but can be surmised by the evolution of lice—see below); the fact that humans are truly erect and bipedal, whereas apes are not; humans are much more omnivorous than apes, which tend to eat more fruit and less meat, and their dentition (shape of teeth) is correspondingly different; and the skulls of apes differs in that they have pronounced brow ridges, slanting foreheads, a flat nose, massive mandibles and prognathous jaws (with the teeth jutting well in front of the nose). In terms of soft tissue, apes do not have lips (outwardly turned tissues in the mouth), and apes cannot talk. This latter point indicates soft tissue changes, since the much lower position of the human larynx permits the complex changes in shape of the throat muscles to generate all the sounds that humans make. Apes cannot make the several dozen individual sounds that humans can, but on the other hand the lowered larynx makes it possible for humans to choke on food. Unfortunately soft tissue like the larynx is not preserved in fossils, though several anthropologists have attempted to infer from the structure of skulls (indications of where jaw muscles inserted) whether or not any of the fossils were able to talk.

There are many more subtle consequences of the differences in lifestyle between and humans and apes. To walk easily and efficiently on two legs requires, for balance and conservation of energy, longer legs. Arms of animals that can swing from trees require strong muscles, leading to a chest that is narrower at the top (to allow for the muscles) than the bottom. Even the spines on the vertebrae serve for attachment of back muscles, giving indication of the size of, and demands on, these muscles. Thus it is possible to extrapolate from fragments of bones to an image of the creature from whence the bones came.

There are also indications of intelligence that, at a higher level, may be recognized. Apes can use primitive tools: Orangutans can strip leaves from twigs and use the resulting sticks as probes to retrieve insects from nests; the preparation of the twig indicates foresight and deliberate intent. Chimpanzees can use rocks to break open large seeds or to throw at a predator or other threat, and they may even protect their feet with shoes of leaves if they walk over thorns or other sharp terrain, but none of these uses leaves identifiable remnants. Early humans learned to split rocks to get sharp edges. Later they learned to hit specific types of rocks to create thin, sharp blades and even to sharpen the blade when it got dull. The use of tools to scrape meat from bones can be identified from the marks left on the bones. Even later humans used fire to create metal tools. All of these activities leave remnants that can be identified, and fire leaves its own traces. More sophisticated tools such as throwing tools (spears, javelins) and fishhooks can be identified by their shape. Creatures that can control fire also have access to far wider ranges of food, since many plant and animal products are indigestible to humans unless they are cooked. Some of the diet may be inferred from the shells and scraps left by the eaters—early humans had a rather sloppy lifestyle, resulting in the collection of middens, or trash heaps, that can be explored—and indigestible materials in fossilized feces give further clues. Finally, art—organized markings, even if their purpose is not understood—is characteristic of all truly modern humans. Anthropologists and evolutionists delve through all of these cues to trace the origin of humans (Fig. 29.6).

Primates have existed for approximately 60 million years. The least evolved primates are small tree-dwelling creatures called lemurs, pottos, lorises, and tarsiers. They are very cute animals because of their forward-facing eyes and opposable thumbs (thumbs that can touch the opposite side of their paws, or hands). Both of these characteristics are adaptations to scampering through trees, the forward-facing eyes giving binocular vision (see Chapters 3 and 4) and the opposable thumb and big toe allowing the animals to grasp branches. Likewise, unlike many other mammals, they rely more on vision than on smell, and they have color vision. All primates retain these characteristics. The so-called New World monkeys are similar but are monkeys (anthropoids, with more human-like body shapes and faces) but

Figure 29.6. (left) Right-tailed lemur, from Madagascar. *(right):* Tarsier from Philippines. These two creatures are prosimians, lower primates. Credits: Ring-tailed Lemur - Wikimedia commons (photographer; Adrian Pingstone), Philippine tarsier - Wikipedia.org

have an interesting adaptation, in that many have tails that can be controlled and used to hold onto branches (prehensile tails). Monkeys like these have been around for about 35 million years. A little over 20 million years ago, in Africa, arose a different version of monkey, one which was also at ease on the ground, could stand and sometimes run on two legs, and whose tail might serve for balance but was not effective for grasping and climbing. All of these monkeys are very social and move around during the day as opposed to night. These are the so-called Old World monkeys. About 15 million years ago, the true apes or hominoids arose. These mostly large monkeys are tailless, and although two apes, the orangutan and gibbon, are almost entirely arboreal, the others, which also include chimpanzees, bonobos (pigmy chimpanzees), and gorillas, are reasonably comfortable on the ground. These "higher" primates have another characteristic that is relevant to the story of the evolution of humans. Almost all female mammals have "estrus" cycles. In an estrus cycle—the word derives from the Latin for "excitement"—the female accepts advances of a male only when she is ovulating, and otherwise actively drives him away, and after ovulation her ovaries do not produce progesterone (an early hormone of pregnancy) unless she has mated and is likely to become pregnant. Female higher primates have menstrual cycles. In a menstrual cycle the female may accept the advances of a male at any time, and after ovulation the ovary produces progesterone for about two weeks after ovulation, preparing the uterus for a potential pregnancy. If pregnancy does not ensue, the tissues in the uterus break down and are released, producing a true "period". These latter characteristics of course apply to humans, linking humans to the higher primates. The continuous sexual receptivity of the higher primate female plays an important behavioral role in keeping the males in proximity, allowing for pairing or, minimally, the organization of troups in which the babies are protected. In most of the apes, males are much larger than females (more about this later.) Thus the question arises, when did humans become distinct from apes, and what evolutionary mechanisms led to the appearance of humans?

The DNA of humans differs from that of chimpanzees and bonobos by less than two percent. It is important to understand that these DNA sequences are measured on current, living, animals, not ancestral animals. Also, genes can differ by being turned on and off at different times, as well as producing different proteins. Since humans in many respects appear to be apes that are sexually mature while retaining many characteristics of infants (prolonged growth of legs, hairlessness, curiosity and capacity for play), much of the difference between humans and apes may reside in the timing of when genes become active. Biochemically we may not be very different but, if we are the equivalent of sexually active juveniles, our behaviors may be very different.

About 30 million years ago, the great forests of eastern Africa were beginning to dry up. India was pushing its way into Asia, and the uplift of the Himalayas in addition to the appearance of a new landmass in the area changed the weather pattern, and the forest gave way to an expanding savannah. By 6 to 7 million years ago, the savannahs were fully established. An ape that could easily move across grasslands would be able to move from one wood to another, and otherwise expand beyond the forests. Thus at about this period we find a type of skull that differs from that of most apes in interesting ways. They have smaller canine, or tearing, teeth, and their faces may have been a bit flatter. The position of the foramen magnum, the hole at the base of the skull where the spinal cord enters the skull, is a bit farther forward than in apes, suggesting that the skull is more balanced on the spine, meaning that the creature is more comfortably balanced in a vertical position. Thus what defines these creatures as being related to the human line is the suggestion that they were more fully bipedal than other apes and that they were eating a more varied diet. Unfortunately, the feet, which are grasping with opposable big toes in apes and flat in humans, are typically lost in the fossil record.

The story begins to get much more interesting about 4 million years ago. Reliable dating of the soils in which they are found indicates that various members of the genus named *Australopithecus* (Southern ape) lived between 4 and 2 million years ago. These creatures were fully bipedal and had teeth much more similar to those of humans. Their brains were still relatively small, about the size of an orange. This is equivalent to the brain of a chimpanzee, which is of similar size to these creatures. "Lucy," a 40% complete skeleton found by the Leaky family in Ethiopia, was an *Australopithecus* who lived 3 ¼ million years ago. Her arms were still relatively long, suggesting that she could easily climb trees, but the pelvis and the remnants of the skull argue for an upright posture and bipedal locomotion. Even more convincing was the discovery of 3.5 million year old bipedal footprints in what is now Tanzania. These footprints were made as a presumptive *Australopithecus* walked across newly-fallen ash from a nearby volcano. The ash subsequently was wet, probably by rain that was "seeded" by the ash in the atmosphere, and solidified into rock. Even more convincing and touching is a detail about these footprints: They start out as two sets, an adult and a child. About half way along, the adult picked up the child, as evidenced by the disappearance of the child's footprints and the deeper impressions made by the prints of the adult.

At approximately this time we encounter some very important fossils, not of humans but of animals found in Ethiopia. What is interesting about these fossil bones, from about 2 ½ million years ago, is that they have marks on them that indicate that the flesh was scraped from the bones by stone tools. Thus we have the first tool use. Thus *Australopithecus* appears to have been a bipedal, upright, tool-using creature.

Between 2.5 and 1.6 million years ago, fossils appear that are similar enough to humans to warrant the genus designation *Homo* (man, or [the] same [as us]). How do they differ from the *Australopithecus* type? They have skulls that can accommodate larger brains, about half the size of modern humans. Associated with these skeletons are well-made and sharp stone tools, enough that the first of these type of fossils were given the name *Homo habilis* ("handy man"). Meanwhile, other hominids persisted along the *Australopithecus* line, continuing with small brains and prognathous (jaw forward) face. This was previously a considerable source of confusion, as long as people imagined a direct lineage from the most primitive anthropoids directly to modern humans. Today we recognize that, similar to other sequences of evolution, there were many branches to the line that led to humans, most of which finally petered out. Thus, contemporaneous with *Homo* were several other hominoids with one or more characteristics approaching those of modern humans, but these were not part of our ancestral line and ultimately the lines died out.

The *Homo* line gave rise, between 1.9 and 1.6 million years ago, to a very interesting new variant, named *Homo ergaster* (working man). This creature now had a brain 70% of modern size (900 cc compared to 1300 cc). Skeletons of *H. ergaster* indicated tall, long-legged individuals with hips clearly structured for straightforward, long strides. Perhaps related to their ability to walk long distances, skeletons of *H. ergaster* are found over much wider areas and in more arid lands. Their teeth were of a more generalized style, indicating a wider variety of food. Their fingers were too short and straight for them to have been good climbers. Their stone tools were sharpened with skill. Unlike the male/female size ratio of 1.5 of *Australopithecus*, in *H. ergaster* the ratio is 1.35, much closer to the ratio for modern humans of 1.2. The decreasing ratio suggests less male-male competition for females and therefore more pairing of partners. (Logically, it would seem more sensible to have females bigger than males, but in the way that the world is constructed, males frequently fight over females and, not only does the larger male often win, the female often prefers the larger male. Thus, where there is competition among males, there is heavy selection in favor of increased size of males.) One driving force may have been a longer period required for infant care. (Human babies take twice as long to reach puberty, and therefore twice as long to increase their knowledge before becoming independent—no jokes please—as do chimpanzees. This slowing of maturation is important and is probably related to the relatively juvenile appearance of humans as apes, and may also be part of the process that gradually lengthened human lifespan to approximately double that of apes.)

A more recent version of the genus *Homo*, *Homo erectus*, is considered by some to be simply a late version of *H. ergaster* and by others to be a different species. It may be nearly a semantic argument, since *H. erectus* survived into much more recent times—from 1.8 million years ago to as recently as 200,000 years ago, but *H. erectus* continued the trend and was the first hominoid to leave Africa. Remains of *H. erectus* have been found in the Republic of Georgia, and in Indonesia.

Finally, in 1856 a most curious hominid skeleton was found in a cave in the valley of the Neander River in Germany. Named the Neander Valley, or Neanderthal, skeleton, it was very human in many respects. For instance, it had a brain size equal to or exceeding that of modern humans (approximately 1300 cc). On the other hand, the Neanderthal people were much more heavily boned than modern humans; they had heavy, massive jaws; and they had pronounced brow ridges (Fig. 29.7). Their hip sockets were a bit different from ours, indicating that they walked with a more waddling gait. However, based on the material found among them, they did many things that were essentially human in style. Since the period in which they existed, 200,000 to 40,000 years ago, was an ice age in Europe, they had to use and control fire, and it is difficult to imagine how they could have coped with the winters unless they had clothing. There is evidence that they constructed wooden homes, or nests if you prefer, on platforms beside lakes in Switzerland. Some of their dead have what appears to be jewelry or other indications of planned burial. They honed stones into very effective hand axes.

However, what may be missing is definitive evidence of artwork. Although for at least one cave modern results suggest otherwise, for the most part these people left no statuettes, drawings, or markings on stones to suggest that their thoughts surpassed the immediate and the practical. Because of these lacks, we cannot be certain that we know or recognize these people. If subsequent research for the Grotte des Fées (Grotto of the Fairies) in southern France confirms the recent findings, we will have to reassess this judgment and reevaluate the issue of why these people disappeared approximately 40,000 years ago.

Late in the period of the Neanderthals, starting according to DNA approximately 200,000 years ago, a different variety of hominid appeared in Africa. Other than the dry Rift Valley of Ethiopia, the African continent is not very conducive to preserving the remains of hominids, and we have little physical evidence of what was going on. What we do know is that, approximately 100,000 years ago, a new type of hominid appeared in the Middle East. This hominid had the full modern brain capacity of 1200–1300 cc, nearly absent brow ridges and a high brow, modern, multi-purpose teeth, a flattened face that had receded behind the nose; and its skeleton was lighter-weight and more delicate than that of the Neanderthals. In short, this was a modern

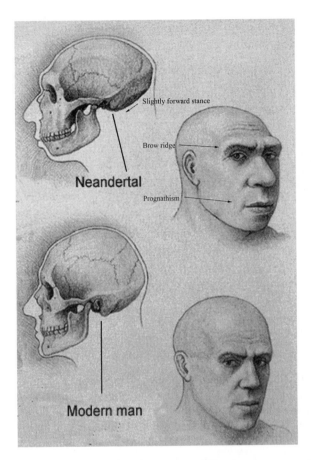

Figure 29.7. Interpretation of appearance of *Homo erectus* compared to *Homo sapiens*. Skin color and relative level of hirsuteness are of course totally speculative. Such images help us to understand that, the similarity of these people would have been enough to confuse us, the differences were such that they would not have moved in our societies without evoking a double-take. Their ability to speak is also in question. Credit: http://sapphire.indstate.edu/~ramanank/heads-sk.gif

human who would not stand out in a crowd today. We do not know the colors of their skins or the form of their hair, though we might speculate, but they were modern humans. Their tools were far superior to those of the Neanderthals, in that they used one rock to shape another and, rather than simply cracking a rock to get a sharp edge, they had learned to break off thin blades, suitable for arrowheads and knives, from flint. Most importantly, they left traces of their passage in the form of odd geometric patterns carved onto bone and stone, curious figurines often in the shape of obese, possibly pregnant, women termed Venuses (Fig 29.8), and paintings on the walls of the caves that they inhabited. When they buried their dead, they surrounded them with tools, stones or shells that appear to have been necklace jewelry, and red-colored dirts that may have been symbolic or may have

Figure 29.8. Left: A controversial Venus apparently over 100,000 years old and therefore at a time at which no Homo sapiens fossils have been found. DNA evidence (see page 392) suggests that Homo sapiens may have appeared by this time. It appears to be a human artifact rather than a remarkable natural formation, and the dating appears to be correct. If so, it would indicate that this species also had a representative (religious?) art. Right: The Venus of Willenburg, typical of the earliest known art (approximately 50,000 years ago). Although Venuses were not the only figures made, they are frequently found. Their purpose is unknown, Credits: Venus, Berekhat-ram - Wikimedia Commons. This image has been released into the public domain by its author, Locutus Borg, Venus, Willendorf - Venus of Willendorf Source first published at de.wikipedia as de:Bild:VenusWillendorf.jpg Date 2004 Author Photo taken by de:Benutzer:Plp at the Naturhistorisches Museum Wien

been used to make them more alive-looking. Though we do not know what they intended by the carvings and the art, it is clear that the important word is "intended": these people thought and could formulate concepts, almost certainly with a sense of future and past. They understood the concept of a symbol, and they may have attempted to control their fate with the use of these symbols. They were modern humans and, though their presumptions were undoubtedly very different from ours, we could have understood how they reasoned. Although most of the caves are now closed to the public because visitors bring in new microorganisms and the lights allow algae to grow, several websites offer innovative tours that are well worth a visit. They are listed in the references section.

These people moved quickly around the world. Though the first exploration as far as Israel does not seem to have survived, by 70,000 years ago a second wave again reached the Middle East. By 50,000 years ago, the first humans had reached Australia, presumably traveling along the coasts past India and island-hopping thereafter—obviously by boat. They also migrated toward the northeast and northwest, appearing in both Europe and central Asia between 40,000 and 30,000 years ago, crossing the Bering Strait during an ice age when there were land bridges and, apparently following the coastlines, reaching southern South America approximately 15,000 years ago. Movement inland in the Americas was a bit slower, so that the earliest non-coastal settlements date from 19,000 to 12,000 years ago. All of these people left skeletons, burial grounds, campsites, weapons, and artwork along their trails, allowing a fairly accurate and well-confirmed record.

Though for a species this speed of expansion is quite remarkable, in terms of what we see in current human behavior it was not a headlong rush. This expansion could have been accomplished if, in each generation, the outermost family decided to move out of sight, out of earshot, or out of the hunting range of its neighbors. One or two miles per generation would have been adequate. What the movement does suggest is a desire to seek more game or more resources or to keep away from perhaps competing tribes, together with a resourceful life style that is able to overcome geographical and geological barriers. We know that these people were resourceful and that they hunted resources; in some areas of Europe there is evidence that they could dig substantial distances into the ground to locate good flint for arrowheads.

There are many worthy questions that are worth asking. Among these questions of course are, what happened to the Neanderthals, where modern humans came from, how we know their origins, and how the world ended up with different races.

The Neanderthal (*H. neanderthalensis*) population had been declining, but the Neanderthals disappeared from Europe at approximately the time that the Cro-Magnon population (*H. sapiens*, named from the cave in France from which they were first clearly distinguished from *H. neanderthalensis*) spread through Europe. Although skeletons of *H. neanderthalensis* do not display wounds indicating attack by humans, the question of whether *H. sapiens* fought with them, made them uncomfortable enough to cause them to retreat, or interbred with them, ultimately diluting them out of existence, is of interest to scientists and most amateurs.

The lack of evidence of battles between *H. neanderthalensis* and *H. sapiens* led most researchers to favor the hypotheses of displacement or interbreeding, but advancing technology gave a much clearer answer. It became possible to collect DNA from first one, then a second specimen of *H. neanderthalensis* and compare the DNA to that of modern humans. More specifically, it was mitochondrial DNA that was analyzed. Mitochondrial DNA, a remnant of the presumed bacterial origin of mitochondria (Chapter 18, page 266), is more conserved than nuclear DNA and, since there is only one nucleus but hundreds of mitochondria in each cell, there is more mitochondrial DNA available.

If we assume that changes in some DNA bases will not markedly influence evolution ("neutral mutations") and that the probability of these mutations appearing is random, one can estimate both the evolutionary distance between two subjects and the age of their last common ancestor. From the specific sequences, one can suggest a probable lineage (page 118, Chapter 9). From the sequences established from the *neanderthalensis* and several samples of *sapiens* mitochondrial DNA, it is clear that the two are similar: of 360 bases, 335 are identical and only 25 are different. But this is not really the surprise. When the same regions of DNA are compared from humans all over the globe, the DNAs differ by no more than 8 bases. All human DNA differs by less than 0.1%. By comparison, humans and chimpanzees differ by 55 bases. What this means is that the evolutionary distance between *H. neanderthalensis* and *H. sapiens* is at least three times the distance that separates the most different modern humans, and half the distance that separates humans from chimps. In other words, Neanderthals were very different from us. Furthermore, the similarity between Neanderthal and European DNA is no greater than the similarity between Neanderthal and Asian, Native American, Australian, or Oceanic DNA, arguing against the hypothesis that Neanderthals interbred with Europeans. The conclusion drawn from this evidence is that Neanderthals did not interbreed with modern humans. Furthermore, the last ancestor shared by Neanderthals and modern humans existed over 200,000 years ago. More recently, the DNA of a second Neanderthal, found in the Caucasus Mountains in Russia, was analyzed. This sample was extremely similar to that of the first Neanderthal, notwithstanding a distance of approximately 1000 miles and a difference in age of as much as 70,000 years; and it, likewise, was not similar to that of that of modern humans. More recent analysis of Y-chromosome DNA has confirmed these initial findings.

Thus the Neanderthals, after a reign on earth approximately three times that of modern humans, disappeared. With, like us, the physiology of a tropical animal, they were capable of withstanding the rigors of an ice age in Europe. They had considerable skills and may have had conceptual thought and buried their dead. Were they driven further, into less hospitable territory, by the more capable, skilled modern humans? Did modern humans bring in diseases that they could not resist? Could they speak? Did they have a religion? We do not know, but the book is still open. What we do know is that they were sufficiently different from modern humans that there apparently was no effort to interbreed, or no success at it.

Concerning the migration of modern humans, certainly we can impute the sequence of events from the appearance of peoples before modern migrations began to mix the races once again. Likewise, we can look for common features to suggest the appearance of the earliest modern humans. For instance, chimpanzees and most groups of humans have straight black hair and moderately pigmented skin. Thus deviations from these patterns are probably more recent innovations. Humans, but not apes, have light colored soles and palms, perhaps used like the tails of white-tailed deer as markers by which adults could signal children or be seen by children as they walk away. Finally, again by DNA analysis, headlice seem to have differentiated from pubic lice approximately 70,000 years ago, suggesting that these humans were sufficiently hairless at that time to make the migration of lice between the two sites a difficult excursion.

Again, DNA analysis both adds precision and adds a surprising twist. First, there is much greater diversity of DNA in Africa than elsewhere, dating back 150,000 years, suggesting that, as the fossil record suggests, modern humans arose there. All human DNA shows similarity to a presumed ancestral DNA, especially in the DNAs that are normally passed on without modification, the mitochondrial DNA

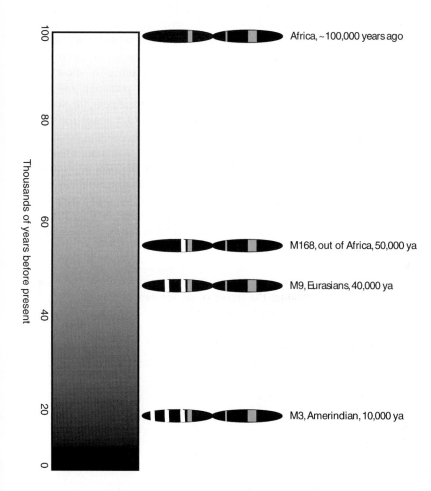

Figure 29.9. Human lineage as traced by Y chromosome markers. All human men, including those in Africa, Eurasia, and the Americas, share a pattern detectable by the use of specific stains (gray markings). These markings therefore are ancestral. The pattern has been stable for approximately 100,000 years. New markers (white) have arisen among the migrants, indicating their paths. The M168 mutation is found in all men whose origin is outside of Africa before the beginning of slave trading. Therefore the migrants from Africa carried this trait, and the ancestral form is African. It is approximately 50,000 years old. The M9 mutation appeared about 40,000 years ago and is common among Eurasians, especially those from the Middle East through Central Asia, marking this migratory route. The M3 mutation is found only in natives of the Americas. It appeared about 10,000 years ago, after this population had become established in the New World

passed on from mother to child and the Y chromosome DNA passed from father to son. Somewhere between 150,000 and 100,000 years ago lived a mitochondrial Eve and a Y chromosome Adam, who gave rise to everyone we now consider to be human. However, they lived in Africa, not the Garden of Eden (Iraq), and—despite the conceits of European artists (Fig. 29.9) more likely looked like the San of southwestern Africa. Note that this does not necessarily mean that there was only one Adam and one Eve, but that only one or very few of the original chromosomes have survived to this day.

The DNA of humans outside of Africa comes from a subset of these types, arguing that a small band of humans left Africa 50-70,000 years ago and spread through Asia, Europe, and Australia by 30-40,000 years ago. Again, the sequence of mutations confirms the migratory patterns suggested by the fossil record and physical similarities (Figs. 29.1 and 29.10).

These sequences have helped us to understand human evolution and underscore some other points as well. The genetic similarity of all humans is remarkable and completely consistent with the history: We are essentially milli-cousins, or second cousins 1000 times removed. We are approximately 2000 generations separated from the first migration from Africa. In fact, there is less difference between any two humans than between chimpanzees from the East Coast and the

Figure 29.10. Contrary to the conceits of European artists (left: Adam and Eve as drawn by Albrecht Durer in 1504) the original humans almost certainly did not look like Europeans. They may have looked more like this San Bushwoman from South Africa, as the San are one of the groups, all in Africa, who seem to have the largest number of ancestral genes. Judging from the apes, it may be more likely that the original human hair was straight. Credits: Durer - Albrecht Dürer (German, 1471–1528) Adam and Eve, Bushmen - http://www.rightlivelihood.org/gallery/2005-first-people-of-the-kalahari.htm See home website: www.survival-international.org

West Coast of Africa, and more variation within a presumed "racial" group than between one group and another. All of this reflects on various attempts to display human evolution as a linear (hierarchical), as opposed to branched, process (see Chapter 5). Second, the wide variation in human appearance argues for evolution in small groups at relatively high selection pressure (Chapter 27). This would be consistent with our impression that humans have always lived in relatively small tribes.

Of the order of 5,000 years ago, a major change occurred in human style. Humans in the Middle East and in Asia learned to control the breeding of certain plants and animals, thus achieving domestication. Shortly thereafter, MesoAmerican and South American peoples marched down the same road. Domestication of plants tied populations to the agricultural fields and provided potentially stable sources of food, as chronicled in the story of Joseph and the Pharaoh (Genesis 39-42) and similar stories in other cultures, while domestication of animals provided sources of milk and meat, as well as the power to undertake large-scale building projects and the speed to move rapidly over great distances. As is described by Jared Diamond, these achievements made possible the rise of cities and perhaps led to selection that made these groups so successful.

Think of what has happened to take us to this stage (Table 29.1 and Fig. 29.11). If the entire history of the earth were encompassed in the space of one hour, the length of time that we have had any understanding of the process would take place in less time than the sound of a single clap of the hands. If the entire history of the

Table 29.1. History of the Earth

Date (years ago)	Event	*Percent of Age of Earth*
-6,000,000,000	Origin of universe	133
-4,500,000,000	Origin of earth	100
-3,500,000,000	Origin of life	77
-1,000,000,000	Multicellular fossils	22
-600,000,000	Origin of animals	13
-500,000,000	Cambrian explosion	11
-350,000,000	Land vertebrates	8
-150,000,000	Origin of mammals; age of dinosaurs	3
-65,000,000	End of dinosaurs	1.4
-5,000,000	Humans, chimps emerge	0.1
-1,600,0000	Genus Homo	0.04
-200,000	Genetic evidence for origin of *Homo sapiens*	0.004
-70,000	*H. sapiens*, jewelry	0.0015
-30,000	First art	.0007
-15,000	Humans reach new world	0.0003
-10,000	Domestication of cattle	0.0002
-5,000	First cities	0.0001
-3,000	First writing	0.00007
-150	First understanding of evolution	0.000008

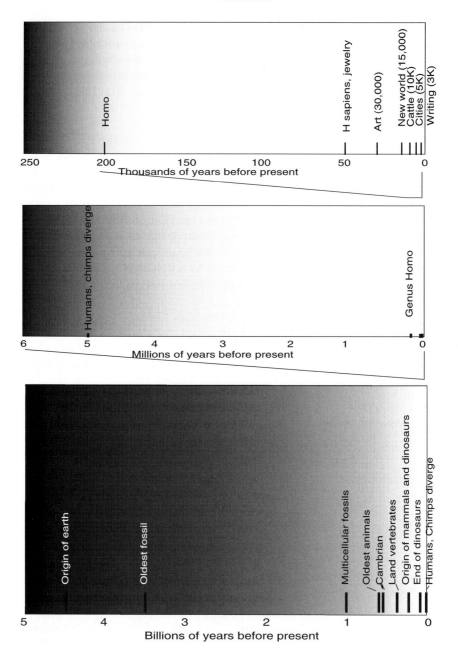

Figure 29.11. Scale of human evolution. Note that the time scales differ from each other by a factor of 1000. Humans have been on this planet for an infinitesimal amount of time since the beginning of the earth. In that time, we have been able to write about our history for a period of less than 3% of our existence, and for 99.9% of that time we have had very little to no idea of how it came about The comment "Genus Homo" refers to the appearance of the large-brained modern versions, H. neanderthalensis rather than the earliest, much smaller-brained H. ergaster, who appeared approximately 1.6 million years ago

earth were encompassed in the space of one day, our understanding would have appeared in the last 25 seconds.

REFERENCES

Shreeve, James, The Greatest Journey, National Geographic March 2006, pp 60–69.
http://www.cbsnews.com/stories/2005/03/18/60minutes/main681558.shtml?CMP=ILC-SearchStories
   (Sea Gypsies see signs in the waves)
http://www.si.edu/resource/faq/nmnh/origin.htm (Paleoamerican origins, from the Smithsonian Institute)
http://cita.chattanooga.org/mtdna.html (mitochondrial DNA and Native American Origins, Chattanooga
   InterTribal Association)
http://www.sciencenews.org/articles/20030823/fob7.asp (The Naked Truth? Lice hint at a recent origin
   of clothing)
http://sapphire.indstate.edu/~ramanank/index.html (story of Neandertals, Indiana State Univ.)
http://www.bonobo.org/whatisabonobo.html (description of bonobos, from an organization devoted to
   their preservation)
http://www.janegoodall.org/chimps/index.html (description of chimpanzees, by Jane Goodall)
http://www.culture.gouv.fr./culture/arcnat/chauvet/en/visite.htm (Chauvet caves (~32,000 years ago): In
   English, very nice)
http://www.culture.fr/culture/arcnat/lascaux/en/index3.html (Lascaux, from the French Government).
http://www.cantabriainter.net/cantabria/lugares/cuevasaltamira.htm (Altamira (~15,000 years ago; In
   Spanish, but very good pictures, from the Cantabria Province of Spain).
http://www.deutsches-museum.de/ausstell/dauer/altamira/e_alta.htm (Altamira; In English or German,
   from the German Museum)
http://vm.kemsu.ru/en/palaeolith/lascaux.html (Russian, French, and Spanish caves ~20,000–15,000
   years ago)
http://witcombe.sbc.edu/willendorf/willendorfdiscovery.html (Venus of Willendorf, 23,000 years ago–
   site is very good, from Sweet Briar College)
http://www.hominids.com/donsmaps/galgenbergvenus.html (Venus of Brassempouy 28,000 BC?)
http://www.hominids.com/donsmaps/ukrainevenus.html (Berekhat Ram, Israel: –800,000 to –233,000
   years ago)

STUDY QUESTIONS

1. Would you consider *Homo neanderthalensis* (Neanderthals) human? Why or why not?
2. Would *Homo neanderthalensis* have all the rights and privileges of citizens of free countries? Or would they be treated as apes? Why or why not?
3. In your opinion, what was the major change that led to the appearance and success of modern humans?
4. What characteristics define progress to becoming human?
5. At what stage do you think that humans learned to use fire? Why?
6. Look at the websites featuring the caves. What do you think that the drawings mean, and why do you think that they were put there?
7. What would have determined which way humans migrated, and where they could and could not go?
8. Is there any contradiction between the DNA evidence for human migration and other evidence, for instance physical appearance of humans or distribution of languages?
9. To what extent is skin color a reliable marker of lineage?

CHAPTER 30

WHEN DID HUMANS ACQUIRE A SOUL?

SCIENCE CANNOT PROVE OR DISPROVE SCRIPTURE, AND SCRIPTURE CANNOT PROVE OR DISPROVE SCIENTIFIC HYPOTHESIS

There are many religions in the world, a few more widely accepted than the others, and each has its own version of the origin of the earth, the heavens, and of humans. For some believers, what scientists describe as their belief poses no problem, as either the religious description is considered to be allegorical or the hypotheses are considered to be in different realms. For some religions, however, notably fundamentalist Protestant Christianity and conservative elements of Catholicism, Judaism, Islam, Hinduism, and other faiths, the scientific view and the religious view are held to be incompatible, and the scientific view even heretical. What are we to make of this?

There are basically three issues to address: **the realm of what science does, the definition of the word "theory," and, most fundamentally, the question of what we choose to accept to structure our world.**

WHAT SCIENCE DOES

Scientists observe, experiment, and analyze the mechanics of how things work. Philosophers and theologians ask why. We can tell you the mechanics of how a clock works and what uses it might be put to, but there are societies on tropical islands for whom the day is simply the period between sunrise and sunset, and the seasons change little if at all. In our sense, they have little concept of time: they eat when they feel hungry, they do not count birthdays, and work (building a house or a boat) is done when it is done. They do not plan for the future. They do not have words such as "want"; if they want, they take. If individuals from these islands were to ask a scientist why one would bother to fractionate a day, the scientist would not be the appropriate person to answer. A philosopher would do better. Similarly, most scientists would say, "I can tell you how I understand the evidence to indicate the mechanics of how we came to have so many species of animals and plants on earth today. If you say that God willed all this to happen, this is fine. I cannot test the hypothesis that God willed evolution to happen. Therefore it is not a subject to which I can respond. I can discuss with you how I think He did it.

As the philosopher-physician Maimonides said, "A miracle is not something that could never happen without the intervention of God. A miracle is something that could always have happened, but did not until God chose to make it happen."

At this point we must accept the fact that there are differences in detail. The Judaic versions of Genesis (there are two, Genesis I, 1-27[13] and Genesis II,

---

[13] **Genesis 1**

The Beginning

[1] In the beginning God created the heavens and the earth.

[2] Now the earth was [a] formless and empty, darkness was over the surface of the deep, and the Spirit of God was hovering over the waters.

[3] And God said, "Let there be light," and there was light.

[4] God saw that the light was good, and He separated the light from the darkness.

[5] God called the light "day," and the darkness he called "night." And there was evening, and there was morning—the first day.

[6] And God said, "Let there be an expanse between the waters to separate water from water."

[7] So God made the expanse and separated the water under the expanse from the water above it. And it was so.

[8] God called the expanse "sky." And there was evening, and there was morning—the second day.

[9] And God said, "Let the water under the sky be gathered to one place, and let dry ground appear." And it was so.

[10] God called the dry ground "land," and the gathered waters he called "seas." And God saw that it was good.

[11] Then God said, "Let the land produce vegetation: seed-bearing plants and trees on the land that bear fruit with seed in it, according to their various kinds." And it was so.

[12] The land produced vegetation: plants bearing seed according to their kinds and trees bearing fruit with seed in it according to their kinds. And God saw that it was good.

[13] And there was evening, and there was morning—the third day.

[14] And God said, "Let there be lights in the expanse of the sky to separate the day from the night, and let them serve as signs to mark seasons and days and years,

[15] and let them be lights in the expanse of the sky to give light on the earth." And it was so.

[16] God made two great lights—the greater light to govern the day and the lesser light to govern the night. He also made the stars.

[17] God set them in the expanse of the sky to give light on the earth,

[18] to govern the day and the night, and to separate light from darkness. And God saw that it was good.

[19] And there was evening, and there was morning—the fourth day.

[20] And God said, "Let the water teem with living creatures, and let birds fly above the earth across the expanse of the sky."

[21] So God created the great creatures of the sea and every living and moving thing with which the water teems, according to their kinds, and every winged bird according to its kind. And God saw that it was good.

[22] God blessed them and said, "Be fruitful and increase in number and fill the water in the seas, and let the birds increase on the earth."

[23] And there was evening, and there was morning—the fifth day.

[24] And God said, "Let the land produce living creatures according to their kinds: livestock, creatures that move along the ground, and wild animals, each according to its kind." And it was so.

[25] God made the wild animals according to their kinds, the livestock according to their kinds, and all the creatures that move along the ground according to their kinds. And God saw that it was good.

[26] Then God said, "Let us make man in our image, in our likeness, and let them rule over the fish of the sea and the birds of the air, over the livestock, over all the earth, [b] and over all the creatures that move along the ground."

[27] So God created man in his own image, in the image of God he created him; male and female he created them.

15-23[14] which differ in the order of creation), accepted in most details by Christians and Muslims, describe Creation in seven days, with what we would call evolution occurring in the last four days. Most scientists would argue that the last four days of creation were in fact the better part of one billion years.

Other than the assumption that one version is allegorical, misleading, or misinterpreted, this disagreement is not easily resolved, and it is inappropriate to attempt to persuade a reader to accept one or the other interpretation. The attitude, however, represents a fundamental difference in how one looks at life, and what one accepts as truth. We all accept as comprehensible what is familiar to us as children, and understand the workings of newer inventions and changes in society in terms of what we previously understood. The same is true for what we accept as the most solid basis for evidence. For humans, sight is the most important sense. If you hear an animal's call in the night, you may not know whether the sound is that of an insect, frog, or bird. However, sight of the creature making the sound will allow you to make what you consider to be certain identification. For a dog, scent takes precedence over sight. When I was a child, my dog would wait for me after school. He would see me from a distance and look inquisitively. If I did not make a sign of recognition, he would wait until I came closer, then come up and sniff me. Only then would he be convinced that I was home. We also know, from optical illusions, that sight can be easily tricked, and juries must frequently contend with witnesses' differing versions of the same incident. In these cases we can force our intellects to take precedence over our senses, but it is not easy.

These allegories bring us to our major point here, the growing strength of rationalism in 17th and 18th C. Europe and North America. In most religions during a major part of their histories, there is only one truth, and this is presented through a gospel or other work of divine origin. However, its meaning is sometimes ambiguous, and the change of society–for instance, increasing urbanization–may render agrarian images difficult to understand or even reveal errors. A priesthood is called upon to interpret the ambiguities.

---

[14] [15] The LORD God took the man and put him in the Garden of Eden to work it and take care of it.

[16] And the LORD God commanded the man, "You are free to eat from any tree in the garden;

[17] but you must not eat from the tree of the knowledge of good and evil, for when you eat of it you will surely die."

[18] The LORD God said, "It is not good for the man to be alone. I will make a helper suitable for him."

[19] Now the LORD God had formed out of the ground all the beasts of the field and all the birds of the air. He brought them to the man to see what he would name them; and whatever the man called each living creature, that was its name.

[20] So the man gave names to all the livestock, the birds of the air and all the beasts of the field. But for Adam [h] no suitable helper was found.

[21] So the LORD God caused the man to fall into a deep sleep; and while he was sleeping, he took one of the man's ribs [i] and closed up the place with flesh.

[22] Then the LORD God made a woman from the rib [j] he had taken out of the man, and he brought her to the man.

[23] The man said, "This is now bone of my bones and flesh of my flesh; she shall be called 'woman, [k]' ' for she was taken out of man."

The concept of the infallibility of prior highly respected sources may even extend to secular documents. For instance, the 2nd Century Hellenistic physician Galen provided a guide for medicine that applied for 1200 years. Even if later anatomists identified obvious errors, the supremacy of Galen was never in doubt. The rationalization was as follows: Most cadavers offered for dissection were those of condemned criminals. As criminals, they were by definition deformed. Surely the anatomy would be as Galen said if one did a dissection of an upstanding member of society.

The challenge to this argument came from many sectors. The great 16th C anatomist, Andreas Vesalius, argued for studying what one saw and mocked following the teachings of Galen in a spectacular fashion that must have deeply offended most of his colleagues (Fig. 30.1). William Harvey, in 16th C England, used a combination of experimentation and deductive reasoning to demonstrate the true function of arteries and veins, thus clearly illustrating errors by Galen.

Two other very important elements were the growing strength of Protestantism and the continuing exploration of the earth. Protestantism, in addition to arguing that each individual could interpret the Scriptures without the intervention of a priesthood, was now an established alternative, and presented several somewhat different interpretations of the meaning of the Scriptures. Explorers, who included interpreters, linguists, and priests, brought back stories of societies with very different stories of creation. The physicists who helped them plan their voyages, by analyzing the nature of motion and force (so that the sailors could design and use sails more effectively) and the motion of the planets (to help them find their way home) were applying these same findings to our planet and beginning to question whether the sun could really stand still, as in Joshua, and whether the earth was truly the center of the universe.

The rise of Protestantism as a religious and political, hence military, force meant that it was no longer possible for states to be the standard-bearers of a single religion. If a state was to govern a large population and not be perpetually at war, it would be obliged minimally to tolerate the existence of followers of a different religion. This movement culminated in the resurrection in Europe, after 1400 years, of a vision of a state as secular, an idea legitimized by the American Constitution and the French Revolution. Competing interpretations of sacred works were available for comparison. The many sources of alternative interpretations meant that it became possible to suggest that a higher standard for truth was analysis and logic. For the first time, people were suggesting that if logic and analysis contradicted Scriptures, it was possible that Scriptures were wrong.

This remains a fundamental difference. Scientists, by virtue of what they do, give precedence to logic and analysis in interpreting the mechanics of how the world works. Fundamentalists of many faiths insist that the exact description of how the world works is given in a holy work, received, as it were, as an email attachment from God, perfect and unalterable, even in translation. Thus if Joshua (Joshua 10: 13-14) says that the sun stopped in the heavens, it did; this is not a figurative statement.

Figure 30.1. Vesalius' insult of his colleagues. Whereas normally the professor gave the lecture while a preceptor performed the messy and smelly work of dissection, in the frontispiece of Vesalius' Anatomy he is illustrated as performing the dissection (here labeled "AV"), while the professor is mocked by having a skeleton placed in his position. In this manner Vesalius not so subtly argued that one must respect facts and observations rather than ordained wisdom handed down from the ancients

The final issue, as is discussed in more detail in Chapter 1, page 11, is the meaning of "theory". In common parlance, "theory" may mean little more than "guess": "Why is the sky blue?" "My theory is that the air is colored blue." This latter statement is identical to "My guess is that the air is colored blue" (and is wrong). To a scientist, "theory" has a much more restricted meaning. "Theory" defines an upper level of a highly structured series of levels of certainty, as defined by Popper. Fundamentalists and rationalists clash over this definition, as in the sense of "Evolution is only a theory." Fundamentalists interpret the word in the popular sense of "guess", while rationalists use it to imply a level of certainty near that of, for instance, being able to calculate the hypotenuse of a right triangle if one knows the length of the other two sides.

REFERENCES

Barbour, Ian G, 2000, When science meets religion, Harper, San Francisco.
Maimonides: Maimonides, *Guide to the Perplexed* (Chicago: University of Chicago Press, 1989), 2:25
For a miracle cannot prove that which is impossible; it is useful only as a confirmation of that which is possible,— http://www.sacred-texts.com/jud/gfp/gfp.htm Part 3, Chapter 24)
http://bibleontheweb.com/Bible.asp Genesis 1 (New International Version) New International Version (NIV) Copyright © 1973, 1978, 1984 by International Bible Society

STUDY QUESTIONS

1. What does science do, and what does it not do?
2. What is a theory to a scientist?
3. Name one authority whose word you would trust. If you did not trust this authority, how would you verify the word of the authority?

CHAPTER 31

THE IMPACT OF EVOLUTIONARY THEORY:
THE EUGENICS SOCIETY AND THE I.Q. TEST

Racism certainly did not arise as a consequence of evolutionary theory. Humans have, as far as we know, always feared or disdained other groups that they have encountered, and usually considered the new groups to be vastly inferior to themselves. We have only to look at our vocabularies: barbaric or barbarian (like the Barbars, the peoples of northwest Africa); vandals (an invading tribe from the East), thugs (a warrior group in India); muggers (another group in India); savages (wild or uncivilized people). Our movies and entertainment rarely describe encounters with people from space as pleasant or agreeable, with peace-loving creatures eager to share their riches. Therefore it should come as no surprise that, long before 1859, writers and scholars were assembling hierarchical lists of humanity, based more on an Aristotelian linear tree of life than on a Linnaeus-style branching tree. There was never any question, of course, that, if Europeans were doing the analyzing and ranking, that Europeans would come to the top of the heap. Among Europeans, obviously, such issues were argued as to whether the rather long-faced northern (Nordic, Slavic, Teutonic) peoples were superior to the rather round-faced (Celtic, Alpine, Mediterranean) peoples. Numerous efforts were undertaken to quantify these alleged superiorities and inferiorities. Such efforts included phrenology (reading the bumps on the skull, which were alleged to indicate certain aspects of character); measurement of total brain size (by filling skulls with shot—small lead pellets—or seeds and weighing the contents); relative size of various parts of the brain; measurement of relative length of arms or legs, or degree of prognathism; and even inferring that shape of hair, color of skin, shape and density of eyebrows, or overall hirsuteness were direct indicators of other features of human worthiness. Drawings were unashamedly distorted to emphasize the more "ape-like" features of the disfavored race. No matter that no ape has the tightly-curled hair of the African and Mediterranean groups: this was obviously a primitive or inferior feature. (Apes have straight hair, shorter but otherwise more like the hair of most of non-African humanity.) Behavioral characteristics derived from being on the unfortunate end of a master-slave relationship, such as servility or stoicism in the face of punishment, were read as innate characteristics. The most well-known of the assumptions of hierarchy was the invention and use of the now-discredited term "Mongolian Idiot" to describe individuals bearing an extra chromosome and afflicted with the abnormality now known as Down's Syndrome. Among the characteristics of this affliction is a deformation of the skin fold of the eyelid, giving a characteristic to the eye quite unlike that of an Asian eye but, to a western eye, similar enough to confirm a presumption. The presumption was that Asians ranked well below Occidentals in the level of development of humanity. They were servile, obsequious, and intellectually far inferior to Occidentals (their great civilizations were known but ignored) and derived from a more primitive state of humanity. Thus the fact that people with Down's Syndrome were typically of low intelligence merely confirmed their relationship to Asians—they were throw-backs to this earlier stage of low intelligence. Another, even more startling, example is to consider why the lightly-pigmented races of humankind should be called "Caucasian," or people deriving a mountain range in southwest Russia. The people who more-or-less belong to this category were first identified as living in a territory stretching from northern India through the Middle East, Africa north of the Sahara, and throughout Europe. J. F. Blumenbach, the father of Anthropology, first used the term in 1795. In 1758 Linnaeus had classified humans into four races: Americans (Native Americans), Europeans, Asians, and Africans. Granted, he described these classifications according to the prejudices of the day. Americans were red, choleric, and upright, ruled by habit; Europeans were white, sanguine, and muscular, ruled by custom; Asians were yellow, melancholy, and stiff, ruled by beliefs; and Africans were black, phlegmatic, and relaxed, ruled by caprice. Who wouldn't prefer the European?

However, Linnaeus did not believe in a strictly hierarchical structure and, in spite of the overtones of his descriptions, did not truly rank humans. Toward the end of the century, though, the general social attitude had changed. Europeans were exploring and exploiting the world, bringing greater and greater riches home; philosophers had pushed the ideas of individual worth, individual rights, and freedom, and these ideas had achieved fruition in the appearance of new and exciting governments in the United States and in France; the motions of the planets and many of the laws of physics were known and exploited to improve human welfare. Likewise chemists were learning to extract riches from the earth and turn them into other valuable products. The wonders of electricity and magnetism were beginning to be known, as were the properties of the air (experiments to understand vacuums and to identify the life-giving functions of oxygen were underway). The attitude among the intellectuals of Enlightenment Europe was of steady, rapid, linear, and inexorable progress. Thus Blumenbach, otherwise an ardent disciple of Linnaeus, was a bit uncomfortable. A four-pronged lineage of humanity could not readily be defined as a hierarchy of progress. Pondering this problem, he realized that there was, in fact, a fifth race of humans, which he called Malay. Now it was clear that there was a symmetrical, elegant hierarchy or pyramid. At the lower level, of course were the Oriental and African lines. Above these were two slightly higher lines, the Americans and the Malays. Finally, the pinnacle, the crowning masterpiece of all humanity, was—ta dah!—the Caucasian group. Truly, this pattern displayed the progress of the species *Homo sapiens*. (We don't really have to go above this level to the future, more perfect stage because through some mysterious process we have achieved perfection and need rise no higher.)

Why Caucasian? Because to Blumenbach the races radiated from a center of origin and (he acknowledged racial mixing) at the center of origin would be the purest and most beautiful representatives of the race. To quote his description of a skull from a woman who had lived in Georgia (now the Republic of Georgia, where the Caucasus Mountains are found): "...really the most beautiful form of skull which...always of itself attracts every eye, however little observant.... In the first place, that stock displays...the most beautiful form of the skull, from which, as a mean and primeval type, the others diverge by most easy gradations...Besides, it is white in color, which we may fairly assume to have been the primitive color of mankind, since...it is very easy for that to degenerate into brown, but very much more difficult for dark to become white."[15]

The late 18th and early 19th Centuries are replete with such writings, in which bald assumptions about the native superiority of Europeans or subgroups of Europeans were defended by mellifluous but circular or otherwise illogical arguments or evidence. Today the hoops and contortions that these writers went through to make their point would be hilarious were they not so painful. Such arguments not only assuaged and flattered European egos, they served very effectively to convince Europeans of the justness and appropriateness of their treatment of natives of other lands. Servility, enslavement, and even slaughter were not crimes against humanity since the subject peoples could really aspire to no higher status and could not meaningfully exploit the resources of their lands. In fact, slavery might even be considered to be a type of beneficence, removing the slaves from the savagery and squalor from which they came. Tribal conflicts were obvious evidence of savagery whereas, of course, European wars were noble and justified.

Such arguments and rationalizations readily justified the European conquests and exploitation of foreign lands and peoples but, with the continued faith in progress and the newly-arising sense of evolution, they came to the support of another fear, miscegenation (mixing of races). There had always been some concern, though not enough, of course, to prevent sailors, explorers, and soldiers from partaking of the pleasures of sexual encounters with women of different races. From the dawn of recorded history and, following the genetics, as far back in human history as we can trace, part of the spoils of war has been the women of the conquered people. However, where economics and social imperatives determined, the children of such encounters were routinely assigned to the subservient, slave or dependent, group. Even so, if miscegenation was not a desirable state of affairs, the problem

---

[15] Beside the remarkable assumptions implied here, to a biologist the final passage is wrong. Pale pigmentation generally indicates a failure to synthesize a pigment, as is described in Chapter 13, pages 186–187. It is much easier to lose a gene responsible for producing an enzyme to synthesize a pigment than it is to gain a gene to make the pigment. Albinos regularly appear among all human races, whereas mutations to darker skin are not seen. Furthermore, since all human races have pale palms and soles, moving to lighter pigmentation requires merely expanding the regions where pigmentation is restricted.

was essentially the result of the licentiousness of the foreign women. Sir Thomas Browne, a physician and philosopher in the mid 17th C, had debunked such myths as those that maintained that beavers castrated themselves to avoid capture; that the legs of badgers were shorter on one side than the other; and that, because Eve was created from the rib of Adam, men had one fewer rib than women. (This latter argument he addressed on two levels: first, by simply counting ribs on skeletons; and, second, by raising the question of how an excision from one man would be passed to his children, since someone who had lost one eye would still have a two-eyed child and those who had lost a limb would still have normal children. In spite of the clarity of his observation, the argument persisted 250 years later—see page 30.) He also debunked the myth that Jews had a peculiar odor, using several interesting arguments: You could not consistently detect an odor; there was no odor in synagogues; no one accused Jews who had converted to Christianity of having an odor; with intermarriage, the children did not have the alleged odor; and with intermarriage, you could not readily define a [genetic] population that you could call Jewish. Yet, even with this coldly analytical style, the image of miscegenation bothered him enough that he assumed that it routinely arose from the desire of Jewish women to enjoy the pleasures of the far more desirable Christian men. Two hundred years later (in 1863) Louis Agassiz, an important figure who is described elsewhere (pages 69ff) could find no better explanation for the appearance of mixed-race (white-black) children than that women of African origin were particularly seductive and eager for sexual intercourse.

To this suspicious, prejudiced, and fear-mongering mix the prospect of evolution brought a new and frightening possibility: If humankind was engaged in a race to perfection, and the front-runners were the light-skinned races of Europe, then interracial marriage (or, more bluntly, sex) threatened to dilute and degrade the obvious superior traits of the dominant population. Societies that permitted the admixture of these poorer qualities would fall behind the purer (Caucasian) societies and would be derelict in their responsibility to continue the progress of humanity. Biracial and multiracial were not only an inconvenience and an embarrassment, they would be like an infection or a bad apple in the barrel, ultimately undermining the whole society. This new fear came to occupy the minds of many. But before one could truly act on this fear, there were of course certain niceties such as laws against cruelty and laws protecting freedom. To contravene these laws, it would be necessary to demonstrate that there really was no cruelty, that the lower races would not really benefit from the full protection of the law and that, in fact, they would benefit from being guided into their proper places (servants, slaves, or other dependent positions) by the paternalistic but beneficent supervision of their superiors.

Alfred Binet is a tragic figure in this story. Working in France at the end of the 19th and the beginning of the 20th C in France, he attempted to solve a problem for the government. Though he explicitly and emphatically stated the limitations of his solution, he opened Pandora's Box. The problem that the government had was that some students did not learn well, and the government wished to provide special education and training for those students. There were potentially three sources of their limitations. Some might simply be of such low intelligence that they could not learn; some might be sufficiently intelligent but, as peasants, lack skills or experience that urban recruits might have, in which case it would be necessary to identify the skills and alter the training; or their lack of literacy might simply make it difficult to give the students new information and for the students to retain important ideas (since they could not write them down as notes). Binet had worked with the great neuroanatomist Paul Broca and had attempted to assess intelligence as a function of brain size and, unlike many of the pretenders of his day, had concluded that with truly unbiased measurements he could establish no correlation between intelligence and brain size. Thus he could not assess the children by measuring their head size. He therefore set out to devise a test that would discriminate among the three possibilities. It would have to test problem-solving ability without depending on written instructions or on experiences available only to certain groups. If he succeeded, the schools could weed out the untrainable and modify training procedures to accommodate different levels of literacy, skills, and experience. His test could produce a score that was indicative of something. He tried to make the test relate to the age of the child by getting mean scores for children of various ages. He assessed the child's relative ability subtracting the child's chronological age from the score the child achieved. All in all, he did an admirable job. Shortly thereafter the German psychologist W. Stern recommended dividing the score by the chronological age. In other words a 10-year old child who scored at the mean for 12-year-olds would score 12/10 or 1.2, which for appearance was multiplied by 100 to produce a score of 120.

Binet warned that the test identified simply the skills of an individual in a specific situation; that the skills and therefore the perceived intelligence could improve; and that the results were not generalizable to entire groups. A single test was very limited in value; one would need numerous tests, testing many functions, over different times. The ONLY function of the test was, in his eyes, to find the best means of teaching a child, never to categorize or restrict a child. In fact, he wrote with considerable anguish and frustration that some teachers assumed that the score, now described as an Intelligence Quotient or I.Q., represented an absolute and immutable assessment of the child's current and future worth: "The scale, properly speaking, does not permit the measure of intelligence because intellectual qualities are not superposable, and therefore cannot be measured as linear surfaces are measured." And, again, arguing against teachers who claim that a student with a low I.Q. can "never" succeed: "Never! What a momentous word. Some recent thinkers seem to have given their moral support to these deplorable verdicts by affirming that an individual's intelligence is a fixed quantity, a quantity that cannot be increased. We must protest and react against this brutal pessimism…"

Binet's writings were translated and published in both England and in the United States, but his meaning sank in the Atlantic and the English Channel. Joseph-Arthur, comte de Gobineau, was a 19th C author who championed the idea that great civilizations fell when their citizens became corrupt enough to mix sexually with the degenerate populations that they had conquered. For the European-American civilization to maintain its hegemony, it must avoid mixing with Asian and African peoples. Gobineau's book was translated in the U.S. in 1856, just before the Civil War, and his translator emphasized in the preface that the U.S., which already had "the Indian" and "the negro" and was now threatened by "the extensive immigration of the Chinese" was triply threatened. Two years before, the translator had co-authored a best-seller entitled "Types of Mankind". The question was certainly a hot-button issue in the U.S.

Henry Herbert Goddard, of good family, brought the test to the U.S., in the process as Stephan Jay Gould notes reifying the IQ score (turning the score into a single object of defined validity, asserting that it is a true, meaningful assessment of ability). He also managed to establish an argument that Binet never claimed, that the I.Q. was hereditary. Goddard ran, and was director of research at, the Vineland Training School for Feeble-Minded Girls and Boys in New Jersey. He eagerly used Binet's tests to assess his wards and, following his general attitude, to determine who was sufficiently unskilled and unreliable that they should not be trusted in society. Note the inversion: Binet devised the tests to determine how best to help students. Goddard wished to use the tests to determine whom to institutionalize. He also considered that the most important determinant of social deviation was low intelligence. Thus if one were to commit people based on their intelligence, one would thereby reduce thievery, murder, and immorality. Goddard fervently believed that most people naturally found the level at which they functioned. Thus manual laborers were, for instance, naturally inferior to students at Princeton, and it would be a mistake to pretend that they could rise higher: "How can there be such a thing as social equality with this wide range of mental capacity?" Furthermore, this level of inheritance was likely inherited as a single gene and easily traceable as Mendelian inheritance, much like the yellow or green color of a pea.

To be fair, Goddard did not really understand the function of polygenic inheritance, the concept of which was elaborated in the 1930's. "Polygenic inheritance" means that many genes influence a character, and it is very difficult to assign a particular effect to a single gene. This is what is meant when physicians or scientists say that a disease "runs in families": your having a relative with the disease increases your chances of having the disease, meaning that you carry a trait that encourages the disease, but many other factors finally determine whether or not you manifest the disease. It is easy enough to see how this works. Suppose that you have a gene that destines you to be tall. However, you also carry a gene that prevents you from absorbing or using a specific vitamin; or a gene that causes you to digest food poorly; or a gene that interferes with the proper deposition of bone; or a gene that forces an abnormally early puberty so that your growth is terminated before you reach your full potential. With any of these other genes acting, you will never see your predestined height. We can put this another way: I have a marvelous gene that will renew all my organs at age 88, giving me the potential for living another vigorous 88 years. Unfortunately I am a soldier and am killed in battle at age 25. Perhaps if Goddard had been less convinced of the direct hereditability of intelligence he might have been more gentle in his assessment.

However, hereditability it was and, to preserve the high level of American society, those of low intelligence should be prevented from contributing to future generations. "It is perfectly clear that no feeble-minded person should ever be allowed to marry or to become a parent" (written in 1914). They should be committed to institutions like his own.

Furthermore, it was most likely that the immigrants coming into the country at this time also harbored large numbers of "defectives". It would also be important to prevent them from coming into the country. He visited Ellis Island and was convinced (by observation) that many morons and feeble-minded were applying for entrance into the country. He therefore trained two women ("The people who are best at this work, and who I believe should do this work, are women. Women seem to have closer observation than men. It was quite impossible for others to see how these two young women could pick out the feeble-minded...") to look for potentially feeble-minded immigrants to whom he would administer the Binet test. The results were astounding. 89% of the Jews, 80% of the Hungarians, 79% of the Italians, and 87% of the Russians were feeble-minded.

The tests were really quite remarkable. The immigrants were pulled from the line, given a pencil, shown a design, and then when the design was removed asked to draw it from memory. In addition to their confusion and fear, many had never held a pencil before. They were asked to state sixty words in three minutes, which most probably could have done if they had understood what the point of the question was. They were asked for the date, notwithstanding the fact that the Russian and Hebrew calendars, to name two, were different from the Julian calendar, and that they had been on a boat for two or more weeks. Goddard was thrilled that through his work in two years he had upped the number of rejections of immigrants by almost sixfold.

He continued assiduously to affirm the inheritability of intelligence and general citizenship. His argument came from his tracing of the lineages of some of the inhabitants of his institution. One was particularly revealing. One group of "paupers and ne'er-do-wells" descended from an affair between an otherwise well-respected young man and a barmaid of presumed low intelligence and morality (his morality was not at issue). The young man later married a good Quaker woman. The children of his marriage all became upstanding citizens. The descendants of the affair produced a startling number of "degenerates". Goddard described them as the descendants of Martin Kallikak, the name being derived from the Greek words kalos (good) and kakos (bad). (Our words such as calisthenics, calligraphy, and cacophony derive from the same roots). The barmaid produced the Kakos line, and the Quaker produced the Kalos line. The conclusions derived from the lineages were impressive: the fruits of the initial affair produced a huge cost to society, in dealing with all the degenerates and immoral descendants. It would have been much more practical to forbid that reproduction (Fig. 31.1).

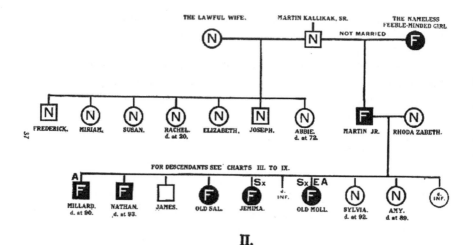

II.

Chart II.

N — Normal.   F — Feeble-minded.   Sx — Sexually immoral.   A — Alcoholic.   I — Insane.   Sy — Syphilitic.   C — Criminalistic.   D — Deaf.
d. inf. — died in infancy.   T — Tuberculous.   Hand points to child in Vineland Institution.   For further explanation see pp. 33–35.

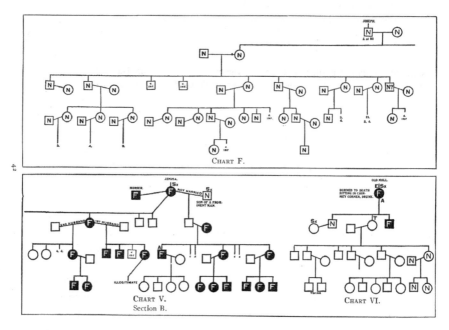

Figure 31.1. Kallikak lineage. The two charts from Goddard 's 1913 monograph. The upper graph follows the two lineages of Martin Kallikak Sr., from "the lawful wife" (left) and "the nameless feeble-minded girl" (right). Males are squares and females are circles. Other indications are listed below the chart, indicating that "the lawful wife" produced all healthy, productive descendents, while "the nameless feeble-minded girl" produced a large number of misfits. The lower chart shows a later generation, indicating that one woman, though feeble-minded, produced only one feeble-minded son while the other woman produced a series of "defective" dependents. Credits: http://psychclassics.yorku.ca/ Goddard/Developed by Christopher D. Green, York U

In his writing his fervor bespeaks a passionately concerned scientist, but in a terrifying way. He followed up the study of the Jukes family published in 1877 by Richard L. Dugsdale. In that paper, Dugsdale compared the genealogy of a disreputable family (the Jukes, a pseudonym) with that of an upright family (that of the stirring Quaker preacher Jonathan Edwards) and concluded that the tendency to be a social misfit was inherited. Goddard improved that dubious genetic analysis by comparing two lineages starting with the same father.

From his illegitimate child, Goddard found 480 descendants in six generations. Most were a problem (Table 31.1).

His whole article reeks of preconceived notions and attitudes. Most of the "Kakos" descendents who appear who appear to be normal are highly suspect and probably misread. In any case, they have sometimes been transferred to other, "upstanding," families and this has been their salvation. (Goddard did not notice his obvious acceptance of environmental factors in this case.). The quotations are spectacular and speak for themselves:

"This is the ghastly story of the descendants of Martin Kallikak Sr., from the nameless feeble-minded girl. "All of the legitimate children of Martin Sr. married into the best families in their state, the descendants of colonial governors, signers of the Declaration of Independence, soldiers and even the founders of a great university. Indeed, in this family and its collateral branches, we find nothing but good representative citizenship. There are doctors, lawyers, judges, educators, traders, landholders, in short, respectable citizens, men and women prominent in every phase of social life. They have scattered over the United States and are prominent in their communities wherever they have gone. Half a dozen towns in New Jersey are named from the families into which Martin's descendants have married. There have been no feeble-minded among them; no illegitimate children; no immoral women; only one man was sexually loose. There has been no epilepsy, no criminals, no keepers of houses of prostitution."
"The foregoing charts and text tell a story as instructive as it is amazing. We have here a family of good English blood of the middle class, settling upon the original land purchased from the proprietors of the state in Colonial times, and throughout four generations maintaining a reputation for honor and respectability of which they are justly proud. Then a scion of this family, in an unguarded moment, steps aside from the paths of rectitude and with the help of a feeble-minded girl, starts a line of mental

Table 31.1. Descendants of Kallikak

Category	Illegitimate line	Illegitimate line; further tracking	Legitimate line
Total	480	1146	496
Normal	46	197	496
Feeble-minded	143	262	0
Undetermined*		581	0
Illegitimate	36		0
"Sexually immoral"	33		1
Alcoholic	24		2
Epileptic	3		0
Died in infancy	82		0
Criminal	3		0
"Kept houses of ill fame"	8		0

* "...we could not decide. They are people we can scarcely recognize as normal; frequently they are not what we could call good members of society."

defectives that is truly appalling. After this mistake, he returns to the traditions of his family, marries a woman of his own quality, and through her carries on a line of respectability equal to that of his ancestors

"Clearly it was not environment that has made that good family. They made their environment; and their own good blood, with the good blood in the families into which they married, told.

"Schools and colleges were not for them, rather a segregation which would have prevented them from falling into evil and from procreating their kind, so avoiding the transmitting of their defects and delinquencies to succeeding generations.

"At times she works hard in the field as a farm hand, so that it cannot be wondered at that her house is neglected and her children unkempt. Her philosophy of life is the philosophy of the animal

"If all of the slum districts of our cities were removed to-morrow and model tenements built in their places, we would still have slums in a week's time, because we have these mentally defective people who can never be taught to live otherwise than as they have been living. Not until we take care of this class and see to it that their lives are guided by intelligent people, shall we remove these sores from our social life.

"There are Kallikak families all about us. They are multiplying at twice the rate of the general population, and not until we recognize this fact, and work on this basis, will we begin to solve these social problems.

"What can we do? For the low-grade idiot, the loathsome unfortunate that may be seen in our institutions, some have proposed the lethal chamber. But humanity is steadily tending away from the possibility of that method, and there is no probability that it will ever be practiced.

"But in view of such conditions as are shown in the defective side of the Kallikak family, we begin to realize that the idiot is not our greatest problem. He is indeed loathsome; he is somewhat difficult to take care of; nevertheless, he lives his life and is done. He does not continue the race with a line of children like himself. Because of his very low-grade condition, he never becomes a parent.

"It is the moron type that makes for us our great problem. And when we face the question, "What is to be done with them – with such people as make up a large proportion of the bad side of the Kallikak family?" we realize that we have a huge problem.

"The real sin of peopling the world with a race of defective degenerates who would probably commit his sin a thousand times over, was doubtless not perceived or realized. It is only after the lapse of six generations that we are able to look back, count up and see the havoc that was wrought by that one thoughtless act.

"When we conclude that had the nameless girl been segregated in an institution, this defective family would not have existed, we of course do not mean that one single act of precaution, in that case, would have solved the problem, but we mean that all such cases, male and female, must be taken care of, before their propagation will cease. The instant we grasp this thought, we realize that we are facing a problem that presents two great difficulties; in the first place the difficulty of knowing who are the feeble-minded people; and, secondly, the difficulty of taking care of them when they are known.

"A large proportion of those who are considered feeble-minded in this study are persons who would not be recognized as such by the untrained observer." [Note that here he takes the responsibility of making the judgment call himself—ral]

"In addition to this, the number would be reduced, in a single generation, from 300,000 (the estimated number in the United States) to 100,000, at least, – and probably even lower. (We have found the hereditary factor in 65 per cent of cases; while others place it as high as 80 per cent.)

"The other method proposed of solving the problem is to take away from these people the power of procreation. The earlier method proposed was unsexing, asexualization, as it is sometimes called, or the removing, from the male and female, the necessary organs for procreation. The operation in the female is that of ovariectomy and in the male of castration.

"There are two great practical difficulties in the way of carrying out this method on any large scale. The first is the strong opposition to this practice on the part of the public generally. It is regarded as mutilation of the human body and as such is opposed vigorously by many people. And while there is no rational basis for this, nevertheless we have, as practical reformers, to recognize the fact that the average man acts not upon reason, but upon sentiment and feeling;

"The question, then, comes right there. Should Martin Jr. have been sterilized! We would thus have saved five feeble-minded individuals and their horrible progeny.

"In considering the question of care, segregation through colonization seems in the present state of our knowledge to be the ideal and perfectly satisfactory method. Sterilization may be accepted as a makeshift, as a help to solve this problem because the conditions have become so intolerable. But this must at present be regarded only as a makeshift and temporary, for before it can be extensively practiced, a great deal must be learned about the effects of the operation and about the laws of human inheritance".

The number of quotes may be excessive, but they are important. Based on a fear of upsetting the wellbeing of the lives of the "good" people—Dugsdale had calculated that New York State had spent $1,308,000 [equivalent to over $26,000,000 today, as calculated from prices in 1913] between 1800 and 1875 on "social degenerates"— Goddard wanted to institutionalize at least 2% of the population and considered, really, that castration would be more effective. He was not a Nazi, and he was not crazy.

There were some problems with Goddard's data. Since he did not have access to many of the people, he would infer their morality "from the similarity of the language describing them to that used in describing persons she has seen". Others were "reputed to be a horse thief," "sexually amoral," or "wanton"—obviously social judgments bestowed on less-favored individuals, and not hard facts.

Even the facts were questionable. Others who have attempted to retrace the lineages note that many of the Kakos line did well for themselves and, in spite of everything, were upstanding citizens, but for some reason were not counted in the evaluations. The surviving photographs of the Kakos line show people who, to a modern eye, do not show the dullness and lack of interest that might betray someone of low intelligence (Fig. 31.2 left). Even worse, it is now reasonably certain that the pictures of the Kakos members who were not in institutions were retouched to make them appear more threatening and alien (Fig. 31.2 right). Nevertheless, Goddard's conviction of the hereditability of intelligence (or feeble-mindedness) and the cost to society of mental defect made him an avid defender of the rising Eugenics Society.

Lewis M. Termin was also totally convinced of existence and value of an intelligence quotient, and he made a major improvement. He simplified the test so that it could be easily administered by less-trained individuals, and he aggressively marketed it as a tool for assessing children under the name of the Stanford-Binet scale (since he was a professor at Stanford). He prepared and marketed two scales: "...either may be administered in thirty minutes. They are simple in application, reliable, and immediately useful in classifying children in Grades 3 to 8 with respect to intellectual ability. Scoring is unusually simple." It was widely successful. Termin considered it to be only fair and reasonable, but even necessary, that society should assess its members and allocate professions and careers based on this assessment.

Of course, the test questions, and their answers, might get some of us—certainly me, and perhaps you as well—into trouble. "An Indian who had come to town for the first time in his life saw a white man riding along the street. As the white man rode by, the Indian said—'The white man is lazy; he walks sitting down.' What

AGE 17.

.H.

MALINDA, DAUGHTER OF "JEMIMA."

Figure 31.2. The institutionalized Deborah Kallikak (left), who in this and other photographs does not look unintelligent. Malinda (right), one of the retarded descendents, certainly looks threatening, but the pictures were tampered with to increase the darkness under her eyes. Credits: Kallikak Deborah - Kallikak family: Kallikaks_deborah2.jpg (53KB, MIME type: image/jpeg) This is a file from the Wikimedia Commons. An image from Henry H. Goddard's The Kallikak Family, 1912. AND http://psychclassics.yorku.ca/Goddard/Developed by Christopher D. Green, Kallikak Malinda - Kallikak family: This is a file from the Wikimedia Commons. An image from Henry H. Goddard's The Kallikak Family, 1912. http://psychclassics.yorku.ca/Goddard/Developed by Christopher D. Green

was the white man riding on that caused the Indian to say, 'He walks sitting down." (Think out your answer before referring to the footnote.)[16] In England, thanks to the efforts of the equally assiduous Cyril Burt, the British developed the "11 plus" exam, in which children were tested at age 11 and, on the basis of this test, allocated toward technical or vocational schools, or towards preparation for college. Before these tests were finally eliminated in the 1960's under the auspices of the Labor Party (roughly at that time equivalent to a party of union members and therefore keenly aware of the consequences of social ostracism) they carefully selected boys and girls on the basis of pure intellectual and immutable standards.

Or maybe not. One question the answer to which depended purely on problem solving and which of course would not have any basis in social experience was the

---

[16] You of course recognized "bicycle" as the correct response. "Horse" is incorrect because an Indian would obviously recognize a horse; "wheel chair"; "on someone's back"; or any other clever or imaginative answer was wrong.

following: "In the following list, which does not fit: dog, cat, car, motorcycle?" Again, try to answer the question before looking at the footnote.[17]

This sort of inanity continues steadily. There were efforts to assess the I.Q.'s of historical figures. When the assessments did not match the predictions, corrections were imposed so that the results came out "right". When looking at correlations with social status, intellectual deprivations in orphanages were considered trivial; schools were considered to be everywhere equal; greater familiarity with another language was not relevant. To his credit, Termin finally became so enmeshed in the entanglements of these rationalizations that, by the end of his life, he conceded that he was measuring environmental, not innate factors. But much damage had been done.

Robert M. Yerkes, a psychologist at Harvard deeply concerned that psychology did not get the respect that it deserved, argued for greater quantitation and emphasis on function. Consequently, addressing for the army the same issues that had been brought to Binet, persuaded the army to test 1.75 million men in World War I. The results were as feared: many recruits were morons, idiots, or feeble-minded. They could not answer questions of general knowledge, independent of culture, such as:

Crisco is a: patent medicine, disinfectant, toothpaste, food product
The number of a Kaffir's legs is: 2, 4, 6, 8.
Christy Mathewson is famous as a: writer, artist, baseball player, comedian.

In looking at drawings to fill in what was missing, it was assumed that they would notice a missing rivet on a pocket knife or a filament in a light bulb. A Sicilian would never put a crucifix in a house rather than completing the drawing of the chimney. And of course, an anxious, sometimes rural, often immigrant, army recruit would not get rattled by the following question and would correctly answer it within ten seconds:

"Attention! Look at 4. When I say 'go' make a figure 1 in the space which is in the circle but not in the triangle or square, and also make a figure 2 in the space which is in the triangle and circle, but not in the square. Go."

The tests were analyzed many ways, including assessing the results by race and social background. The facts that many recruits scored 0; that the scores correlated heavily with geography; that recent immigrants did far more poorly; that Blacks from the North did far better than blacks from the South; and other indicators of heavy bias were not considered but rather rationalized out of existence. For instance, the differences between Northern and Southern Blacks were explained by the argument that only the brightest Blacks were smart enough to move to

---

[17] If my experience is any guide, welcome to vocational school. The correct answer is "cat" because it is the only one that does not require a license. The 11+ exam, incidentally, was instituted largely on the influence of Sir Cyril Burt, whose studies of the similarity of intelligence in identical twins was the primary basis for the belief that intelligence was highly heritable. After his death in 1947, reexamination of his data led to the conclusion that almost all of it was fabricated.

the North. The argument continued with a student of Yerkes, C.C. Brigham, who published a book, *A Study of American Intelligence* in 1923. To him, the results were completely consistent with what one knew about the races. It was known that Jews had scored (in English, at Ellis Island) among the lowest of the immigrants, but by 1923 there were many Jews of undoubted abilities who were prominent in the U.S. The reason that they were prominent was because Jews were so routinely poor performers that the occasional success was truly startling, rather as we would notice a man who stood 5"7" but was surrounded by pygmies.

   It might all sound laughable, but it led to two of the worst laws in American history as well as to the rise of Nazism in Europe. The country was already, following the First World War, fairly xenophobic, and hordes of immigrants were arriving, mostly from southern and eastern Europe. Francis Galton, a cousin of Darwin and an important player in the development of statistics, had argued for the hereditability of intelligence and described a philosophy he called eugenics, or good breeding. According to this philosophy, societies should strive to assure that only the fittest (note that he considered this word to be equivalent to "best") members procreate and leave young to the next generation. In this manner the human species will continuously improve. Brigham was a believer in eugenics and an active member in the fledgling Eugenics Society, which espoused the antipodal philosophy that the poorest members should not breed, or should be excluded from the country. Even though he, like Termin, ultimately realized that he had talked himself into a circle, his influence was considerable. The British Eugenics Society was formed in 1907. The American Eugenics Society was founded in 1922, and the Human Betterment Foundation, in 1928. Their founding members included lawyers, bankers, economists, and professors and chancellors of universities. The Eugenics Society, including the psychologists as well as prominent geneticists such as Thaddeus Hunt Morgan (the developer of *Drosophila* genetics) appeared before congress arguing for the protection of the country, resulting in the passage of the Johnson Immigration Restriction Act of 1924. This act reduced the then current immigration, which already had been somewhat curtailed. The law of 1921 allowed, per year and per country, an entry of immigrants equal to 3% of those already present from that country. The 1924 act dropped the number to 2% and used as its baseline the numbers present at 1890, thereby substantially reducing the numbers from southern and eastern Europe. As Calvin Coolidge commented, "America must be kept American". The laws were not revised until the mid 1960's, once again allowing immigration from southern and eastern Europe, as well as from Asia and Latin America. Many of your classmates will recognize the history of their families in these laws.

   The other truly horrifying result was a collection of laws based on the assumption that inherited defects caused a huge financial drain on society and that it therefore was in society's interest to guarantee that these defects not be propagated. In other words, those deemed likely to generate cost for society should be sterilized. The issue finally reached the Supreme Court in 1927, with an issue that a young, apparently feeble-minded, woman had given birth to a likewise feeble-minded child. The woman's mother was also of dubious intelligence, and the state of Virginia requested the right to castrate her. In deciding for the state, the otherwise highly respected Oliver Wendell Holmes, Jr. wrote,

"We have seen more than once that the public welfare may call upon the best citizens for their lives. It would be strange if it could not call upon those who already sap the strength of the state for these lesser sacrifices...Three generations of imbeciles are enough."

It really was almost like the right of cities to require the neutering of free-running dogs. It does not take much extrapolation to move from this attitude to the conclusion that whole races should be annihilated. We were not that far from a Nazi attitude. The last known sterilization forced by these laws took place in the U.S. in the 1960's. Another horrifying aspect was that these policies were not promulgated by obvious kooks, white supremacists, or jingoists. Many prominent scientists, including leaders in psychology and some of the best biologists, argued for these laws. In more recent times prominent scientists, though not biologists, such as Arthur Jensen, Richard Herrnstein, and Charles Murray, have attempted to resuscitate these arguments. It matters little that intelligence—whatever it is—is so polygenic that its heritability is either nonexistent or extremely hard to measure, or that, according to the evolutionist Richard Lewontin, all the differences among races amount to 6.3% of the total variation seen among humans. There is a direct, if inadvertent and unintended, trail from Darwin to Hitler. What matters is not that we should not explore these ideas, because exploration, curiosity, and understanding are what make us human, but rather a more general admonition. As scientists we can easily overlook the social attitudes that create a sense of obviousness, as has been seen in the case of chimpanzee and gorilla social structures (page 431). For scientists and non-scientists alike, we need to remember that information is not morality. Access to information does not make the scientist a seer, and a society has a right to judge the moral and social value of that information.

REFERENCES

New York Times, Feb. 8 2003, Bad Seed or Bad Science: The Story of the Notorious Jukes Family Available as http://www.wehaitians.com/bad%20seed%20or%20bad%20science.html

Goddard Henry Herbert (1913) The Kallikak Family:A Study in the Heredity of Feeble-Mindedness Available as Classics in the History of Psychology An internet resource developed by Christopher D. Green, York University, Toronto, Ontario http://psychclassics.yorku.ca/Goddard/

http://en.wikipedia.org/wiki/Jukes_and_Kallikaks (Jukes and Kallikaks, from Wikipedia)

http://www.criminology.fsu.edu/crimtheory/week4.htm (A long, good essay on biological and social factors in criminology, from Florida State University)

Estabrook Arthur H. 1916 The Jukes in 1915 Carnegie Institution of Washington (available as http://www.disabilitymuseum.org/lib/docs/759.htm http://www.disabilitymuseum.org

Gould, Stephen Jay, 1981, 1996, The mismeasure of man, W.W. Norton and Co., New York, N.Y.

Eldredge, Niles, 2001 (2000) The triumph of evolution and the failure of creationism, Henry Holt & Co., New York.

STUDY QUESTIONS

1. Under what laws did your ancestors come to the United States? What restrictions were placed on immigrants from those countries? To what extent were the laws designed with Eugenics goals in mind?
2. What is the I.Q. test? What does it measure, and how does it measure it?
3. What is polygenic inheritance?
4. What factors could affect performance on an I.Q. test?
5. Before reading this chapter, what was your understanding of the value, accuracy, and importance of the I.Q. test?

CHAPTER 32

EVALUATING POPULATION MEASUREMENTS: BELL CURVES, STATISTICS, AND PROBABILITY

We discussed the bell curve in Chapter 9, where we indicated that it described a normal distribution of a variable in a population. The bell curve describes a trait that can be measured but which varies continuously in a population, such as weight or height. It does not describe variables that are essentially discrete or discontinuous. For instance, there is no continuous distribution along a gradient between "female-like" and "male-like". Almost all individuals are clustered in one or two categories. On a more complex level, because of the historically late meeting of various groups of humans, in some societies skin color might be discontinuous while in other societies, such as in Brazil or Hawaii, it might tend towards a continuous distribution. It is important to realize that bell curves do not describe all situations, and that the crudest versions of statistical calculations may not apply. This is the most mathematical of restrictions that we encounter. Others are more fundamental, and they relate to the most common misuses of scientific reasoning, testimonials and false associations of logic. You can find examples of these nearly every day, and indeed a study problem at the end of the chapter is to find and analyze them.

TESTIMONIALS

The testimonial is the single example, most commonly in first person: "I followed Dr. J's diet plan, and I lost 50 pounds!" It is important to remember that the function of such a testimonial is to give a single example to entice the listener to generalize from that example to a general rule, from which the listener can deduce the effect on himself or herself: Testimonial-giver X lost 50 pounds→all people who follow Dr. J's plan lose 50 lbs→if I follow Dr. J's plan, I will lose 50 pounds. This is the logic of a false syllogism. A true syllogism allows one to deduce an individual truth from a general truth: If all girls are pretty, and if all children named Mary are girls, then Mary is pretty. A syllogism, basically an issue of logic, works only in this fashion, and reaches only one conclusion. It does not work backwards: If a flower is pretty, it does not follow that a flower is a girl. In other words, the logical flow does not of itself establish the "if and only if" structure of a true scientific experiment.

First, the assumption may be false. The statement "if all girls are pretty" may be false in fact even if the structure is acceptable. If the answer to "if all girls are pretty" is "no," then it is not necessary that Mary be pretty. She might be, but if some girls are not pretty, then Mary might be one of those girls.

The problem with a testimonial is that it attempts to create the syllogism where none exists. In large populations, many things are possible, with or without cause. Someone will win the lottery, but the fact that this person chose the numerical version of his son's birthday as numbers does not create the prediction that choosing birthdays will win again. Likewise, a certain number of people will be involved in traffic accidents. Many of these will have proximate causes, but the fact that you were driving on the priority road when someone pulled out from a stop sign without seeing you does not necessarily make you a less worthy driver. You may or may not have been less attentive than you should have been. Likewise, the dieter giving the testimonial may have been enormously motivated by any of several social or financial consequences of his obesity, or a serious medical problem. In this circumstance almost any effort to lose weight would probably have worked. The causal relationship between Dr. J's diet and his weight loss is not proven, and it will not necessarily work for you or anyone else. Other examples include non-prescription medicines for conditions that spontaneously resolve themselves. The old joke is that this treatment will cure a cold in seven days. Without treatment, the cold can last a week!

STATISTICAL MEASUREMENTS

The only valid predictor is the statistical measurement, done with suitable controls. If 1000 people follow Dr. J's diet and are compared to 1000 people who make no effort to lose weight and 1000 people who try another procedure (exercise or another diet) and, for instance, 200 of those on Dr. J's diet lose 10 or more pounds, compared to 20 who lose weight while doing nothing and 30 who lose weight doing something else, then one can conclude that the diet has some benefit. Note, however, that there is still a catch: 800 people who followed Dr. J's diet nevertheless did not lose weight. It will not work for everyone. In more dismal terms, if 90% of patients survive a difficult operation, the loss of 10% does not necessarily indicate failure or malpractice on the part of the surgeon.

Thus the testimonial relates one instance to one other instance. It does not establish any greater causality than my winning the lottery by choosing numbers based on the license plate of a car stopped ahead of me at a red light. In addition to the issue of bell curves discussed in Chapter 9, there is the other extremely difficult problem when dealing with large populations: since we cannot truly control the situation, there are huge numbers of variables that potentially can affect results. Suppose we tried to compare rates of heart disease among Mexican-Americans in Los Angeles and among people of Swedish descent in Minnesota. Using your hands and feet, you can readily count off the numerous likely differences in genetic background, diet, probability of smoking, exposure to sunlight, exposure to childhood diseases, exposure to airborne carcinogens, amount of exercise, level of education, probability that they have lived in the same location all their lives—and we have not begun to consider the age and sex of the subjects. If there is any social component, for instance if the data are gathered by interview, there may be other differences. It is well known that different ethnic groups give different responses depending on circumstances such as the race of the interviewer. Even if one relies on more solid data, such as hospital admissions, some groups are more likely than others to present themselves to formal medicine at an early stage of a disease. Thus, as was discussed in Chapter 9, correlation is NOT causality.

IMPLICATIONS OF CONFUSING CORRELATION AND CAUSALITY

The reason that this discussion is important to us is that it is precisely this type of misuse of data that is so often used to argue against the theory of evolution or, at a more destructive level, to justify a particularly heinous or cruel attitude toward fellow human beings. One can readily argue that there is a direct line from Darwin to Hitler, and that most of the cruelest political activities of the 20th C were based on somewhat innocent to intentional distortions of the meaning of evolutionary theory. To avoid any potential confusion or misinterpretation of the intent of this chapter, we will state the following as the basis of our current belief:

- The modern human species evolved once in biological history and consists of a group highly variable in phenotype but genetically extremely close, much closer than the range in many other species.
- Although there are geographic differences among humans, the designation of race is far more political or social than biological. For instance, it might be of medical interest to know that a person carries genes derived from Africa or the Middle East (in which case the person might carry the sickle-cell gene (Chapter 15, pp 247–255) but for other purposes, though designated for census and even self-designated as Afro-American, the person might well have much less than 50% of his or her genes derived from African populations.
- The range of variability within a group such as what we describe as a race is much greater than the difference between races.
- In tests such as those described below, purporting to measure intelligence, many aspects of the social situation substantially affect the results, and in any event it is not obvious that the tests measure a vital biological characteristic.
- In any event, in a democratic society one judges individuals, and we do not allow any informed or uninformed opinion of a group to influence the manner in which we interact with an individual. If there were a statistical argument that children born in [choose the month of your birth] were, on average, of slightly lower intelligence than children born in [choose the month six months later], would you feel that it was fair that people discriminate against you?

We have to deal with the issue that humans tend not to display the most noble instincts when encountering each other. Most ancestral histories, whether Biblical, Greek, Asian, African, or other, recount mostly long episodes of war, and the history

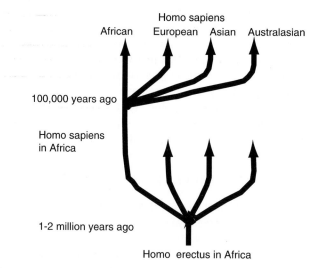

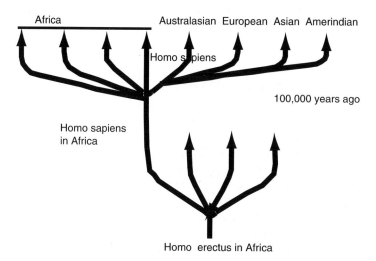

Figure 32.1. Even modern textbooks tend to drift into subtle implications. The figure above, taken from a textbook printed in 2002, manages to suggest that the African lineage is the most direct line from (and therefore closest to) the ancestral lineage. The lower figure gives a more accurate picture, in that the *Homo sapiens* lineage appeared in Africa about 100,000 years ago and generated many branches, most of which remained in Africa. About 50,000 years ago one branch left Africa and gave rise to the Australasian (Melanesians and Australian Aborigines) and European branches. The European branch gave rise to the Asian branch, which gave rise to the Amerindian branch. (The 2006 edition of the book has a more representative figure.) Credits: Redrawn from textbook figures

of exploration of the world, the development of nations, and the development of empires, relates far more to opportunities for plunder than to any more noble or innocent cause. Similarly, humans have not in general viewed with pleasure the encounter with other humans who were noticeably different. In fact, it is not unreasonable to comment that, when two humans find a difference between themselves and a third human, they are likely to use the difference as a basis for discrimination. This discrimination has little memory. Many individuals encountered the prejudice of the currently reigning population in spite of proud, even magnificent, achievements of their cultures. For instance, Mayans, Aztecs, and Incas had very sophisticated, impressive civilizations; most of the Middle East, including Egypt, Syria, Iraq, Persia, Israel, and Greece were elegant when Europe was barbarian; and the same is true for China and some parts of Africa.

Because of the human proclivity to fear and demean strangers, the first understanding of the rudiments of evolutionary theory led almost immediately to a grouping and hierarchical ranking of presumed human groups. There were many efforts to place different groups of humans along this scale, and the scale was almost always linear rather than branched. (Fig. 32.1) Most of this effort took place in Europe and North America, where evolutionary theory had created a stir, and, needless to say, the hierarchical rankings always placed Europeans and, especially,

Figure 32.2. Chimpanzee. Note that, contrary to popular impression, the skin color of our closest relatives is not necessarily dark. The color of the skin, as opposed to the hair, varies quite a bit. Credits: http://en.wikipedia.org/wiki/Image:South_Djoum_Chimp.jpg

Table 32.1.

Characteristic	Bonobo	Human
Skin color	Variable	Variable
Hair color	Black	Black, Brown, Red, blond
Length of legs	Short	Long
Length of arms	Long	Short
Brow ridge	Pronounced	Very slight (out of Africa) to absent (Africa)
Hair Style	Straight	Straight, wavy, curly, tight curls
Male Beard	None	All races; variable
Hirsuteness	Very hirsute	Moderate to almost none
Lips	None	Pronounced
Light color palms and soles	No	Yes (all races)
Nose	No nose projection	Modest to pronounced

northwestern Europeans, at the very top of the scale. These rankings were unapologetically prejudiced.

Other offshoots of these presumptions included the often-illustrated interpretation that dark-skinned people were more closely related to apes than the lighter-skinned people, though by most skeletal and other criteria African populations are more differentiated from apes than are the lighter-skinned peoples, and our closest relatives among the apes have light, not dark, skins (Fig. 32.2; Table 32.1).

The Eugenics Society contributed other misconceptions. Based on a misunderstanding of the manner in which genes spread in a society, the members worried about the capacity of "unfit" people to breed. The misunderstanding was the assumption that a harmful gene will spread throughout a population. It will not. Genes will spread only if they are selected for—that is, if the carrier is more successful in leaving children to the next generation.

THE RISE OF NAZISM

By far the greatest excess of the 20th C was the rise of Nazism. Based on the Malthusian argument of struggle among competing groups, the Nietzschian concept of the superior being, and Darwinian selection, the theory of Nazi racial policy was that the best humans should be selectively bred and those who were not beneficial to the gene pool should be prevented from breeding by confinement, sterilization, and execution. The undesirables included Jews, Blacks, homosexuals, Gypsies (members of an itinerant population coming from the area of northern India—the name "Gypsy" refers to the erroneous Medieval assumption that they were from Egypt), non-representational artists, and several others deemed by Hitler and his staff to undercut what they perceived to be their highest culture. Although there have been many other massacres in history, this one derived its primary philosophy on the most brutal (and inaccurate) interpretation of Darwinian Theory.

BIAS IN SCIENCE

It is easy to blame society and leaders for misinterpreting and misusing the findings of a politically neutral science, but the explanation is not that easy. First, science is never totally politically neutral. As certain facts make no sense until we have a logic to explain them, scientists are humans and operate within the assumptions of the society. Thus early primatologists, who were all male, observed among apes the structure of male social hierarchy and interpreted the behavior of the females as totally dependent on the social ranking of their male partners. It was not until women primatologists, notably Diane Fossey and Jane Goodall, joined the research that the social structure of the females was noted and recognized to be important. Similarly, Konrad Lorenz, who was a brilliant observer of behavior but also someone who was willing to help draft Nazi policy, promulgated the concept of the "alpha male," the one who is the natural leader and dominant member of the troop. Field observations, combined with a more questioning or at least different political viewpoint, revealed that, while one male might be dominant (or, depending on your viewpoint, a bully) in a zoo setting, in the field one male might be the alpha defender against attack, another the alpha fruit-seeker, another the alpha chooser of the nesting site, etc. In the case of the immigration issues and the Eugenics Society, although ultimately science proved the interpretations wrong, well-meaning, intelligent scientists sometimes led the way. It is the responsibility of scientists to assess the social implications of what they do and say, and to consider the possibility that expertise in the laboratory is not expertise in social policy. Similarly, it is the responsibility of those who are not doing the science to remember the social imperatives and to be sufficiently aware of the fallibility of science to resist making or asking for bad law in the name of science.

REFERENCES

Hatton, John and Plouffe, Paul B, 1997, Science and its ways of knowing. Prentice Hall, Upper Saddle River, NJ.

Carey, Stephen S., 2004, A beginner's guide to scientific method, 3rd Ed., Thomson Wadsworth, Belmont, CA

Piel, Gerard, 2001, The age of science. Basic Books (Perseus Books Group), New York.

Wilson, Edward O. 1995, Naturalist, Warner Books, New York.

Wilson, Edward O, 1996, In search of nature, Island Press/Shearwater Books, Washington, D.C.

STUDY GUIDE

1. Note a testimonial advertisement on television or in a newspaper, or a political endorsement based on a personal experience. To what extent does it meet or fail the requirements of a valid syllogism?

2. Identify any medical report in a newspaper or on television that notes the results of a study in which the sample size is described ("small," "very large," or an actual number. Locate the original article and read the description of how one population was compared to another. Were all possible sources of bias eliminated?

3. How many instances can you identify in which something is accepted as scientific truth, but the opinion reflects the beliefs of the time? Hint: All historical times are valid, not just the modern period.

4. Observe the behavior of any domestic or caged (zoo) animal toward others of its species. Can you identify any aspect of the behavior that is likely to be a result of current situation, rather than its status in the wild? How do you know?

5. Domesticated animals are frequently highly selected for characteristics such as docility and tolerance of crowding.
   a. What does this sentence mean?
   b. How might this selection affect what we interpret to be their natural behavior?

CHAPTER 33

CONCLUSIONS—WHERE DO WE GO FROM HERE?

The scientific mode of thinking can be described as a type of philosophy—a mode and structure of analysis. Its basis is the assumption that an analytical interpretation of the evidence of the senses is the best means of understanding our world. It does not rely heavily on the sensual or emotional side of human experience (passion) as a guide to interpretation, because it has no means of weighing, experimenting with, or falsifying the meaning of passion, trances, or other emotive experiences, and it considers to be irrelevant, evidence based on immeasurable factors such as faith, communication with the dead, extrasensory perception, telekinesis, or other considerations beyond human experience. Note that the operative words are "considers irrelevant": the scientific approach does not reject out of hand such evidence; rather, the scientist states that he or she cannot evaluate such evidence and therefore cannot incorporate it into a logic of the workings of the world. For science is about the mechanics of the world, how the world is put together and how it functions. Science does not consider why the world is here.

Humans have always used the rules and logic of evidence, even in the most adamantly faith-based procedures known to mankind. What were miracles but evidence of the existence of a superior being? And trials by ordeal in all faiths were an effort to establish evidence. They ranged from the African ordeal bean, in which someone accused of a crime was made to eat a bean containing a deadly neuropoison to torture of the accused in a court of Puritans, or the Inquisition. In each case the survival of the accused was evidence of innocence, and death was evidence of guilt. The rule was still evidence, but the logic included assumptions of untestable forces ranging from the power of God to unknown forms of energy. As long as they remain untestable, they are beyond the reach of science and the scientific approach. They may exist—before the existence of the microscope, bacteria were inconceivable, and before the discovery of radioactivity, the idea that a rock could explode and release enormous amounts of energy was unimaginable. Scientists simply say that we know of no forms of energy and no mechanisms by which ghosts, for example, could exist and come to haunt the earth. We can attempt to detect their existence, by setting up numerous detection devices and, above all, attempting to reproduce the conditions in which they appear. If we fail in spite of our best efforts to capture an unequivocal and measurable sign of their existence, they remain an unproven hypothesis, currently falsified by evidence supporting the opposing hypothesis that ghosts do not exist. The evidence supporting the hypothesis that ghosts do not exist is weak, since it consists entirely of negative evidence: the ghost was not recorded by a camera, motion sensor, heat sensor, magnetic field detector (such as a metal detector), microphone, or any of the numerous other means we have of detecting distortions in the environment. Any well-planned and executed experiment that detected a ghost would in a single step overturn the hypothesis that ghosts do not exist, but we would then have to move to the next step of logic—how do they exist? What is their source of energy? Of what are they composed? Science merely tells us where to place our money in a bet, and in this case the best wager is that ghosts do not exist.

It is also very important to remember that morality is a human trait but that science is amoral. By "amoral" we mean that science does not have morals, that it is neutral to morality. Science is not "immoral," or against human codes of morality. It is amoral, in the sense that the value of any human action or judgment is a human decision for which science can provide evidence but not interpretation. A scientist can state when the genome of a new human being is created and at what point the nervous system is developed to the level at which we can presume that an infant feels pain or has a thought, but the value of that information, meaning whether or not the state or the church assumes interest and responsibility for that life, is a value judgment made by societies, and the conclusion has varied from society to society and throughout history. Likewise, we can provide evidence that evolution has occurred and our analysis of this evidence can inform our predictions as to what will happen if we raise the carbon dioxide level in the atmosphere (global warming) or what we will lose if we destroy great ecosystems such as the tropical rainforests. We can likewise interpret how genes will and will not spread in our population, or calculate how many people this planet can hold. But we cannot make decisions for a human society. The society must assess its own values, and in this endeavor all participants have a say. Sometimes societies make very bad decisions, and sometimes they make excellent ones. The role of the scientist is to tell us how it works and therefore to predict the consequences of specific actions. Hopefully you, the citizenry, will be sufficiently well-informed to understand the importance of evidence, logic, and falsification, and you will evaluate the data, and make moral and compassionate decisions on that basis. If you can do this, then we as scientists have succeeded in our mission.

REFERENCES

http://www.unesco.org/science/wcs/eng/declaration_e.htm
American Association for the Advancement of Science, 1993, Benchmarks for Science Literacy (Benchmarks for Science Literacy, Project 2061) (Paperback)
Chalmers, Alan 1999,What Is This Thing Called Science: An Assessment of the Nature and Status of Science and Its Methods (Paperback)

Rothman, Milton A. 2003 Science Gap: Dispelling The Myths And Understanding The Reality of Science, Prometheus Books, New York

Tambiah, Stanley J, 1990, Magic, Science and Religion and the Scope of Rationality (Lewis Henry Morgan Lectures) (Paperback) Cambridge University Press, Cambridge, UK

STUDY QUESTIONS

1. Compare the concepts of "truth," "evidence," and "logic" in science and in other fields.
2. To what extent can scientific facts be considered to be absolute? To what extent can the interpretation of those facts be considered to be absolute?
3. What major scientific subjects will have the most political or moral impact in the future?
4. Suppose it were well established that people born in the month that you were born had a medical problem that would cost insurance companies so much that the cost of everyone's insurance policy would be increased 10%. What would you do? Would your response be the same if the problematic month were something other than your birth month?
5. List the three most important ideas that you have learned from this book; give the evidence that backs the idea; and explain why you consider it to be so important.

INDEX